APPLICATI REFERENCE

PRACTICAL ANTENNA DESIGN

140-150 MHz
VHF TRANSCEIVERS

APPLICATI™ REFERENCE

PRACTICAL ANTENNA DESIGN

140-150 MHz
VHF TRANSCEIVERS

APPLICATI™
MEDIA GROUP
www.applicati.com

APPLICATI REFERENCE
PRACTICAL ANTENNA DESIGN
140-150 MHz VHF TRANSCEIVERS

Published by

APPLICATI,™ LLC
www.applicati.com

First Year of Publication, 2008
Published in the USA

While every precaution has been taken in the preparation of this book,
the publisher assumes no responsibility for errors or omissions, or for damage
resulting from the use of the information contained herein.

Printed in the USA

10 9 8 7 6 5 4 3 2 1 0

DEDICATION

This book is dedicated to

Rene, Rick & Roel

ACKNOWLEDGEMENT

Many thanks to:

Bernd Hohl, for the computer and laser printer

that were used in making the first drafts of this book,

and

Le Van Tam, for the DTP program.

CONTENT

INTRODUCTION

This book is one in a series designed to help anyone who wants to construct antennas for radio transceivers, but who has only a basic knowledge of radio-communications technology. The approach applied in this book is similar to the do-it-yourself methods of trade books. Theories are kept to a minimum. Detailed illustrations are extensively used throughout the whole process of antenna construction, to simplify the otherwise difficult-to-comprehend technical jargon.

The antenna designs presented here are specifically cut to the dimensions necessary for proper operation in the 140-150 MHz VHF band. Each chapter deals with a particular design, and an extra chapter at the end is added to help the individual assembling the antenna in converting the given antenna dimensions for other frequencies. However, the formulas for conversion give only generalized information, and much of the fine-tuning of the new dimensions is left to the actual experimentation of the constructor. A highly detailed set of no-guesswork antenna dimensions for other frequency bands are described in other books in this series, also written by the author.

The choice of a certain design for a particular application is left to the decision of the constructor. In selecting a design, certain factors like portability, ruggedness, compactness, signal gain versus size, weight, wind loading, and availability of materials must be taken into account, in order to obtain an optimum performance from a particular antenna.

The author assumes that the constructor has already some experience in basic construction techniques related to radio communications equipment installation, like soldering VHF connectors to coaxial cables, making a pig tail, cutting aluminum tubes, and using an SWR meter. Obviously, knowledge in operating a VHF transceiver is the most important.

Here is one rule of thumb in installing VHF antennas: If you use an RG-58/U coaxial cable to feed the antenna, do not use more than 20 meters or 60 feet. More than this length results in much of the signal (almost half) being lost in the cable, and will substantially degrade your antenna's performance. If it is unavoidable to extend to this length, use the larger RG-8/U cable instead. Although this cable is about four times more expensive than RG-58/U cable, this is the only way you can avoid signal loss in the cable.

It is the author's hope that this book will provide adequate information to anyone wishing to build their own antennas for VHF transceivers.

1 GROUND PLANE ANTENNA

Model FA-2

Reliable communication in radio systems depends upon the overall effectiveness of both the base station and the mobile unit antennas. The radiation pattern of the transmitted signal is extremely important, since it must be transmitted and received in densely populated areas as well as over long distances.

If you are situated in the center of a town or a city, an omnidirectional pattern is best suited. Omni-pattern is also the best choice when you do not know the exact direction or location of the station you are communicating with. Directive pattern is practical only if you know exactly which direction must the signal be beamed to, in order to maximize the transfer of RF energy. However, antennas with directive patterns are more complex in design, and will be discussed in later chapters.

Generally, antennas for VHF bands are mounted as high off the ground as practical to overcome the limitations of line-of-sight transmission and reception. An artificial ground must then be used, since the antenna is well above the ground in this case. This is not a problem in automobiles, since this artificial ground is provided by either the metal roof or body of the car. For tower installations, however, some means must be provided to simulate this artificial ground. This is accomplished by the ground plane radials, which are usually made of thin metal rods or tubes each cut to quarter wavelength long and mounted at the base of the antenna. The rods sometimes bend downward at an angle of about 45 degrees below the horizontal. This angle is important to maintain the correct impedance match of the system.

The ease of construction and low cost of a ground plane antenna makes it an ideal choice for VHF operators. The unit described in this chapter uses bronze rods for the radiating element because of their availability, and because a bronze rod is the easiest to connect to the center pin of the coaxial connector.

The ground plane radials are made of cheaper aluminum tubes. Obviously, the antenna is not easy to disassemble once completed, so its use is commonly confined to fixed installations requiring little maintenance.

The operational frequency bandwidth of FA-2 is from 140 MHz up to 150 MHz, exhibiting an SWR response of less than 1.5:1 over the entire bandwidth. It has a gain of 1 dB (unity gain) compared to a real dipole. Its signal pattern is omnidirectional.

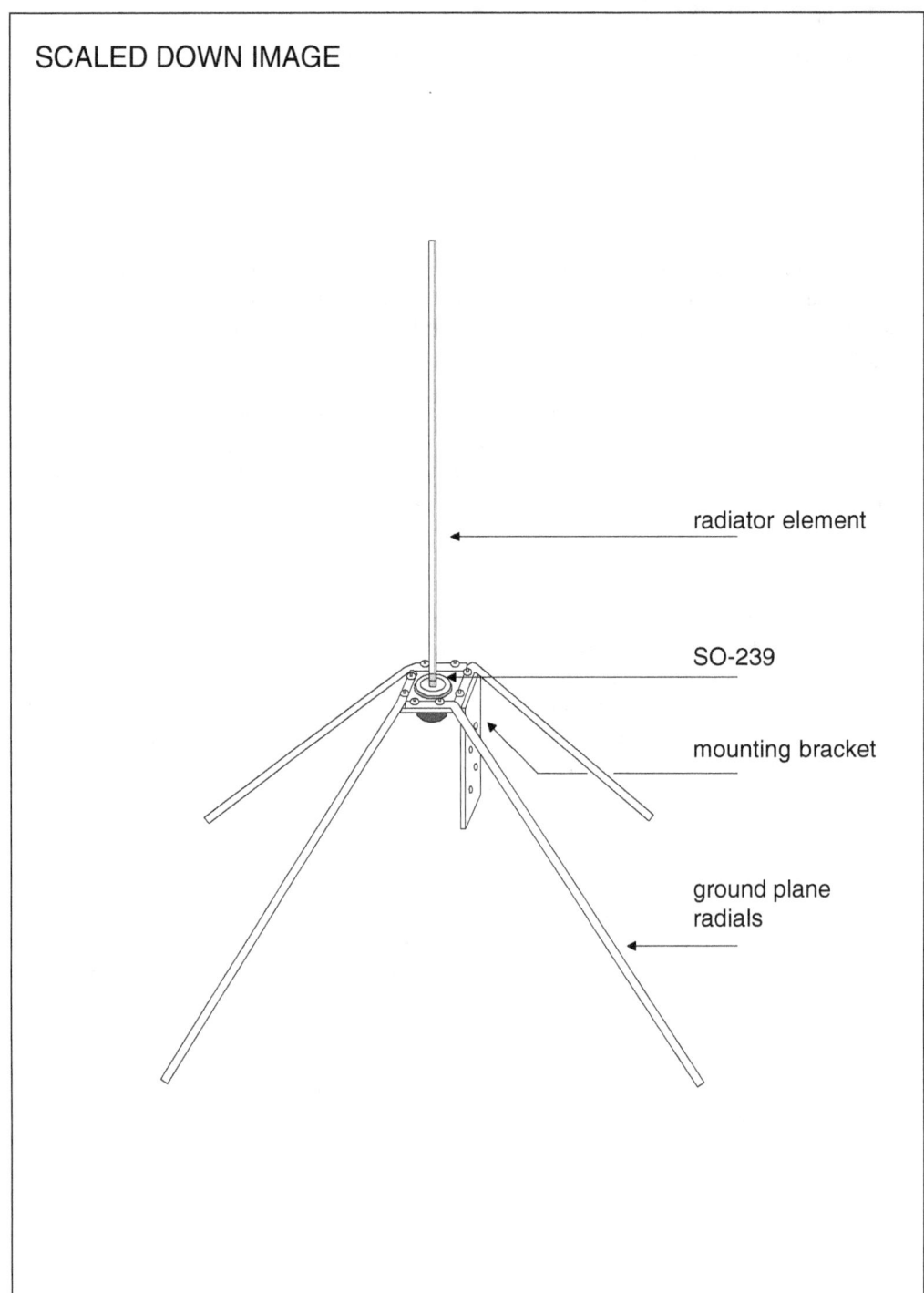

Figure 1.1   Ground plane antenna Model FA-2

Materials List

Quantity	Specification/Description	Dimensions
4	Aluminum Tubes	3/8 id x 20" each
1	Brass Rod - the brass rod for acetylene welding is recommended	1/8" diameter
1	SO-239 VHF female connector without flange	
8	Stove bolts - brass or stainless	1/8" x 3/4"
8	Lockwashers - brass,stainless or GI	1/8" id
8	Hex nuts - brass, stainless or GI	1/8" id
1	Aluminum plate gauge 14 or 16	2" x 6"
2	U-bolts with accompanying hex nuts and lockwashers	

*id - inside diameter

Construction

First of all, construct the antenna mount. It is made from a 1/8" thick aluminum plate cut to 2" x 6". Drill a hole in the plate big enough for the SO-239 VHF connector to insert into (about 5/8" or 15.8 mm). Drill the hole at the point about 1" away from one end (see Figure 1.2).

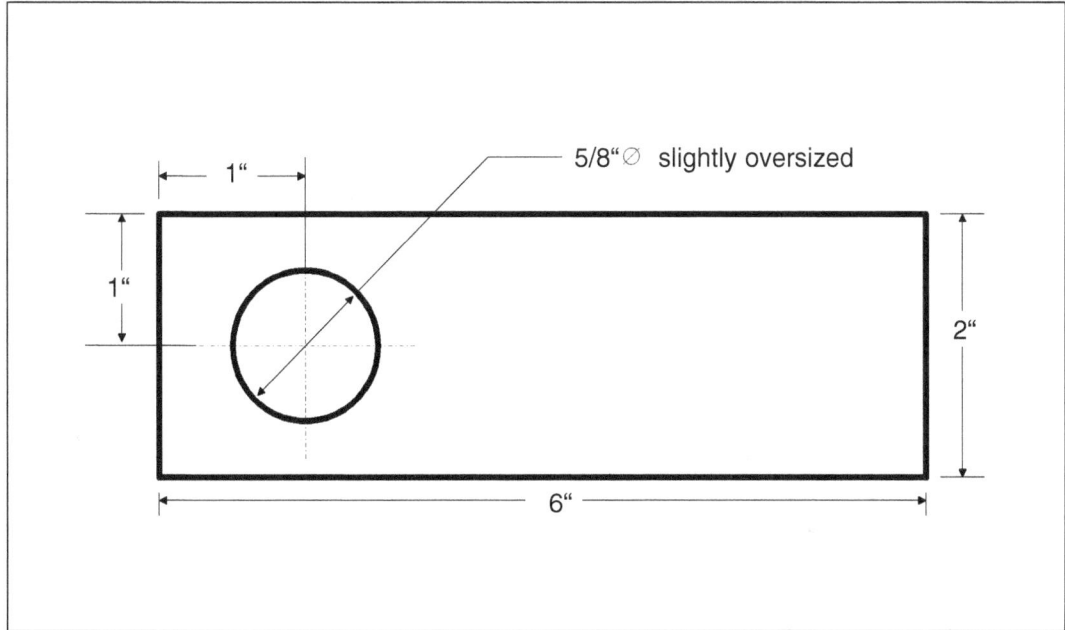

Figure 1.2   Antenna mount and hole dimensions.

Next drill four holes at the other end of the plate, following Figure 1.3 for the proper dimensions. Make sure that the distance between one pair of holes perpendicular to the length of the metal sheet must be the same with the distance of both ends of the U-bolt that will be inserted into it.

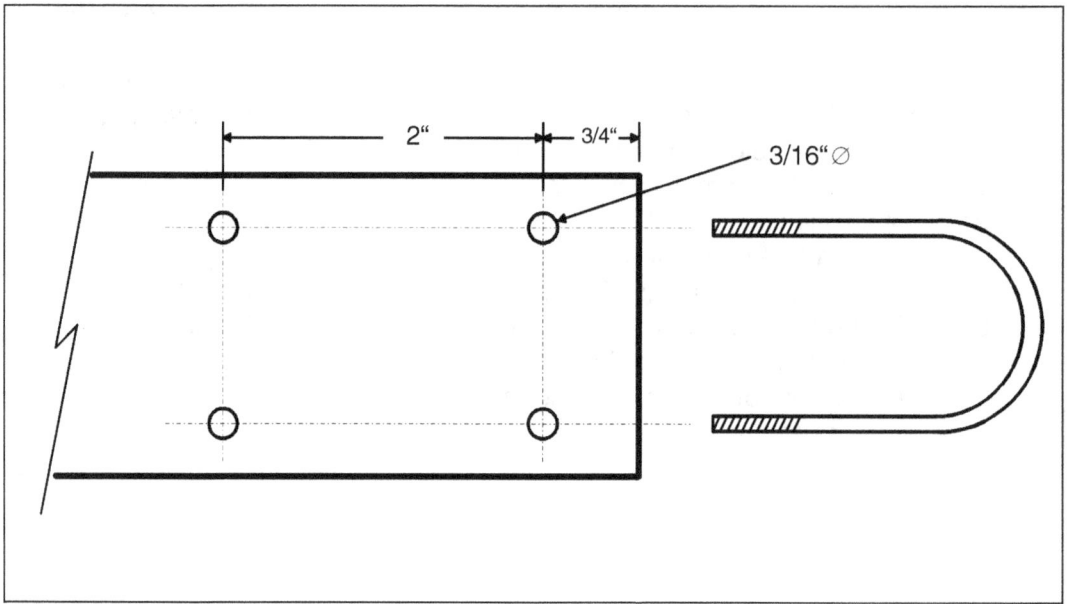

Figure 1.3   Hole dimensions for the U-bolts.

Next, drill eight holes (1/8" diameter) around the large hole, following Figure 1.4 for the proper dimensions.

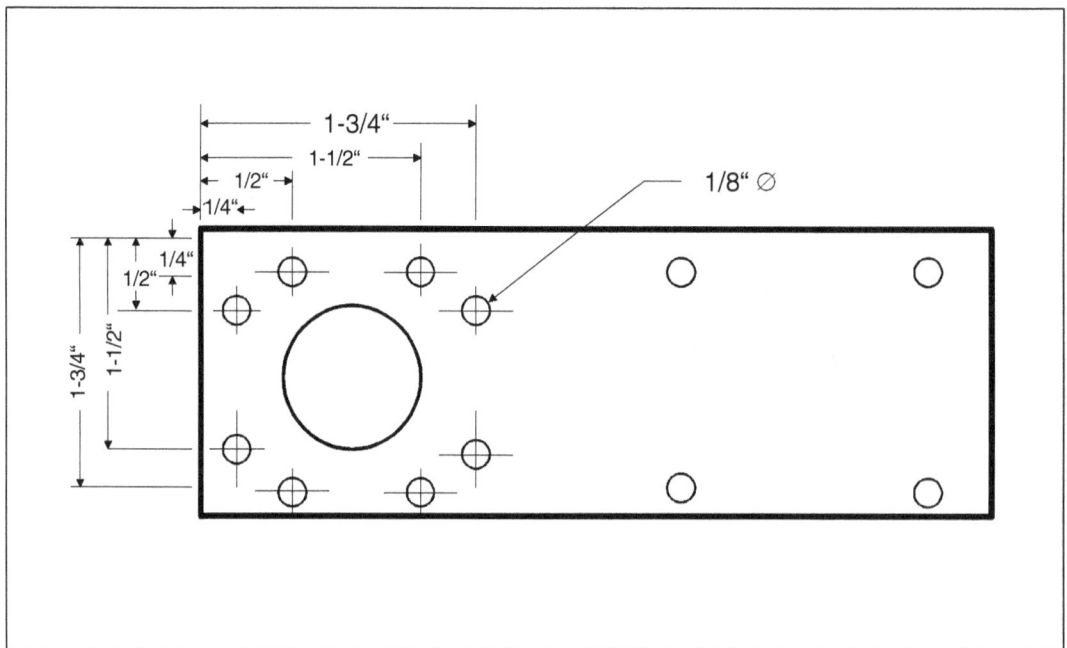

Figure 1.4   Hole dimensions for the radial elements around large hole.

Bend the aluminum plate down to a 90° angle (see Figure 1.5). Follow the illustration for the exact point to bend.

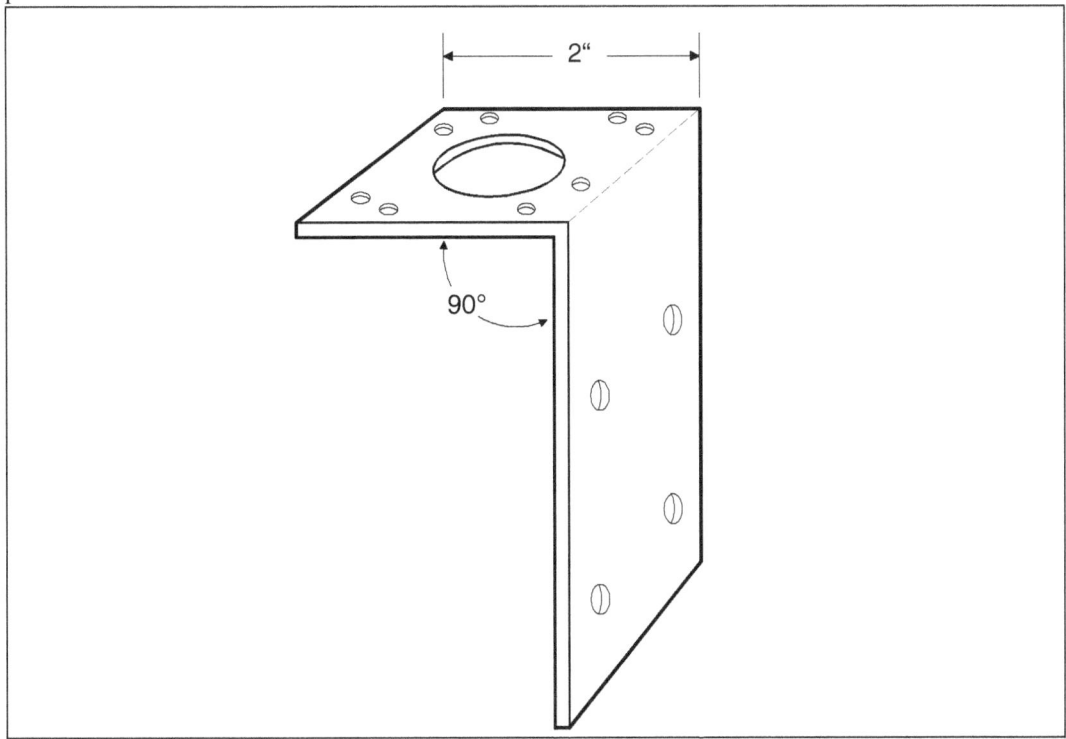

Figure 1.5   Bending the aluminum mounting plate.

Insert the SO-239 VHF connector facing downwards into the mounting plate and fix it permanently with its nut (see Figure 1.6). Discard the grounding ring/lug.

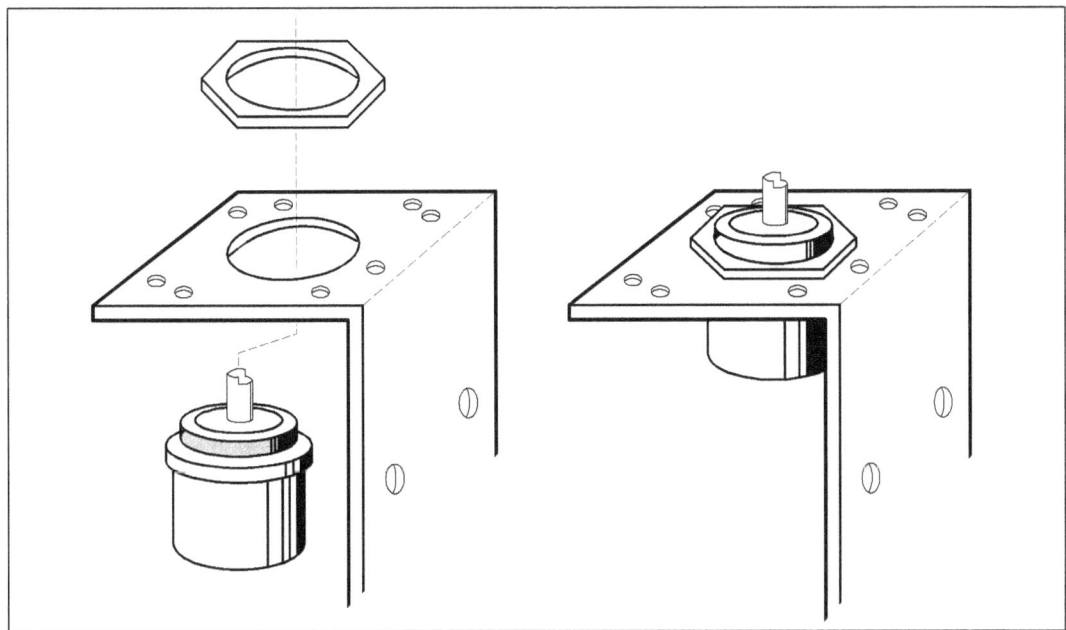

Figure 1.6   Mounting the SO-239 into the plate.

Cut the brass rod to a length of 19" (48.26 cm) and insert one of its end into the center pin of the SO-239 connector (see Figure 1.7). The brass rod may or may not fit into the center pin immediately, so you may need to file away a small portion at the end of the rod to reduce it to a smaller diameter.

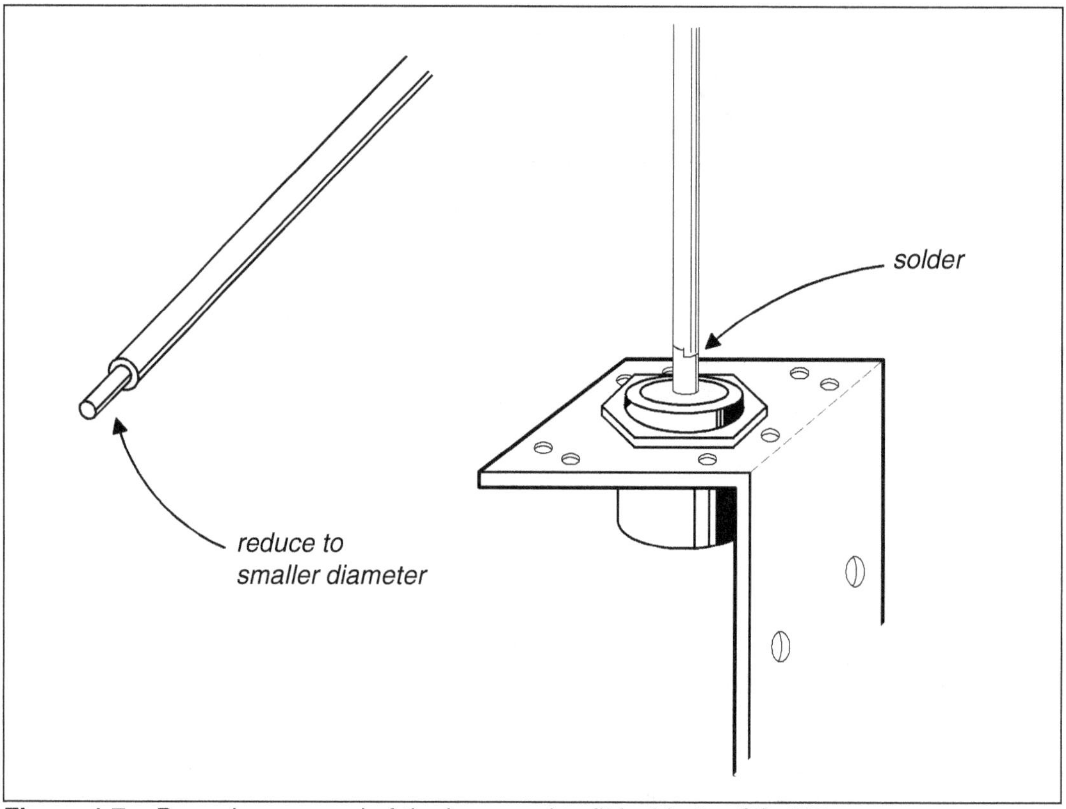

solder

reduce to
smaller diameter

Figure 1.7   Preparing one end of the brass rod to fit inside the SO-239.

Cut four aluminum tubes to a length of 20" each, and drill two holes (1/8" diameter) at one end (see Figure 1.8).

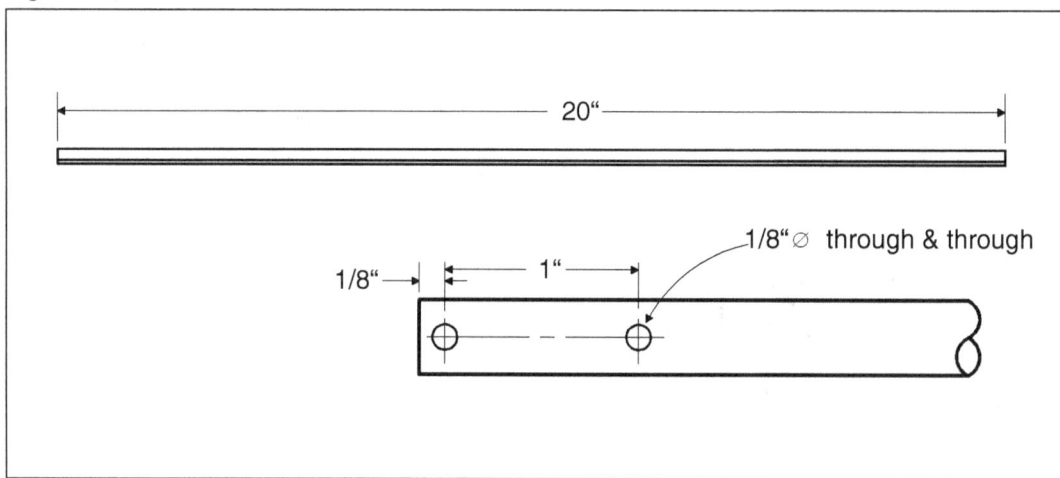

20"

1/8"⌀ through & through

1/8"→

1"

Figure 1.8   Preparing the tubes.

Bend the aluminum tubes to a 45 degree angle at the point 1 inch away from the end with two holes. The direction of the bend must be parallel with the axis of the drilled hole (see Figure 1.9).

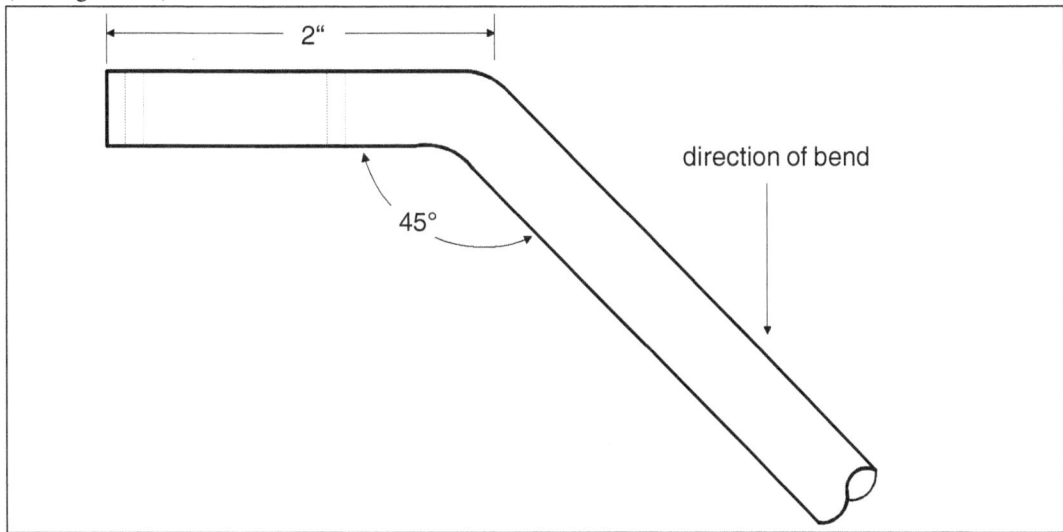

Figure 1.9 Bending the tubes.

Mount the four aluminum tubes into the angled plate by bolting each element with 1/8" x 3/4" stove bolts (see Figure 1.10). The stove bolts must be made of rust-resistant material, such as stainless steel, brass, or GI.

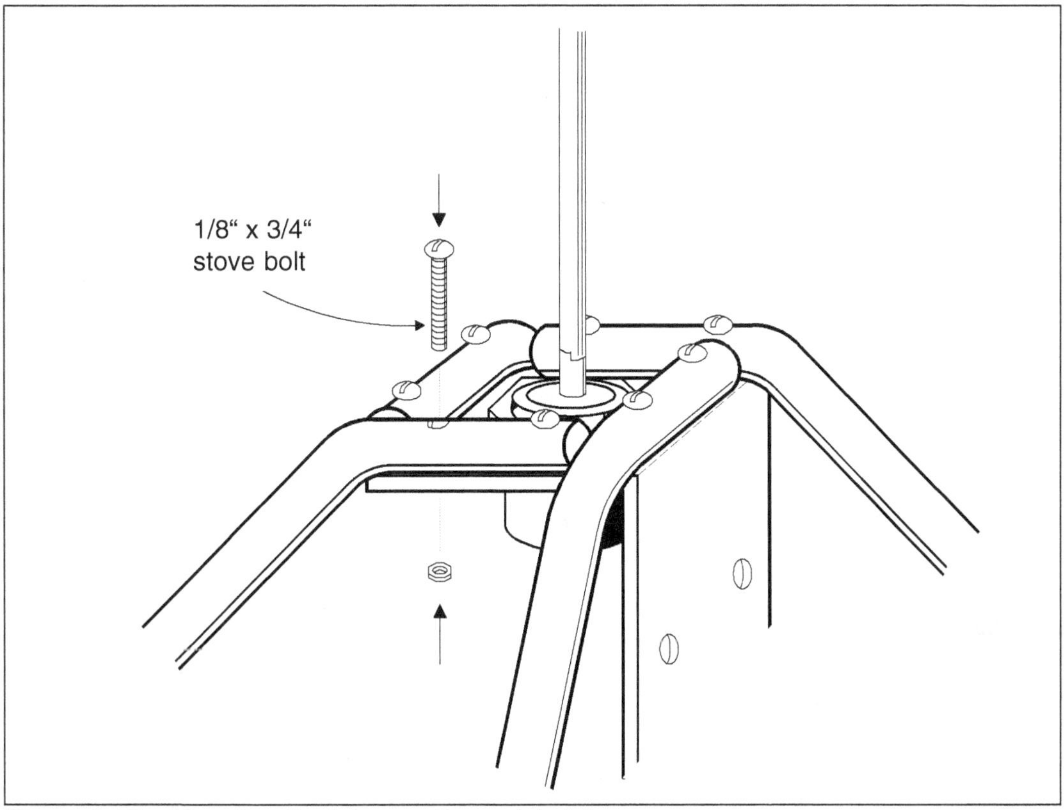

Figure 1.10 Mounting the tubes on the metal plate.

Finally, you can mount the antenna to the mast using the two U-bolts.

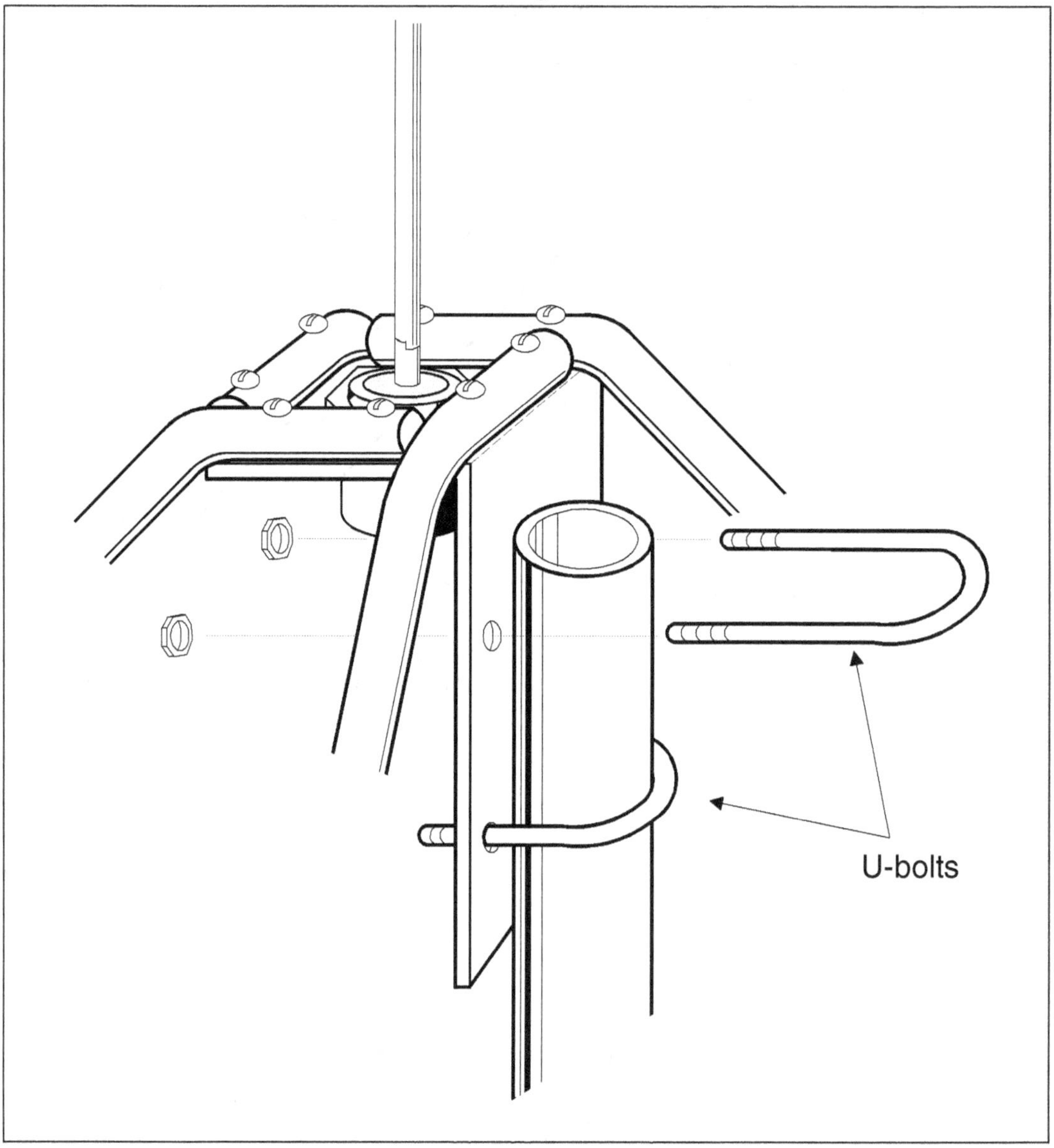

U-bolts

Figure 1.11   Mounting the antenna to the mast.

REVIEW QUESTIONS

1. Why is the radiation pattern of an antenna important in radio communication?
2. What is the advantage of an omnidirectional pattern over a directive pattern?
3. When is a directive pattern practical?
4. Why are VHF antennas installed as high off the ground as possible?
5. Why must the ground plane radials bend downward below the horizontal?
6. What is the function of the ground plane radials.

2 GROUND PLANE ANTENNA

Model FQ-2

The antenna model FQ-2 is an extension from the basic configuration of a ground plane. This unit features quick-detach elements to facilitate easy and fast disassembly of the antenna. The total size of the antenna is much reduced when disassembled, and becomes convenient to carry and transport. It is also a lot easier and faster to construct compared to the FA-2 design.

This particular version of the ground plane was evolved in an emergency situation where there were very few tools available. The place was aboard a fishing boat, and there was no drilling tool around, so a ground plane design was created which did not require any drilling of holes.

If you plan to use a ground plane in mobile operations, then this design is recommended. It can be easily inserted inside your backpack while you are travelling. Assembly or disassembly takes only a couple of minutes. The antenna elements are made of durable bronze materials, so they can survive the stresses caused by the regular mounting and dismounting of the antenna. If you have accidentally bent an element, just straighten it, and it is again functional. A slight bend or kink in the elements has no negative effects on the performance of the antenna. It is so durable that you have to intentionally cut it to pieces to destroy it. Experience has proven its reliability in the rugged life of mobile operations.

The electrical characteristics of this antenna are the same as those of the model FA-2. The only difference between the two models is the mechanical construction and type of materials used.

SCALED DOWN IMAGE

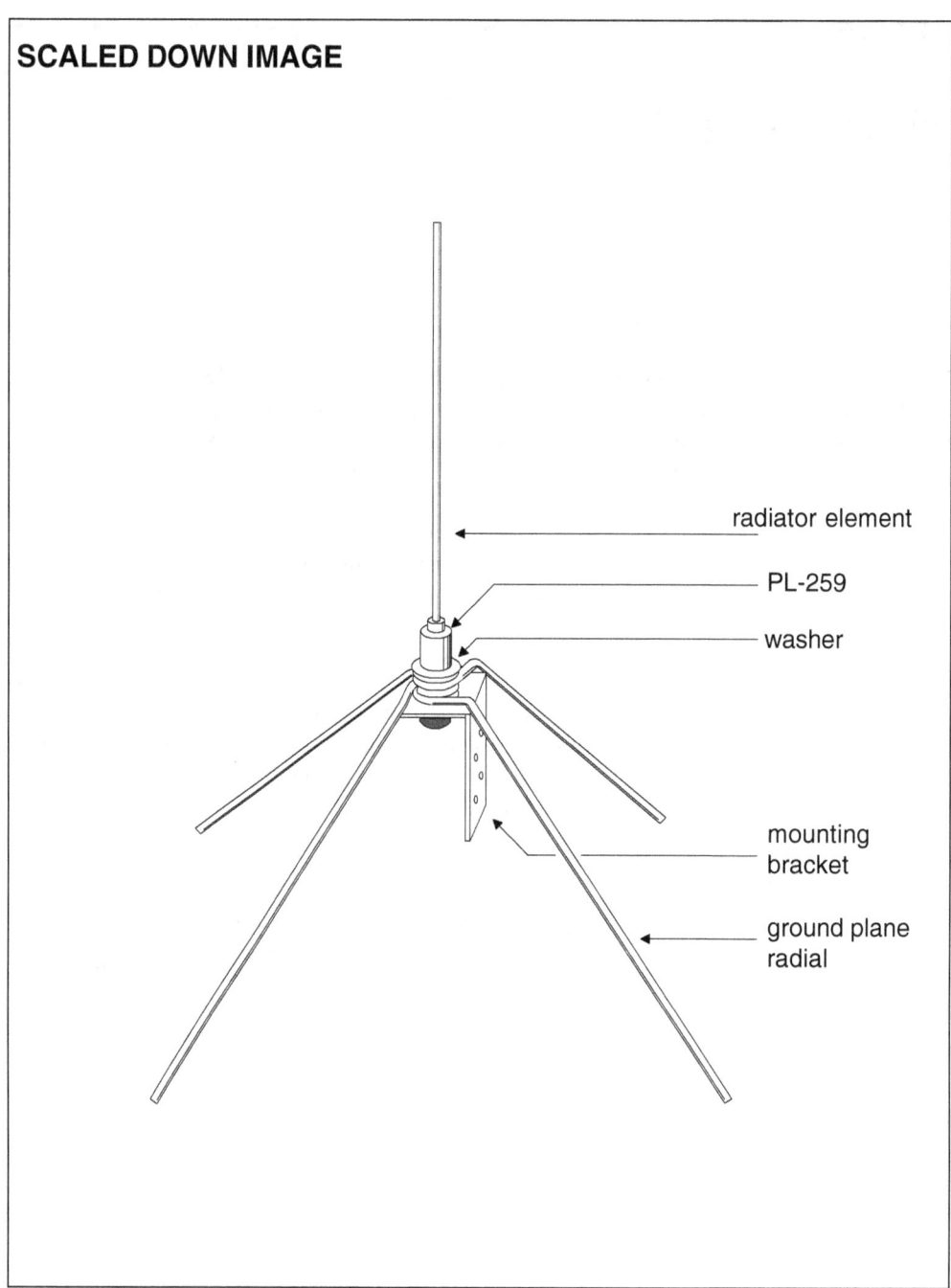

radiator element

PL-259

washer

mounting bracket

ground plane radial

Figure 2.1   Quick detach ground plane model FQ-2

Materials List

Quantity	Specification/Description	Dimensions
5	Brass rods - the brass rod used for acetylene welding is recommended	1/8" diameter
2	PL-259 VHF male connectors	
1	PL-258 VHF straight connector	
1	Aluminum plate gauge 14 or 16	2" x 6"
2	U-bolts with accompanying hex nuts and lockwashers	
1	Plain washer GI or stainless steel	
1	short length of coaxial cable	2" long

Construction

Reduce one end of a brass rod to a smaller diameter enough to be inserted into the center pin of the PL-259 connector. Also file a notch at its end, as shown in Figure 2.2A. Insert the rod into the PL-259, and solder it to the center pin (see Figure 2.2B). File away any excess solder that is bulging out of the center pin.

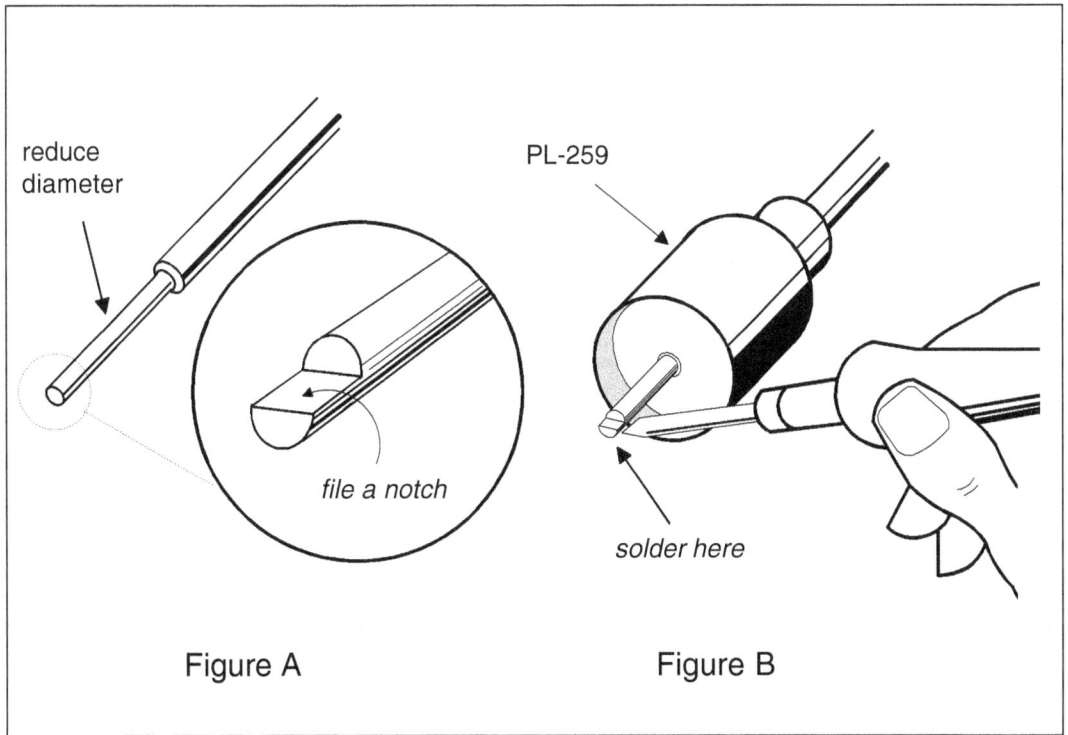

Figure A   Figure B

Figure 2.2   Preparing the end of a brass rod, and soldering it to the PL-259.

After soldering the rod into the PL-259, cut it to a length of 19" (48.26 cm), following Figure 2.3.

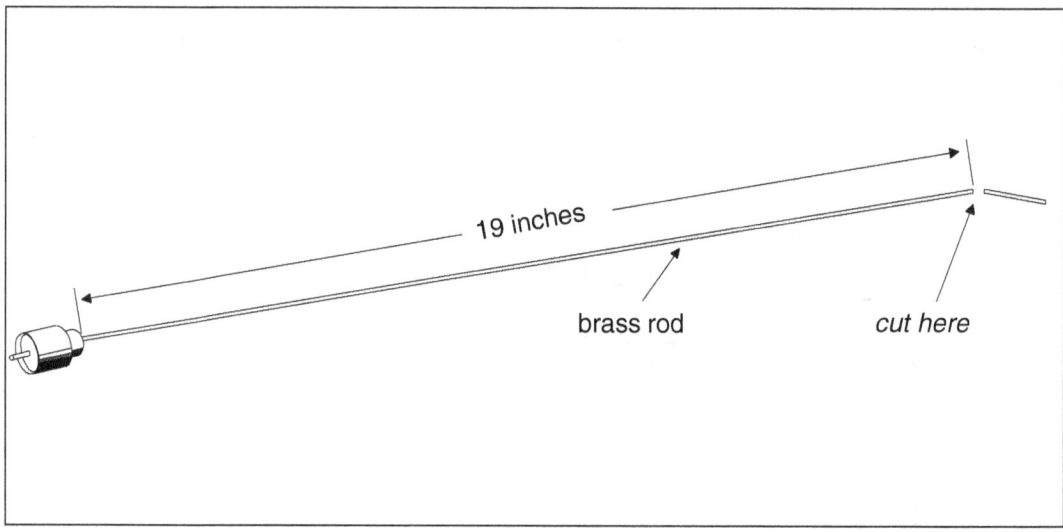

Figure 2.3   Trimming the rod to its proper length.

Cut a small piece of coaxial cable (about 2 inches), and remove its inner conductor and braid (shield). You will only need the vinyl outer jacket. Insert the vinyl jacket into the rod all the way inside the PL-259 (see Figure 2.4). Cut any protruding portion of the jacket. The vinyl jacket serves as an insulator between the brass radiator rod and the body of the PL-259.

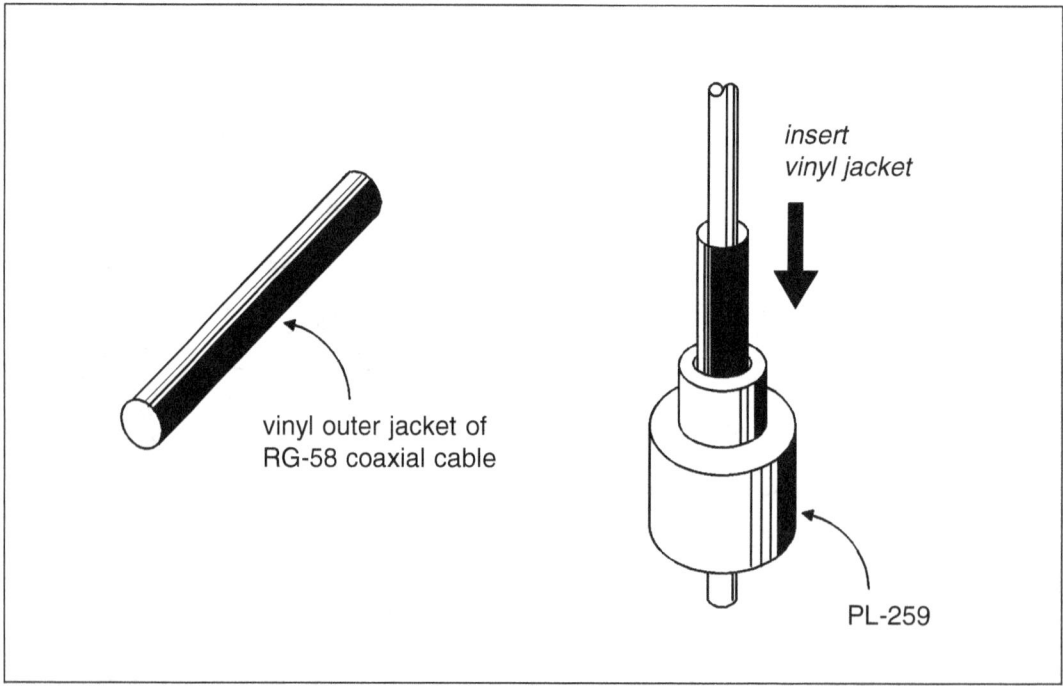

Figure 2.4   Inserting the vinyl insulator into the PL-259.

Prepare a small amount of epoxy glue and place it over the protruding portion of the vinyl insulator. The epoxy glue must cover the gap between the rod and the PL-259 to prevent the seepage of rainwater or other moisture inside the connector (Figure 2.5).

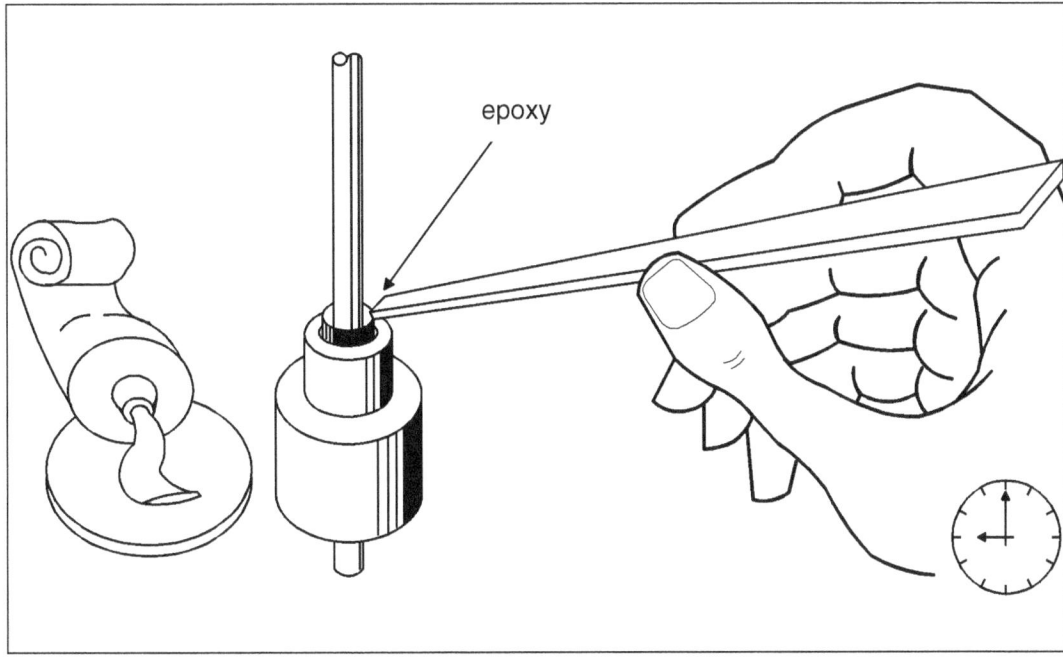

Figure 2.5   Sealing the gap with epoxy glue.

Next, prepare the radial elements. Bend one end of each brass rod to an eyehook shape as shown in the following illustration (Figure 2.6). The diameter of the eyehook formed must be dimensioned in such a way that the straight connector can be easily inserted into or pulled out of it.

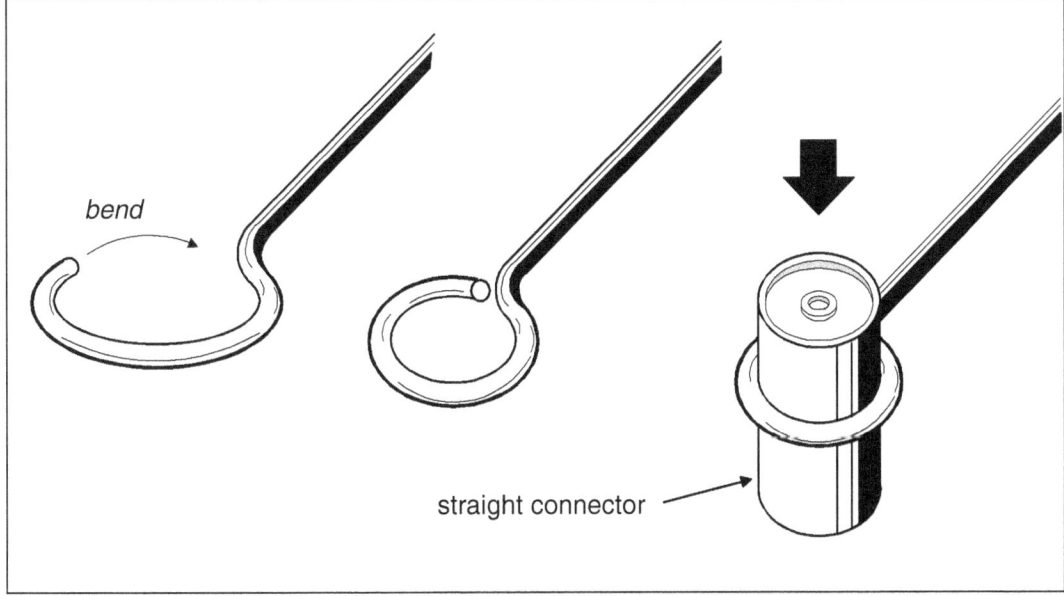

Figure 2.6   Shaping one end of the brass rods to an eyehook form.

After bending one end of all four brass rods into the necessary shape, measure 19 inches from the point where the rod starts to bend into the eyehook form. Mark the measured point at the other end, and cut the brass rod at this point (see Figure 2.7).

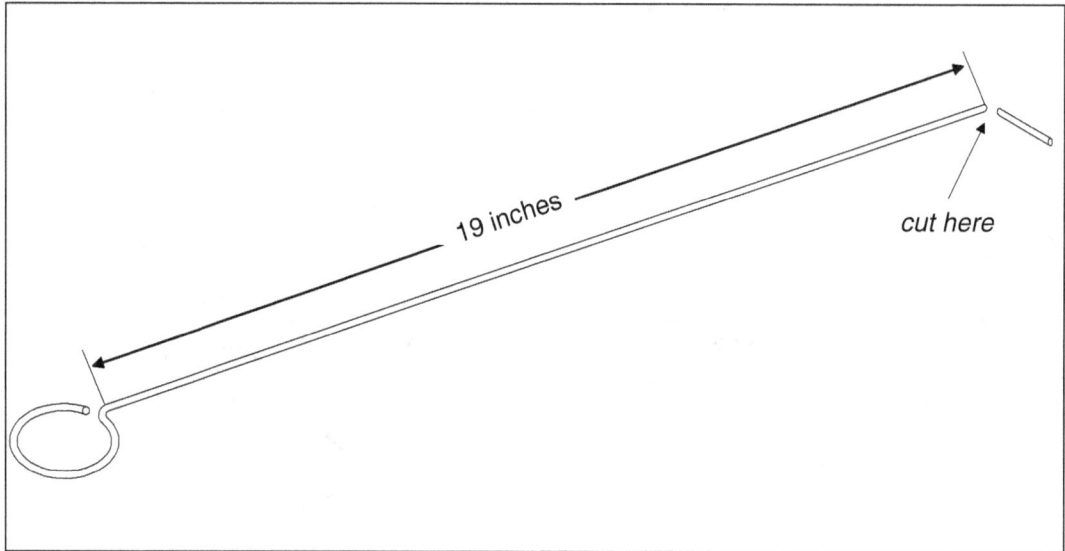

Figure 2.7   Trimming the radial rod to its proper length.

Next, bend the brass rods to a 45 degree angle (see Figure 2.8). Bend the rods at the point 1-1/4 inches away from the center of the eyehook form. The direction of the bend must be perpendicular to the plane of the eyehook end.

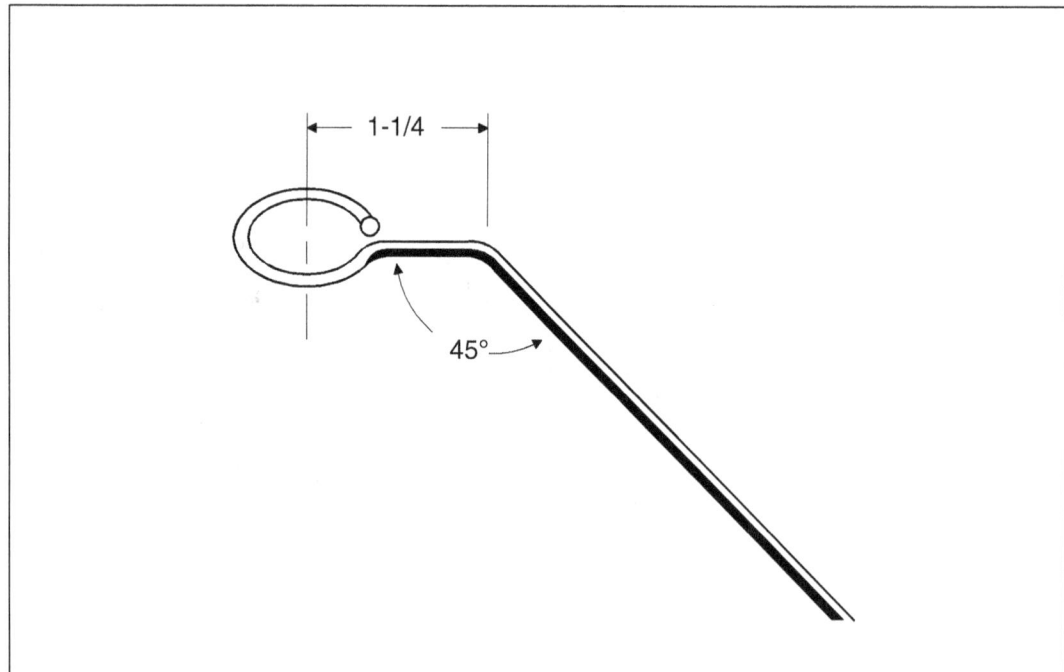

Figure 2.8   Bending a brass radial rod.

Solder the RG-58/U coaxial cable to the remaining PL-259 following the illustrated steps (see Figures 2.9 and Figure 2.10).

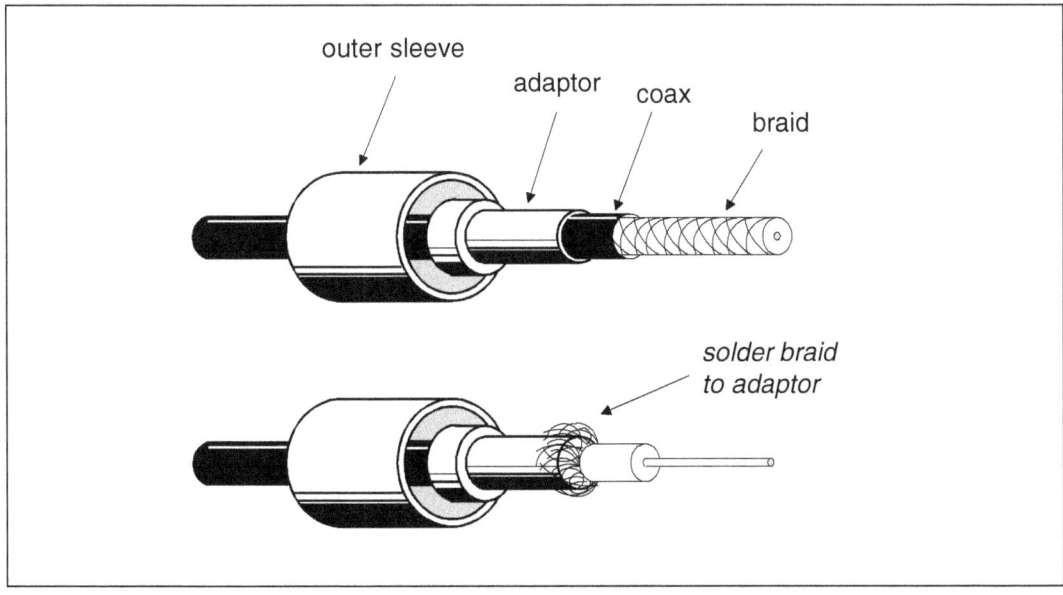

Figure 2.9   Connecting the coaxial cable to the PL-259.

IMPORTANT:
Check the coaxial cable for a possible short after soldering it to the PL-259 connector.

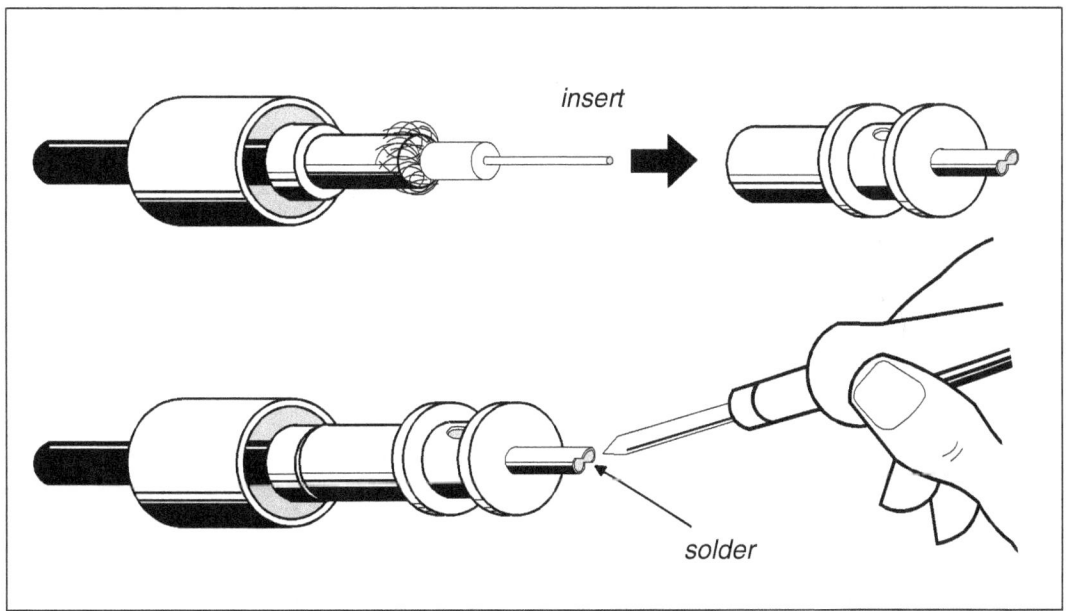

Figure 2.10   Assembling the PL-259.

The mounting bracket for the Model FQ-2 is similar to that used for the Model FA-2. The only difference between the two is that the eight small holes around the 5/8" size hole are absent in the bracket for the Model FQ-2 (see Figure 2.11).

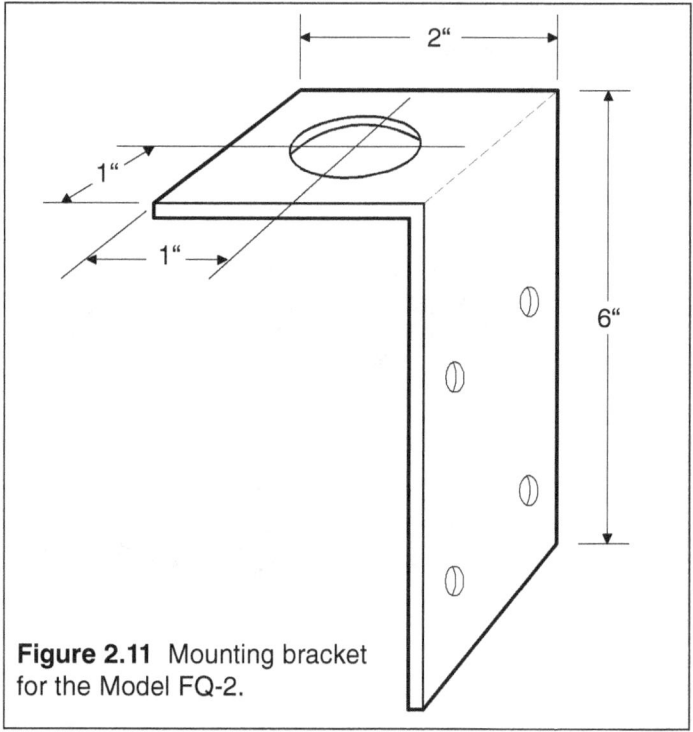

Figure 2.11 Mounting bracket for the Model FQ-2.

PL-259

straight connector

5/8"∅ washer

Assembly

Attach the radiator element into the straight connector. Next, attach the plain washer (5/8" diameter) into the straight connector (see Figure 2.12).

Figure 2.12 Assembling the radiator portion.

Insert the straight connector into the eyehook ends of the brass radial elements. The other ends of the elements must be sloping downwards (see Figure 2.13).

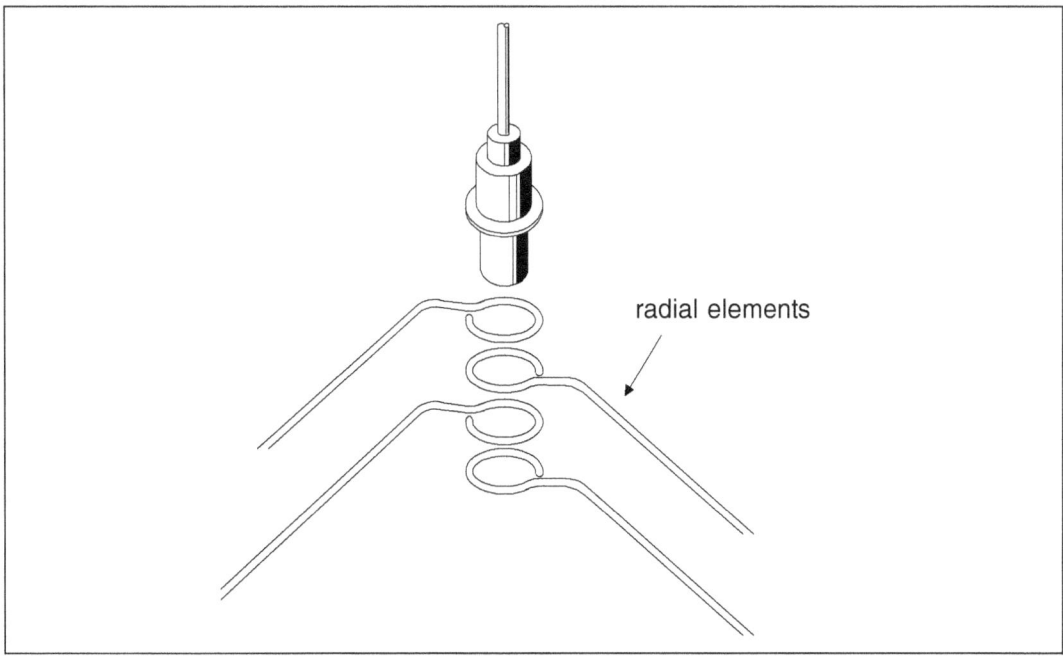

radial elements

Figure 2.13   Assembling the radial elements into the antenna base portion.

Insert the remaining portion of the straight connector into the mounting bracket you made earlier, sandwiching the radial elements between the bracket and the plain washer (see Figure 2.14). Secure the whole assembly by connecting the other PL-259 connector into the protruding part of the straight connector.

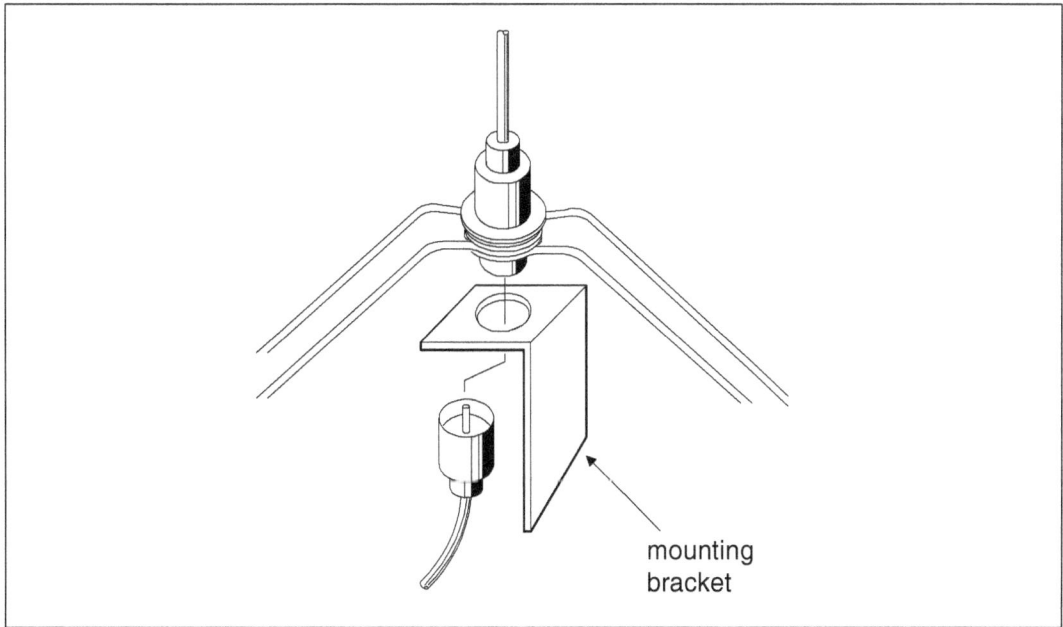

mounting bracket

Figure 2.14   Assembling the antenna into the mounting bracket.

INSTALLATION

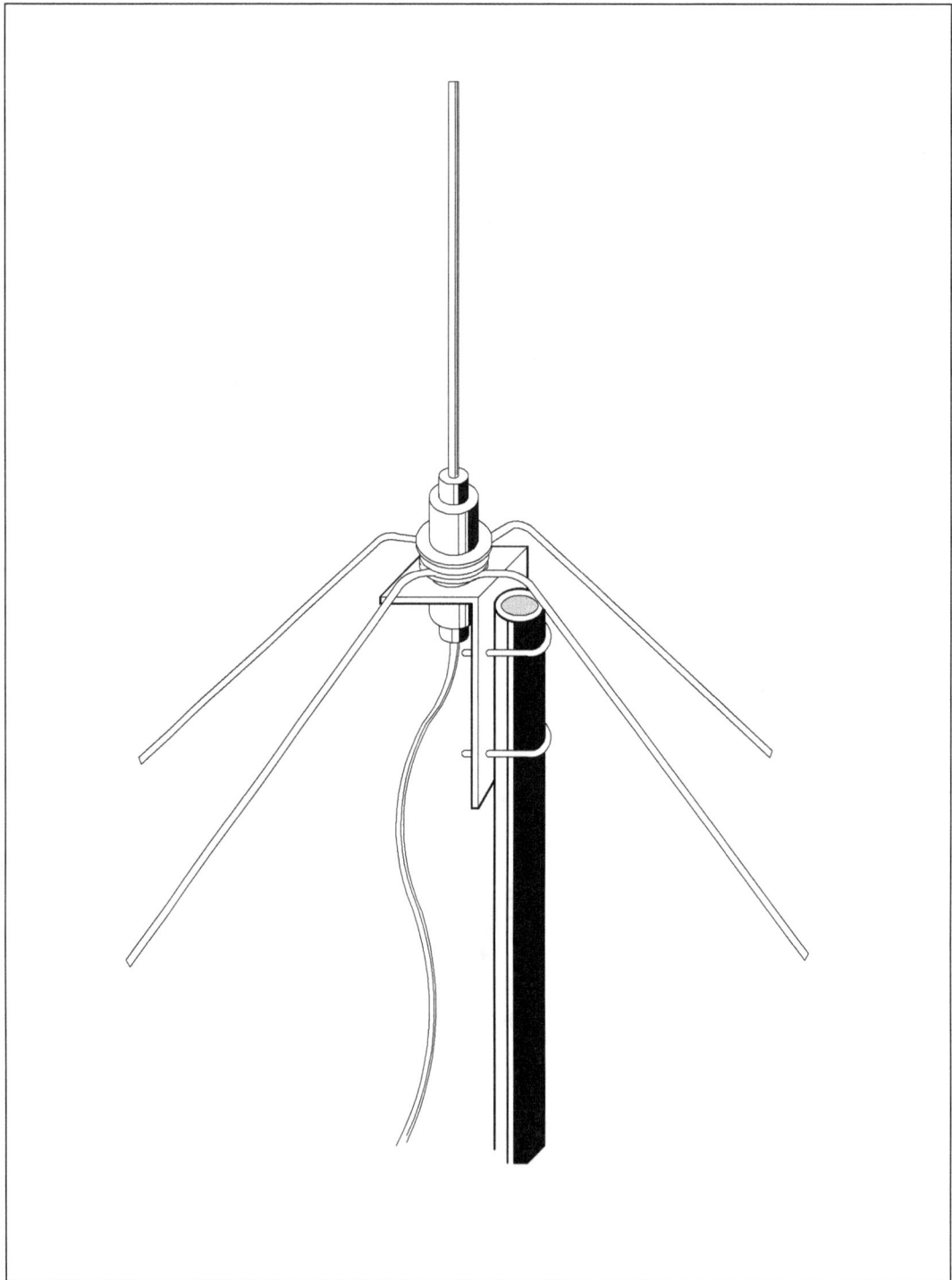

Figure 2.16   Mounting the ground plane antenna Model FQ-2.

Spread the radial elements around with equal spaces between them, and tighten the PL-259 to fix the assemby firmly (see Figure 2.17).

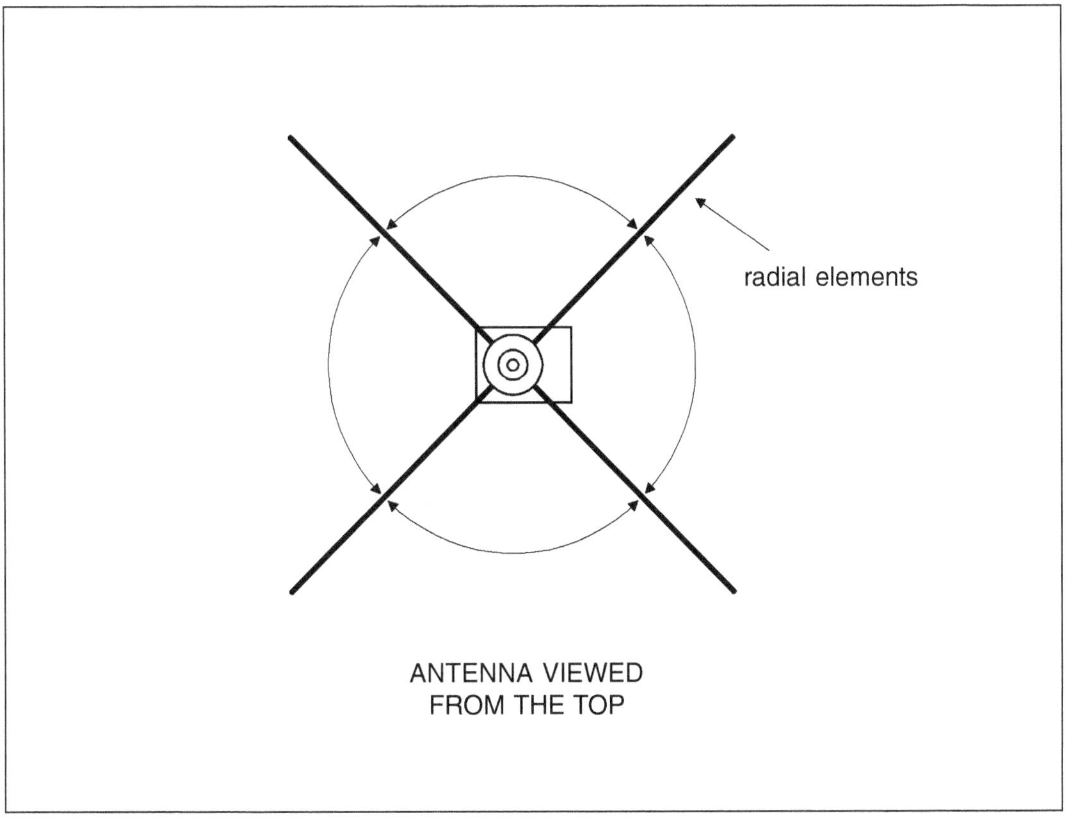

radial elements

ANTENNA VIEWED
FROM THE TOP

Figure 2.17   Spreading the radials.

REVIEW QUESTIONS

1. What are the advantages of using a ground plane antenna with detachable elements?

2. What is the function of the vinyl insulator that is inserted between the radiator rod and the PL-259?

3. Why must the measurement of radiator length start from the point where it emerges from the PL-259, and not from the end of the center pin?

4. Why was the eyehook form of the radials' ends used?

3 GROUND PLANE ANTENNA

Model FC- 2

The antenna is a vital link in the chain of radio communications and numerous designs have come off the drafting boards in a never-ending search for improved performance. Experience shows that one major factor influencing the overall design of antennas is the particulars of the situation where it will be used. For example, the situation around fixed installations allows the antenna to be constructed with durable and heavy materials to make it mechanically strong.

High power gain can also be easily attained by stacking a number of identical antenna. However, in mobile operation, the situation drastically changes and using antennas designed primarily for fixed installations becomes impractical. Mobile operation imposes limitations on the design of an antenna regarding its weight, size, ruggedness, ease of assembly and disassembly, and power gain. The mobile operator has to choose a type of antenna which is highly portable and at the same time functionally efficient in mobile operations.

The ground plane antenna described in this chapter is another development from the FQ-2 model. It is actually the same antenna but "compacted" further to make its total size smaller and more portable when disassembled. This antenna was designed by a mobile radio operator several months after constructing his first ground plane antenna similar to Model FQ-2. Perhaps being unhappy about the bronze rods protruding out of his small knapsack, he cut each rod in half and devised an ingenious way of connecting the elements together during assembly. That is how the FC-2 antenna was developed.

The electrical characteristics of this antenna are similar to those of the Model FQ-2. It also retains the mechanical durability of the earlier full-length version. Being more compact, it has become very popular among mobile radio operators.

SCALED DOWN IMAGE

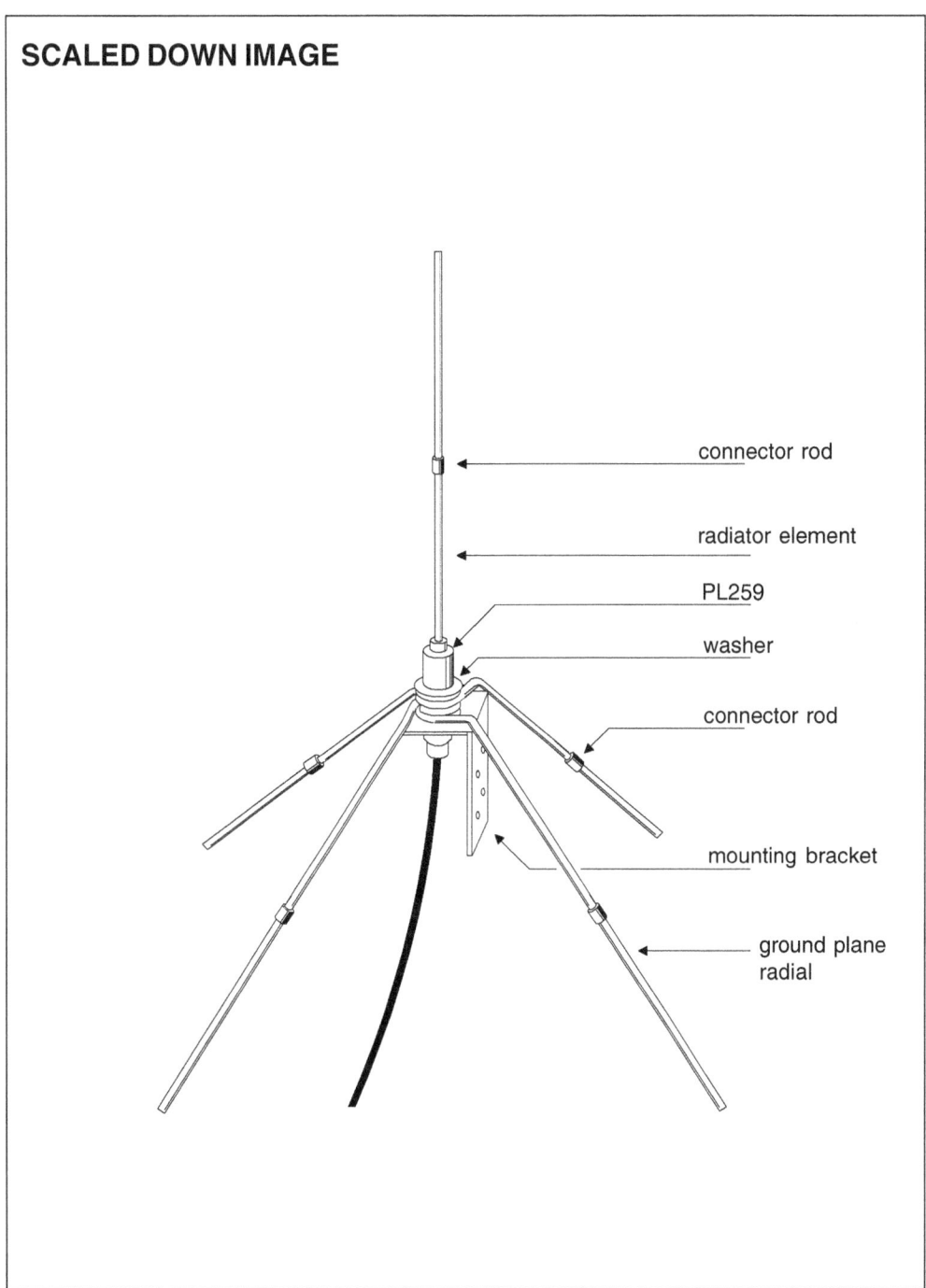

connector rod

radiator element

PL259

washer

connector rod

mounting bracket

ground plane
radial

Figure 3.1 Highly portable ground plane antenna model FC-2.

Materials needed

The materials needed for the Model FC-2 are the same as those needed for the Model FQ-2. Refer to the preceding chapter for the exact description of materials. The only difference between the two is the additional 3/16 inch diameter brass rod, which is used to interconnect the detachable elements of the Model FC-2.

The compact detachable elements of the Model FC-2 permit it to be carried inside a pack or bag for mobile operations.

Construction

Fabricate a complete Model FQ-2 antenna following the construction methods described in the preceding chapter. After you have constructed the Model FQ-2, disassemble it and cut the radiator and each radial element into two equal lengths (see Figure 3.2).

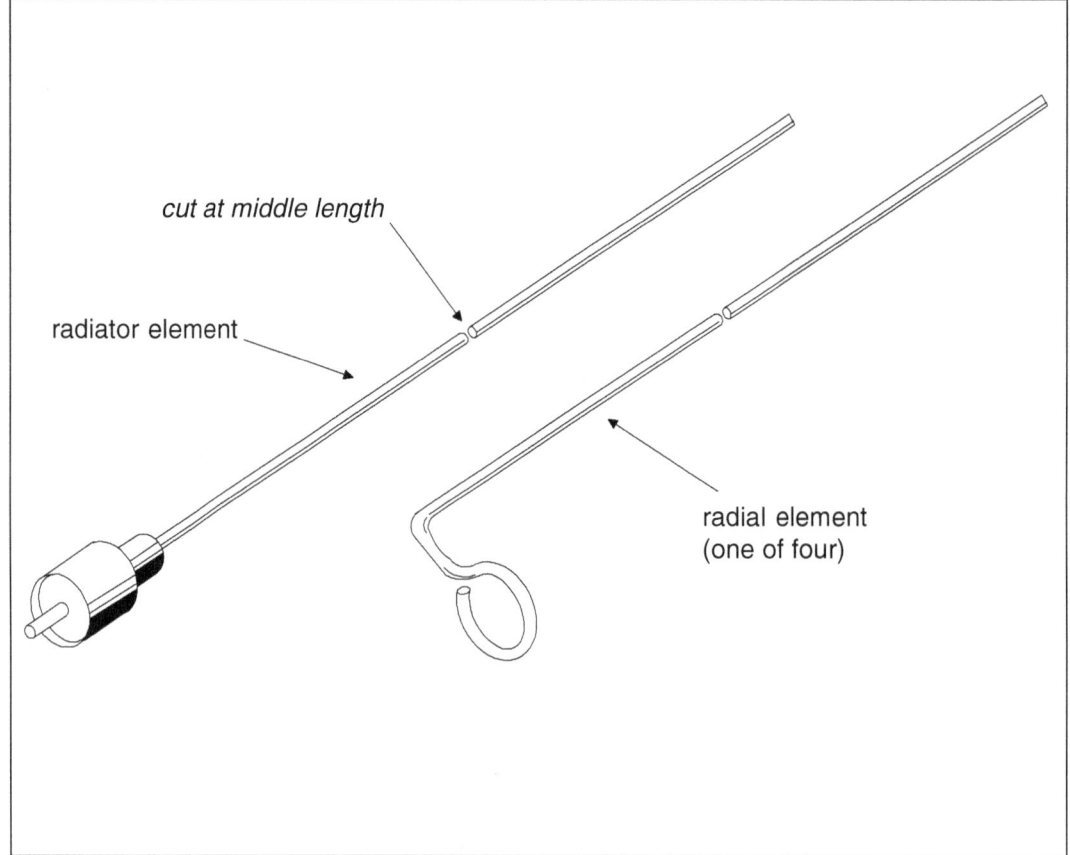

cut at middle length

radiator element

radial element
(one of four)

Figure 3.2   Cutting the elements into two equal lengths.

Next, take the 3/16" diameter brass rod and cut five 3/4 inch pieces from it. These short pieces of brass rod will be used to connect the two equal lengths of each element (see Figure 3.3).

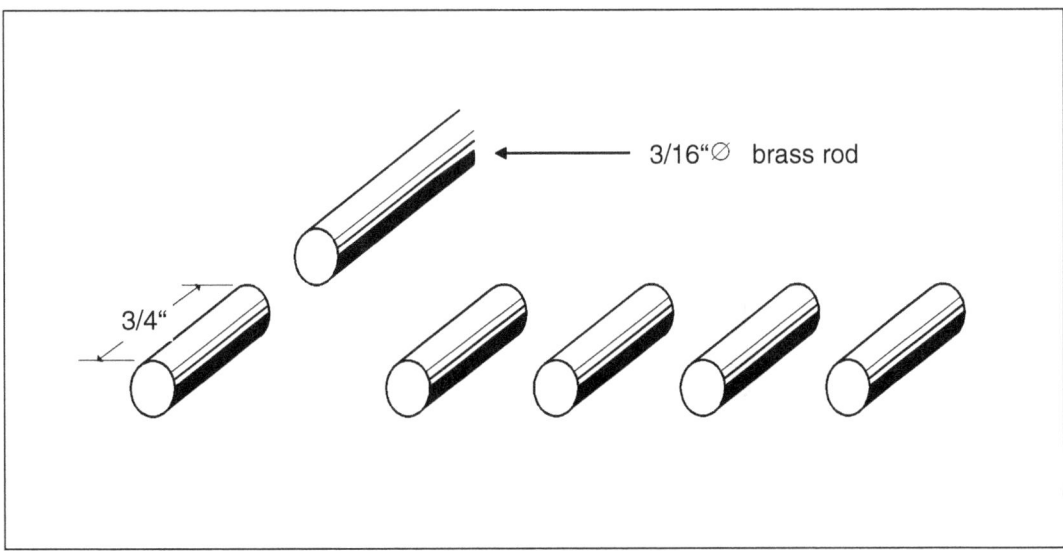

3/16"⌀ brass rod

3/4"

Figure 3.3   Preparing the connecting rods.

Drill a hole with about 1/8 inches diameter at one end of each connector rod. The hole must be about half the length of the connector rod depth (see Figure 3.4).

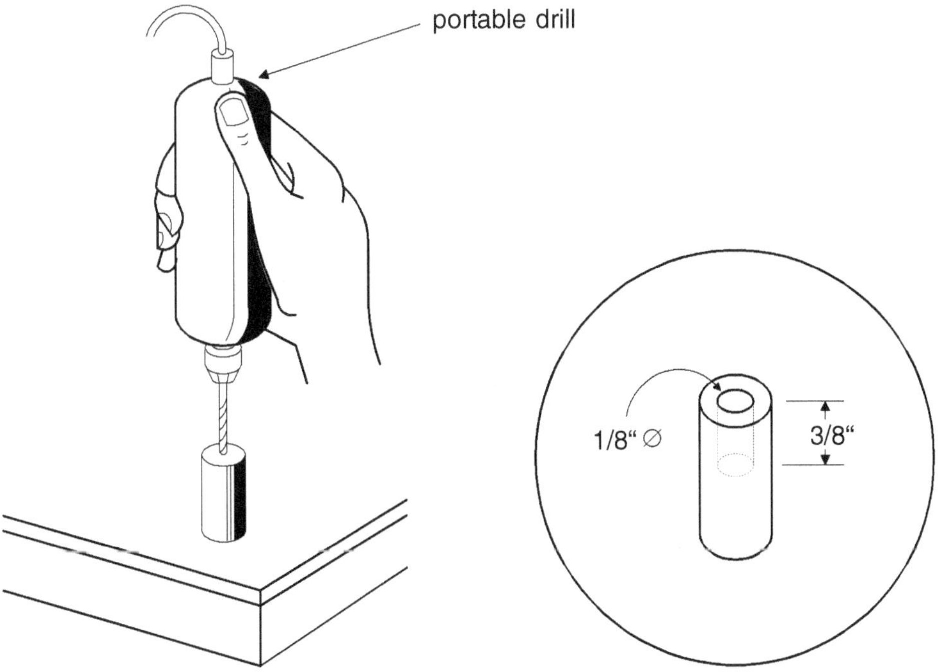

portable drill

1/8" ⌀   3/8"

Figure 3.4   Drilling holes in the connector rods.

Next, drill another hole with about 3/32 inches diameter at the other end of each connector rod, with the same depth as the first hole (see Figure 3.5). Repeat the procedure for all five connector rods.

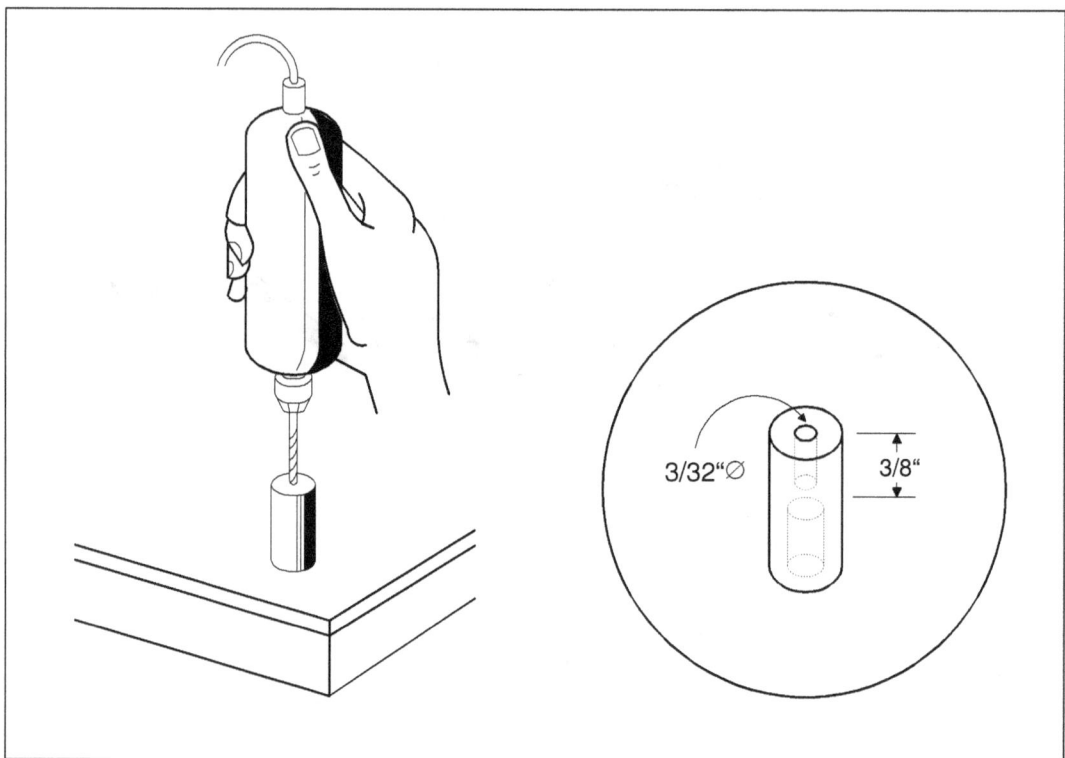

Figure 3.5   Drilling a 3/32" hole in the other end of the connector rods.

Next, secure the connector rod in a table vise, and make a thread *inside the smaller hole* (the 3/32-inch hole to be sure) with a 1/8" gauge NF hand tap (see Figure 3.6). Repeat the same procedure for the remaining connector rods.

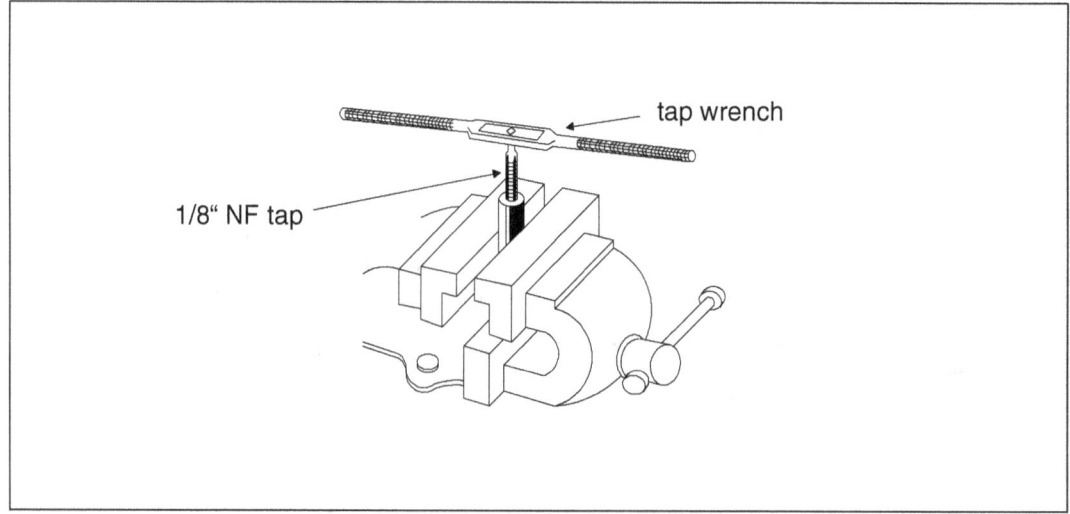

Figure 3.6   Making a thread in the smaller hole.

Insert the inner half rod of the radial element (the half part with the eyehook end) into the larger hole of the connector rod (with a 1/8" diameter hole unthreaded) and solder the two parts together. Do the same with the other radial elements (see Figure 3.7).

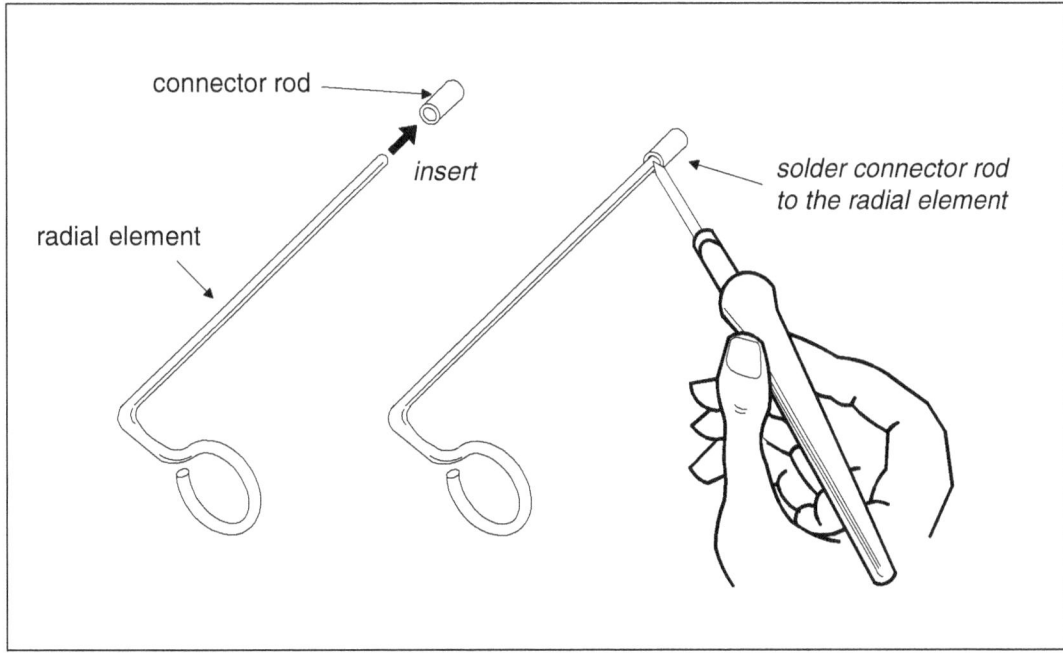

Figure 3.7   Coupling the connector rods to the radial elements.

Insert and solder the top half of the radiator element into the larger half of the remaining connector rod (see Figure 3.8). *Note: The purpose of this arrangement is to avoid the mistake of connecting the top half of the radiator to any of the radial elements.*

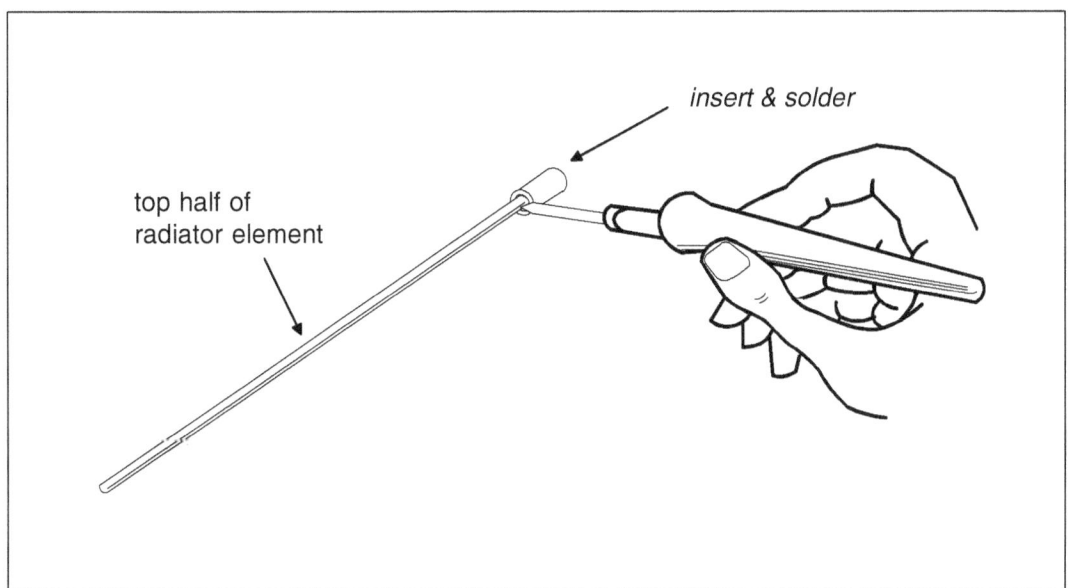

Figure 3.8   Coupling the top half of the radiator element to a connector rod.

The next step is to make a thread around one end of the outer half of the radial element. Use a manual threading die to make the thread. Secure the rod firmly in a table vise while threading. The thread must be at least 3/8 inches long. See Figure 3.9.

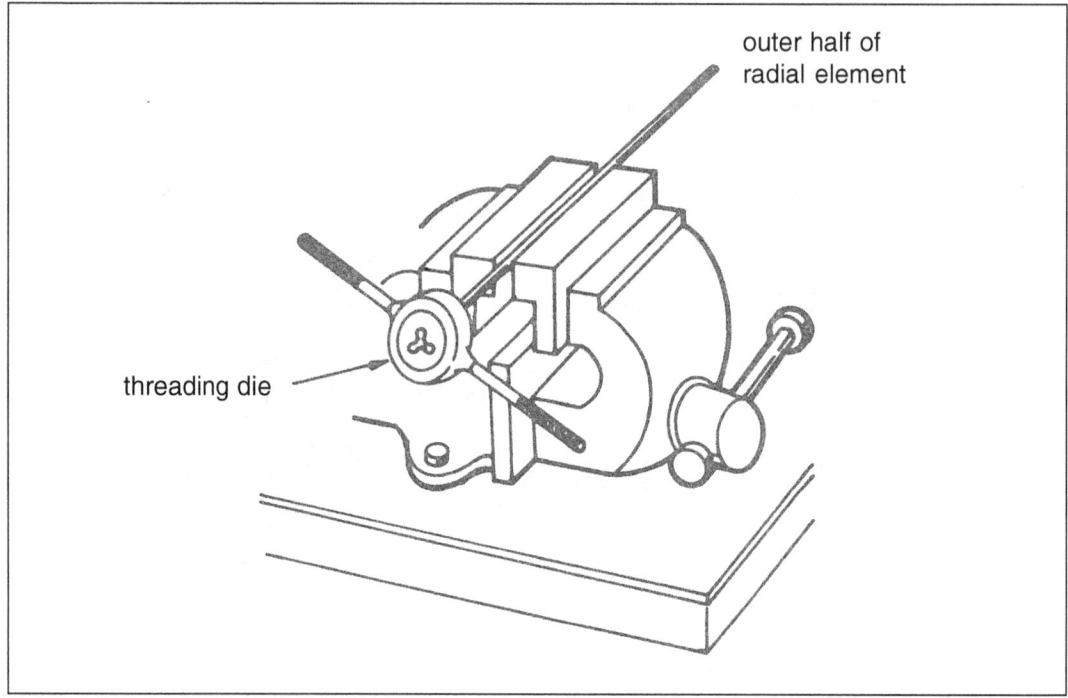

Figure 3.9   Making a thread at one end of the outer radial element.

After you have succesfully made the threads, screw each outer half into its respective connector rod (see Figure 3.10).

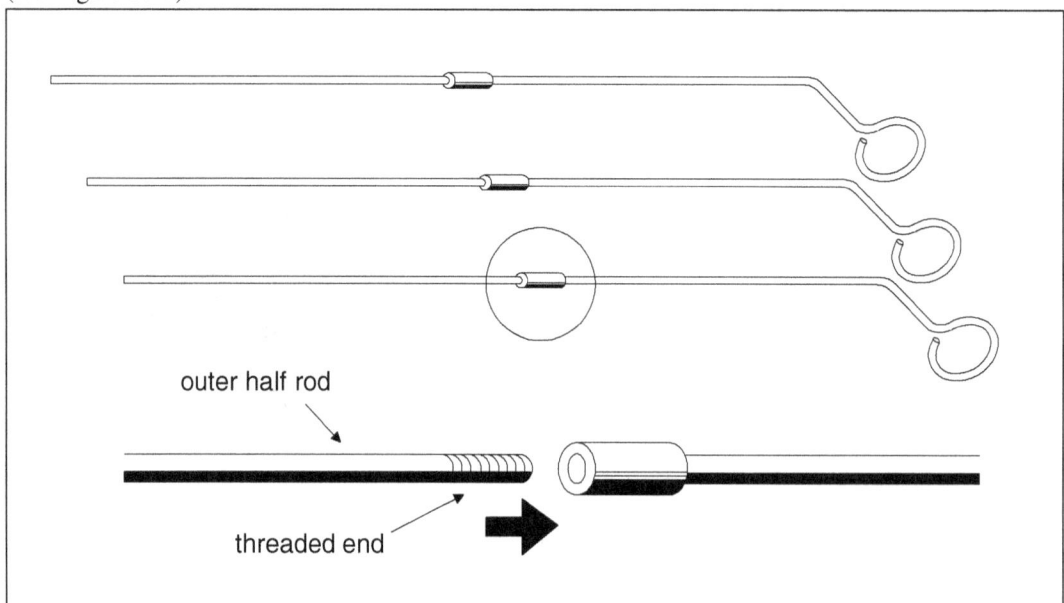

Figure 3.10   Assembling the radial elements.

Next, make a thread at the end of the lower half of the radiator element similar to what you have done to the radial elements (see Figure 3.11). Join the two halves of the radiator element together.

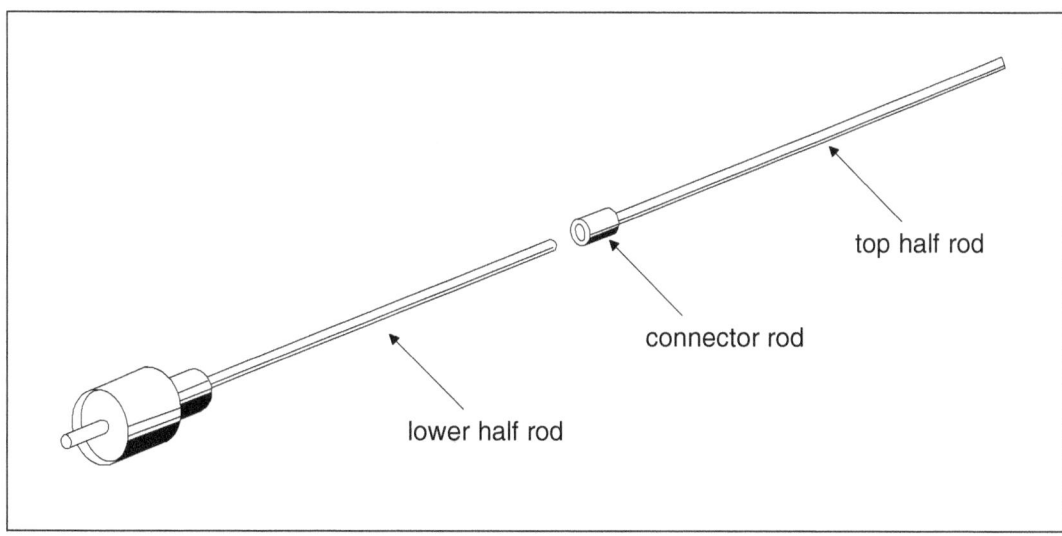

Figure 3.11   Assembling the radiator element together.

The final assembly of the Model FC-2 is similar to the Model FQ-2.

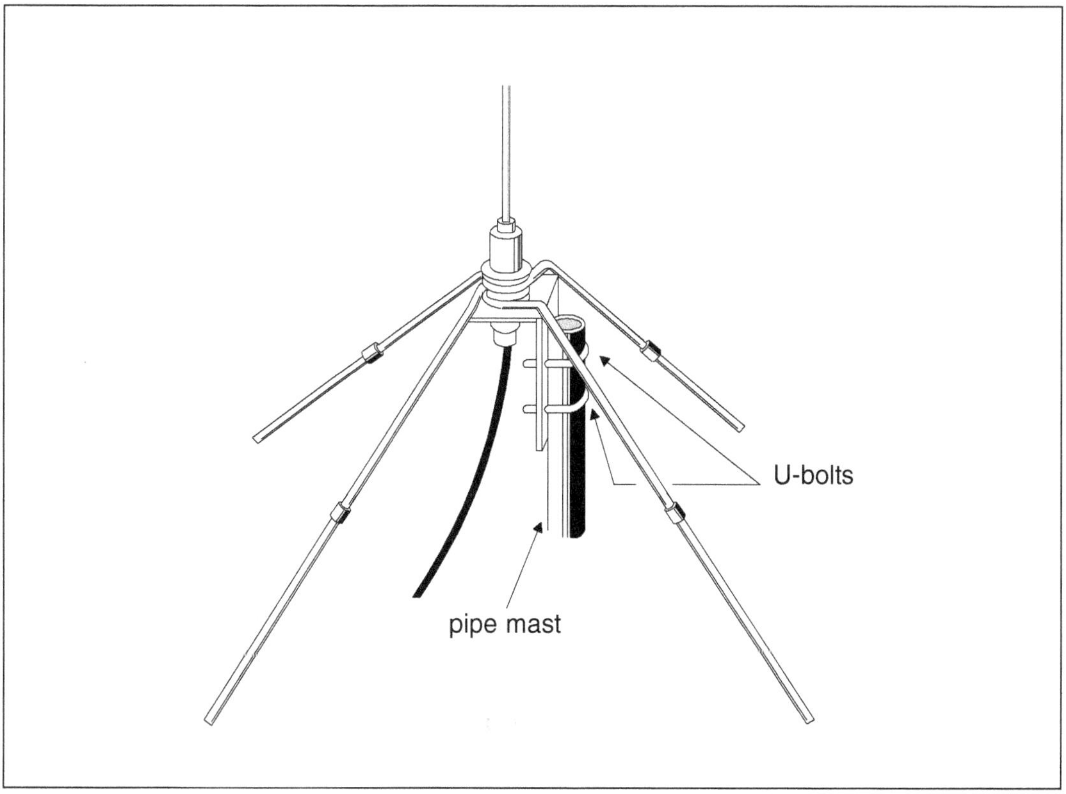

Figure 3.12   Final assembly and mounting of FC-2.

Mobile Installation

In mobile installations, the aluminum mounting bracket is not necessary and may be discarded and substituted with a 5/8" id* plain washer to hold the radial elements assembly. The antenna is then mounted by tying a rope at its base and hanging it under a tree or a makeshift post (see Figure 3.13a). An alternative method of hanging the FC-2 is to bend the tip of the radiator element into a small hook, allowing a nylon rope to be tied to this hook to hang the antenna (See Figure 3.13b).

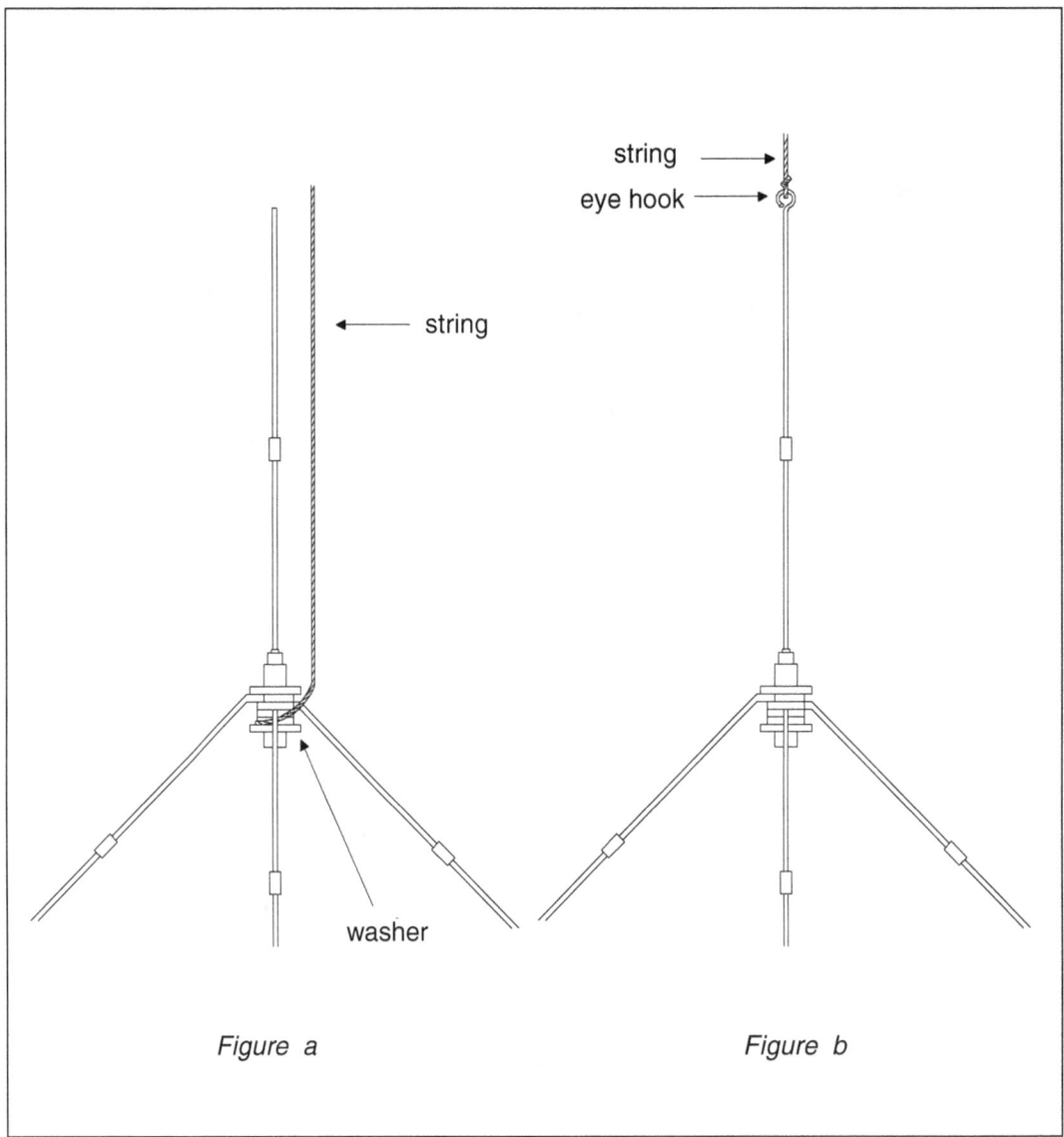

Figure a   Figure b

Figure 3.13   Mobile operation installation techniques.

REVIEW QUESTIONS

1. Why do different operating situations need different antenna designs?

2. What are the important characteristics that must be considered in designing an antenna for mobile applications?

3. Is it good to hang the antenna with a metal wire?

4. What is the reason for cutting each antenna element in half?

Table 3.1   **Conversion table - english foot to meter**

engl. foot(')	0"	1"	2"	3"	4"	5"	6"	7"	8"	9"	10"	11"
0	0.000	0.0254	0.0508	0.0762	0.1016	0.1270	0.1524	0.1778	0.2032	0.2286	0.2540	0.2794m
1' (= 12")	0.305	0.330	0.356	0.381	0.406	0.432	0.457	0.483	0.508	0.533	0.559	0.584m
2'(= 24")	0.610	0.635	0.660	0.686	0.711	0.737	0.762	0.787	0.813	0.838	0.864	0.889m
3'(= 36")	0.914	0.940	0.965	0.991	1.016	1.041	1.067	1.092	1.118	1.143	1.168	1.194m
4'(= 48")	1.219	1.245	1.270	1.295	1.321	1.346	1.372	1.397	1.422	1.448	1.473	1.499m
5'(= 60")	1.524	1.549	1.575	1.600	1.626	1.651	1.676	1.702	1.727	1.753	1.778	1.803m
6'(= 72")	1.829	1.854	1.880	1.905	1.930	1.956	1.981	2.007	2.032	2.057	2.083	2.108m
7(= 84")	2.134	2.159	2.184	2.210	2.235	2.261	2.286	2.311	2.337	2.362	2.388	2.413m
8'(= 96")	2.438	2.464	2.489	2.515	2.540	2.565	2.591	2.616	2.642	2.667	2.692	2.717m
9' (= 108")	2.743	2.769	2.794	2.819	2.845	2.870	2.896	2.921	2.946	2.972	2.997	3.023m
10' (= 120")	3.048	3.073	3.099	3.124	3150	3.175	3.200	3.226	3.251	3.277	3.302	3.327m
11' (= 132")	3.353	3.378	3.404	3.429	3.454	3.480	3.505	3.531	3.556	3.581	3.607	3.632m
12' (= 144")	3.658	3.683	3.708	3.734	3.759	3.785	3.810	3.835	3.861	3.886	3.912	3.937m
13' (= 156")	3.962	3.988	4.013	4.039	4.064	4.089	4.115	4.140	4.166	4.191	4.216	4.242m
14' (= 168")	4.267	4.293	4.318	4.343	4.369	4.394	4.420	4.445	4.470	4.496	4.521	4.547m
15'(= 180")	4.572	4.597	4.623	4.648	4.674	4.699	4.724	4.750	4.775	4.801	4.826	4.851m
16'(= 192")	4.877	4.902	4.928	4.953	4.978	5.004	5.029	5.055	5.080	5.105	5.131	5.156m
17(=204")	5.182	5.207	5.232	5.258	5.283	5.309	5.334	5.359	5.385	5.410	5.436	5.461m
18'(=216")	5.486	5.512	5.537	5.563	5.588	5.613	5.639	5.664	5.690	5.715.	5.740	5.766m
19' (= 228")	5.791	5.817	5.842	5.867	5.893	5.918	5.944	5.969	5.994	6.020	6.045	6.071m
20' (= 240")	6.096	6.121	6.147	6.172	6198	6.223	6.248	6.274	6.299	6.325	6.350	6.375m
21' (= 252")	6.401	6.426	6.452	6.477	6.502	6.528	6.553	6.579	6.604	6.629	6.655	6.680m
22' (= 264")	6.706	6.731	6.756	6.782	6.807	6.833	6.858	6.883	6.909	6.934	6.960	6.985m
23' (= 276")	7.010	7.036	7.061	7.087	7.112	7.137	7.163	7.188	7.214	7.239	7.264	7.290m
24' (= 288")	7.315.	7.341	7.366	7.391	7.417	7.442	7.468	7.493	7.518	7.544	7.569	7.595m
25; (= 300")	7.620	7.645	7.671	7.696	7.722	7.747	7.722	7.798	7.823	7.849	7.874	7.899m
26' (= 312")	7.925	7.950	7.976	8.001	8.026	8.052	8.077	8.103	8.128	8.153	8.179	8.204m
27 (= 324")	8.230	8.255	8.280	8.306	8.331	8.357	8.382	8.407	8.433	8.458	8.484	8.509m
28' (= 336")	8.534	8.560	8.585	8.611	8.636	8.661	8.687	8.712	8.738	8.763	8.788	8.814m
29' (= 348")	8.839	8.865	8.890	8.915	8.941	8.966	8.992	9.017	9.042	9.068	9.093	9.119m
30' (= 360")	9.144	9.169	9.195	9.220	9.246	9.271	9.296	9.322	9.347	9.373	9.398	9.423m

1' = 0.3048 m;   1" = 0.0254 m;   1' = 12"

4 J-FED HALF WAVE ANTENNA

Model JF-2

This antenna is specifically designed to satisfy the need for a simple but effective vertical antenna that does not require any grounding system. It is one version of a monopole antenna that carries its 'ground' along with it. The unit is composed of a half wave radiating element and a quarter wave length matching section. The combination of these two elements provides the transformer action that matches the impedance.

Although it is actually a quarter wave antenna, its radiation pattern and characteristics are very similar to those of a half wave vertical antenna. It also exhibits a slight gain compared to a quarter wave ground plane antenna.

This antenna radiates its signal in an omnidirectional pattern like most vertical antennas do. Its operational bandwidth is 140 - 150 MHz, and exhibits an SWR response of less than 1.5:1 over the entire band.

The unit described in this chapter is designed for fixed installation. If you intend to use it for mobile operation, then it would be better if you were to modify the design to adapt it to the rugged environment it may encounter. Aluminum tubes in general are thin and soft, and thus will easily crack if handled roughly; so you must substitute it with brass, bronze, or copper tubing. These materials are more expensive, but they are more durable. They are also more resistant to corrosion.

The elements must be cut in two or three sections, and some means must be provided to join the pieces of tubing together in assembly (similar to the FC-2 technique). You must also devise a method of mounting the antenna in a much simpler fashion than the one described here. Hanging the antenna under a tree or post will do, but there might be some other way that you can think of. One word of caution though: Never use any metallic material to mount the antenna. All points in the antenna element are electrically active, so it must be insulated from ground.

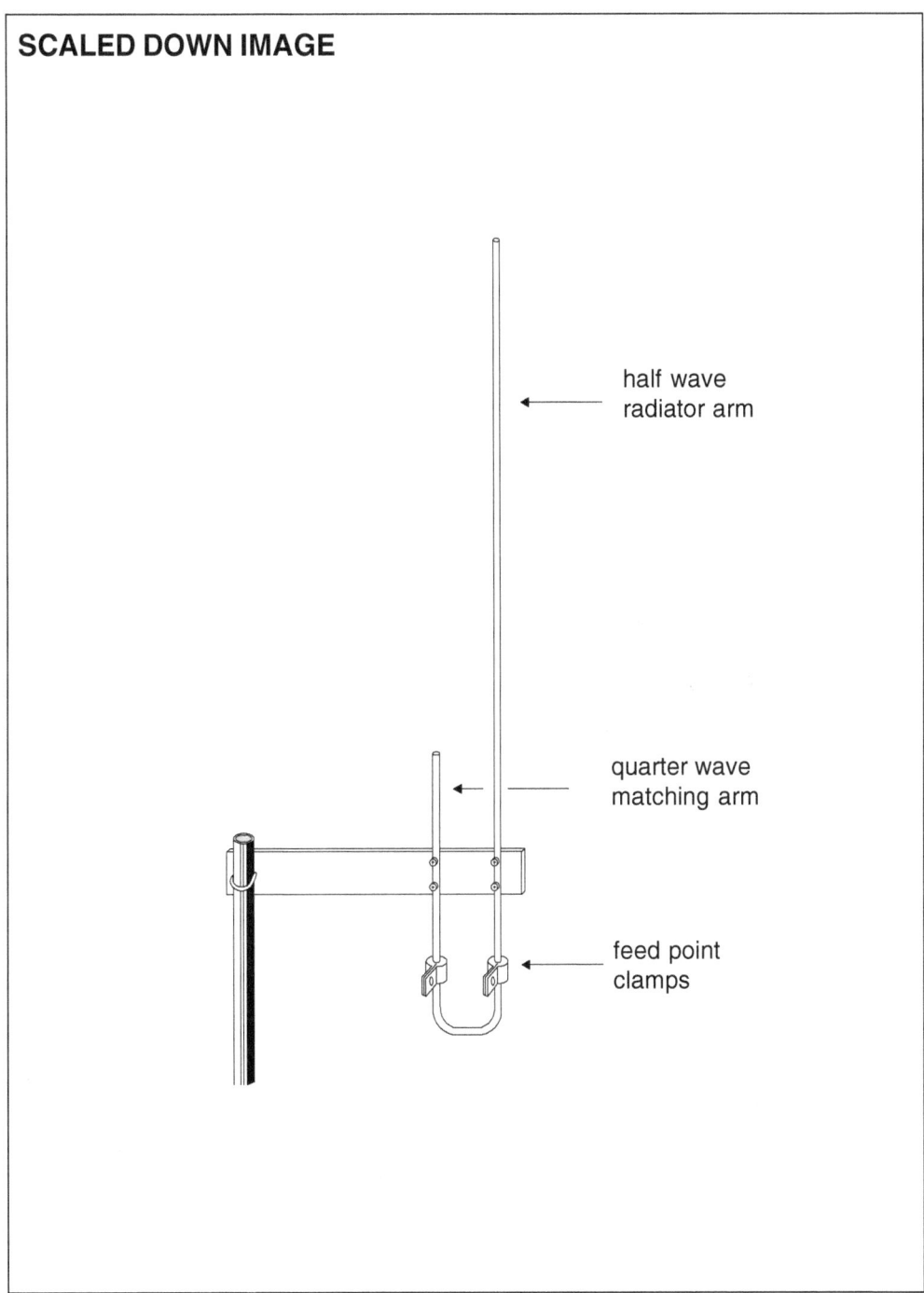

SCALED DOWN IMAGE

half wave
radiator arm

quarter wave
matching arm

feed point
clamps

Figure 4.1   J-fed monopole half wave antenna Model JF-2

Materials List

Quantity	Specification/Description	Dimensions
1	Aluminum or Brass tube 3/8" od*	1" long
2	Aluminum strips - see text to make a strip out of a short length of aluminum tube	1/2" x 1-1/2"
1	Plastic plate 1/2" thick see text for details	3" x 12"
1	U-bolt with accompanying hex nuts and lockwashers	
4	Stove bolts - brass or GI with 1/8" x 1" accompanying hex nuts and lock washers	
2	Stove bolts - brass or GI	1/8" x 3/8"
2	Eye terminals - vinyl insulated	
4	Plain washers - 1/8" id**	
1	Hose clamp - enough to hold 1" diameter tube	
Miscellaneous:	Epoxy glue	

* od- outside diameter

** id- inside diameter

Construction

Cut the tube to a length of 81 inches using a suitable tube cutter. Next, starting from one end, measure about 55 inches. Starting at this point, bend the tube to a U-shape. The two 'arms' of the bent tube must be spaced 2-1/8" apart from each other (see Figure 4.2 on the next page).

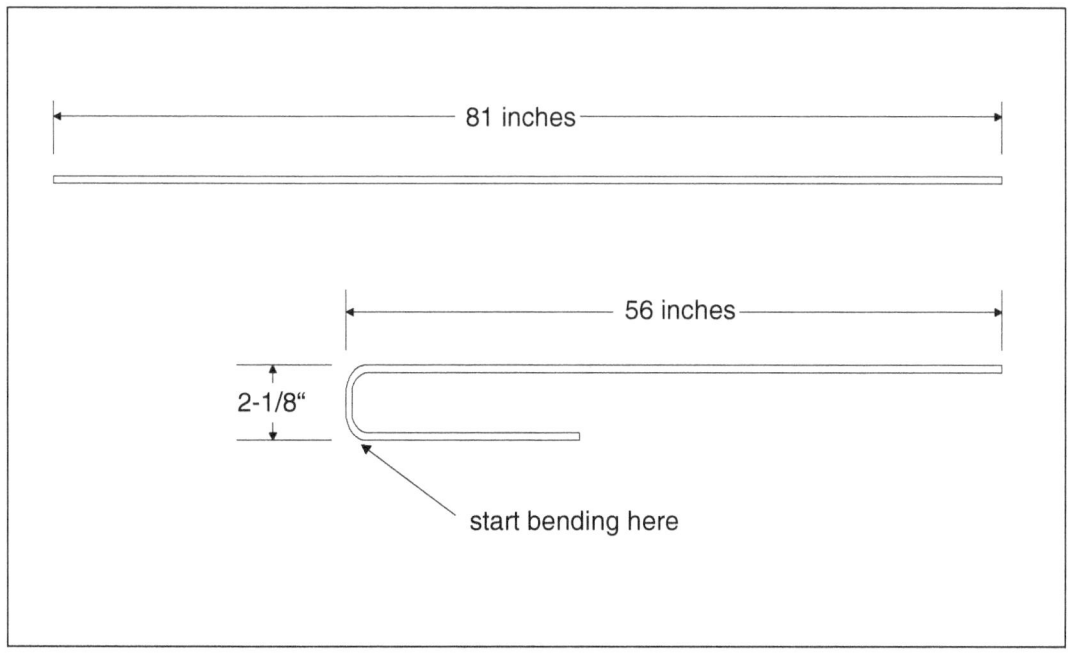

Figure 4.2 Bending one end of the tube. Preparing the end of a brass rod, and soldering it to PL-259.

Trim each arm of the tube to its proper length, measuring from the extreme edge of the bend (see Figure 4.3). This method is employed to give an allowance for possible errors in bending the tube.

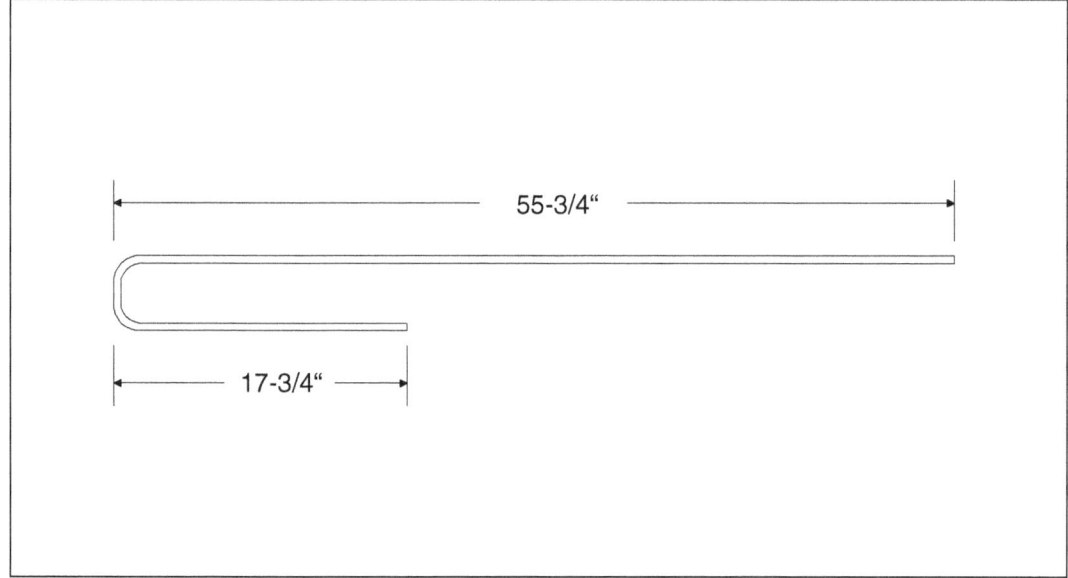

Figure 4.3 Trimming the tube to its exact length.

Drill four holes near the bend of the tube (see Figure 4.4). Each hole must be 1/8" in diameter.

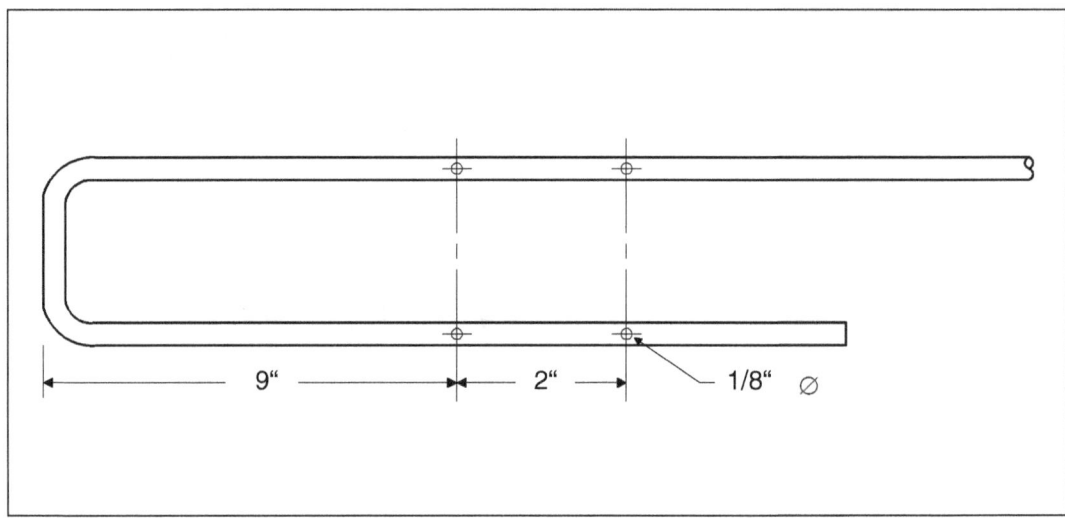

Figure 4.4   Drilling holes in the tube.

After drilling the holes, seal off both ends of the tube with an epoxy glue to prevent rainwater from seeping inside (see Figure 4.5). First, insert a substantial volume of cotton inside, to act as a stopper for the epoxy. Then, follow it up with epoxy glue, leveling it to the edge of the tube. Let the epoxy set and dry before proceeding.

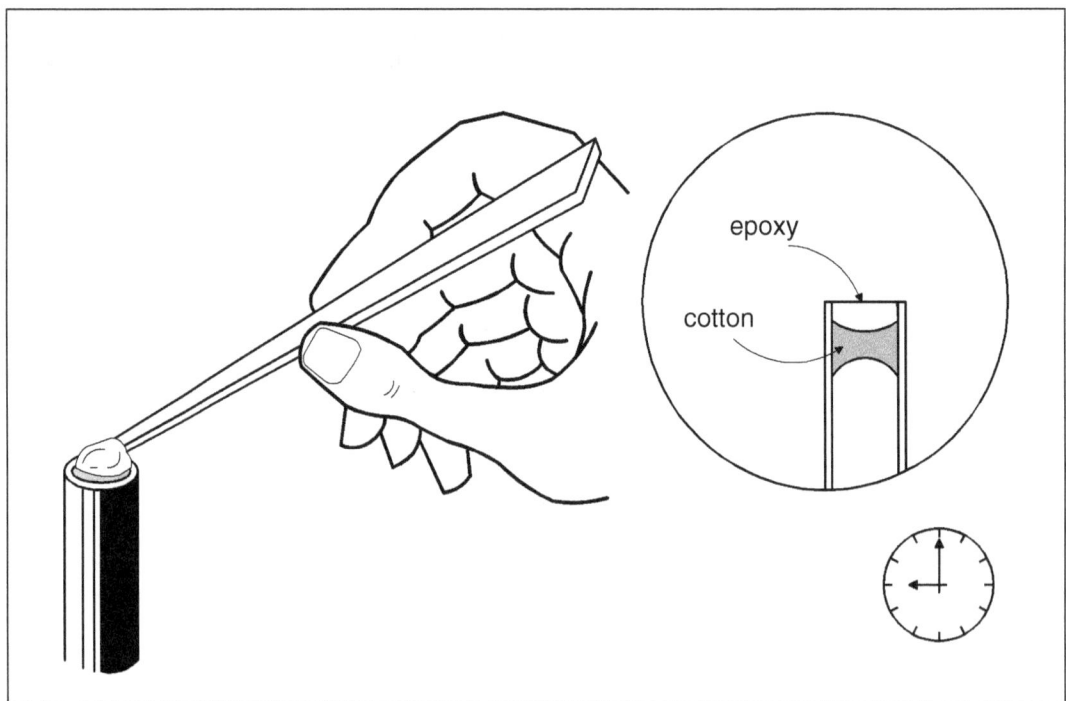

Figure 4.5   Sealing off the open ends of the tube with epoxy glue.

While you are waiting for the epoxy glue to dry, prepare the plastic plate for the antenna mount. Drill holes in the plastic plate following the dimensions shown in Figure 4.6. The larger hole (3/16" diameter) is intended for the U-bolts, so their dimensions must coincide with the actual U-bolt used.

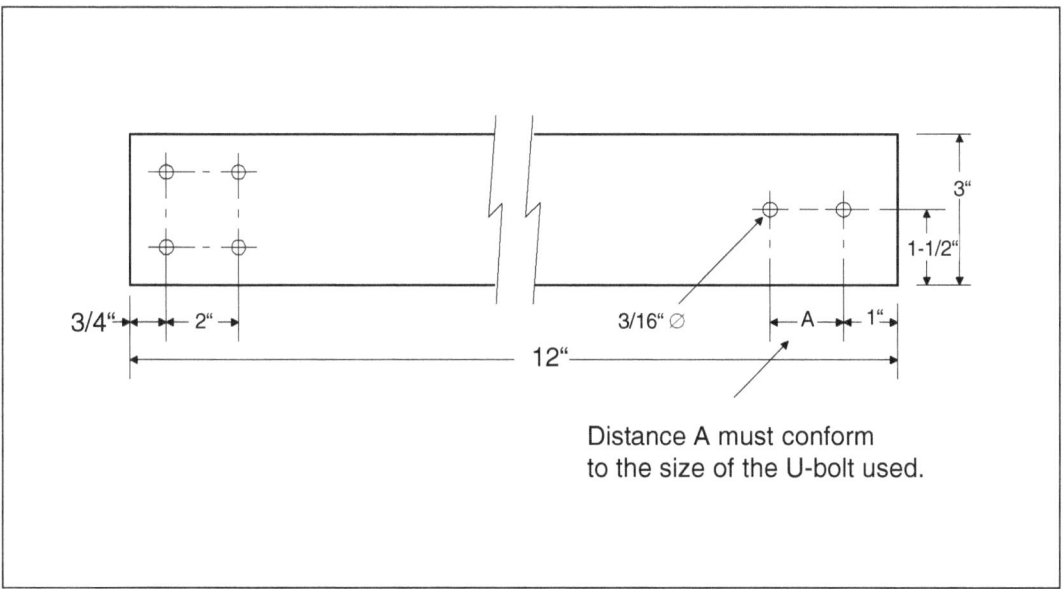

Figure 4.6   Preparing the plastic mounting plate.

Fabricate a metal strip out of a short length of aluminum tube (about 5 inches long) by pressing it in a table vise until the tube is flattened (see Figure 4.7).

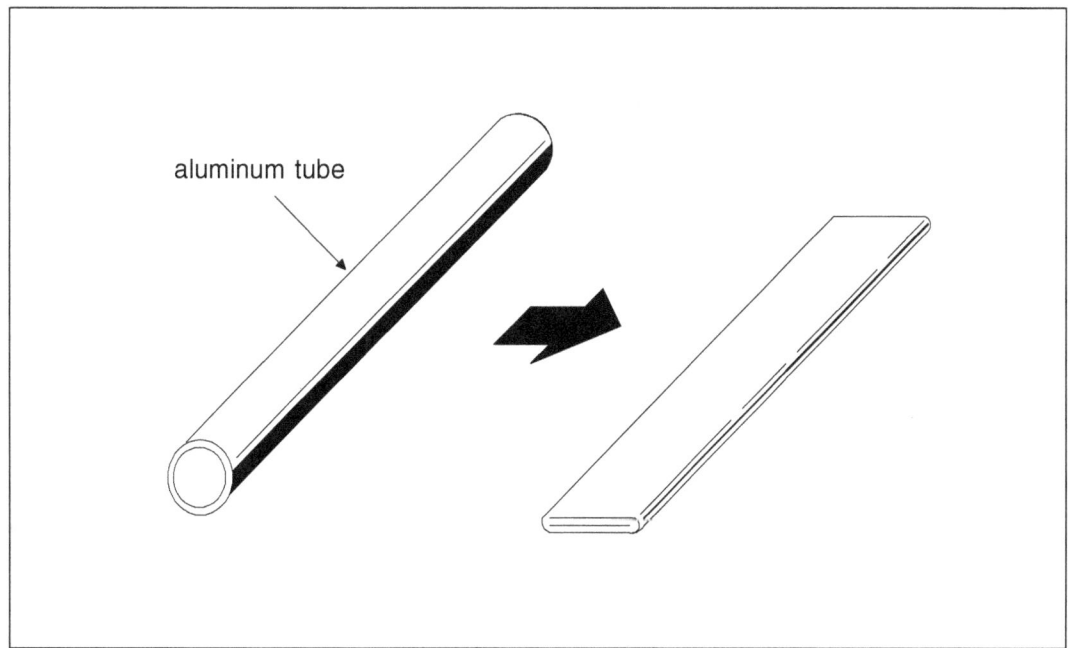

Figure 4.7   Fabricating a metal strip out of a short aluminum tube.

Out of this strip cut two short pieces (about 1-1/2" long). Bend the two strips into a form of a clamp to fit tightly around the antenna tubing (see Figure 4.8). These clamps serve as the feed point terminals of the antenna.

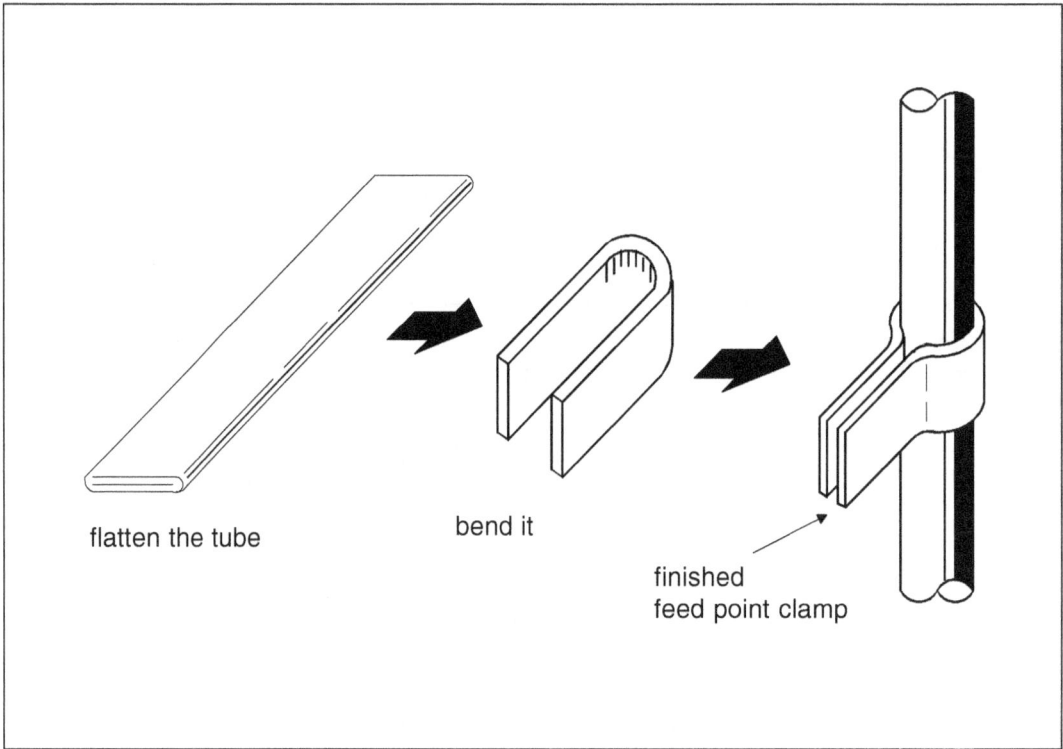

flatten the tube

bend it

finished
feed point clamp

Figure 4.8 Fabricating the clamps.

Next, drill a hole about 1/8" diameter through the flattened end of each feed point clamp (see Figure 4.9).

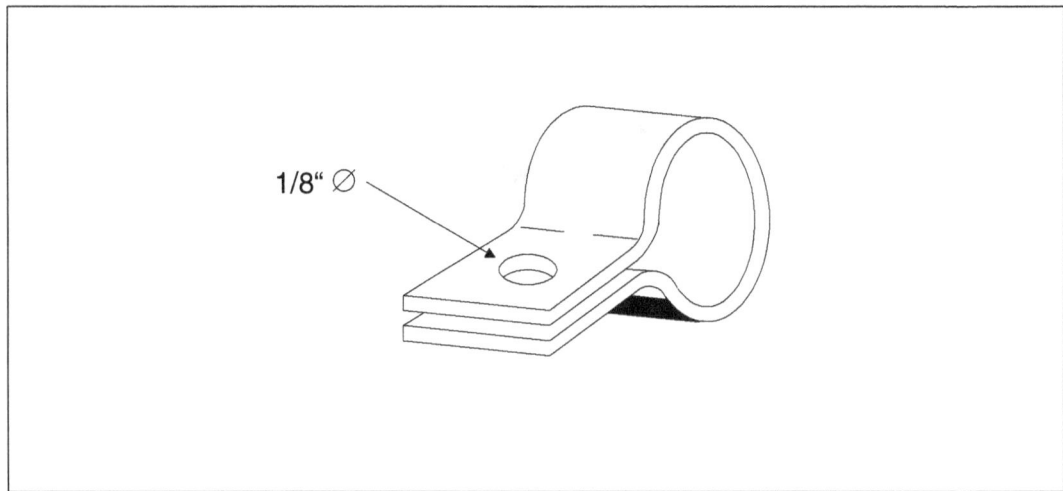

1/8" ⌀

Figure 4.9 A feedpoint clamp with a drilled 1/8" diameter hole.

Assembly

First, attach the J-shaped tube to the plastic mounting plate with 1/8" x 3/4" stove bolts made of corrosion-proof materials such as brass or stainless steel. Do not forget to include the necessary lock washers in the attachment (see Figure 4.10). Be careful in tightening the nut, because the tube is hollow inside, and it might collapse damaging the tube. Apply torque to the nuts just enough to hold the tube rigid.

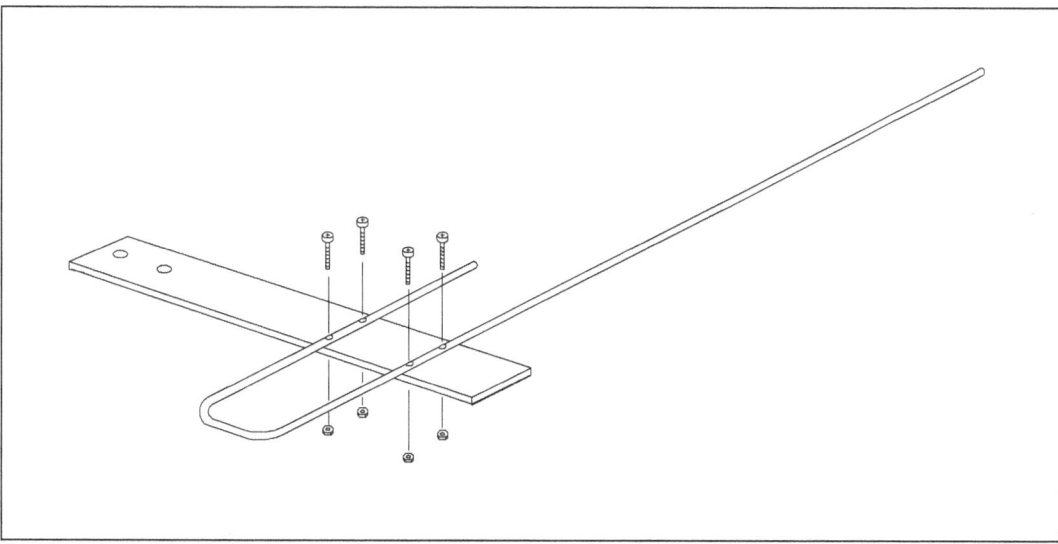

Figure 4.10   Securing the J-shaped tube to the mounting plate.

Attach the feed point clamps into both arms of the tube. Attach one clamp on the shorter arm of the tube and attach the other clamp on the longer arm (see Figure 4.11).

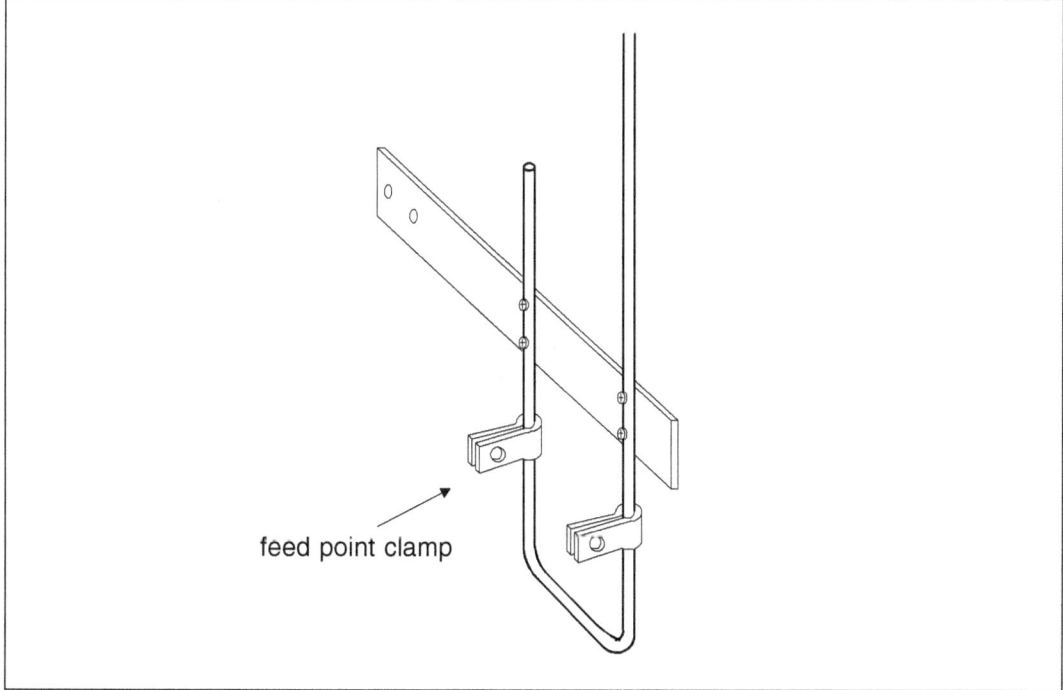

feed point clamp

Figure 4.11   Feed point clamps attached to the antenna.

Attach a plain washer and an eye terminal into a stove bolt (1/8" x 3/16"), and then insert the bolt into the hole in the feed point clamp, sandwiching the eye terminal inbetween (see Figure 4.12). Place a lock washer and a hex nut at the other end of the bolt, and then tighten the clamp lightly. Repeat the same procedure for the other clamp.

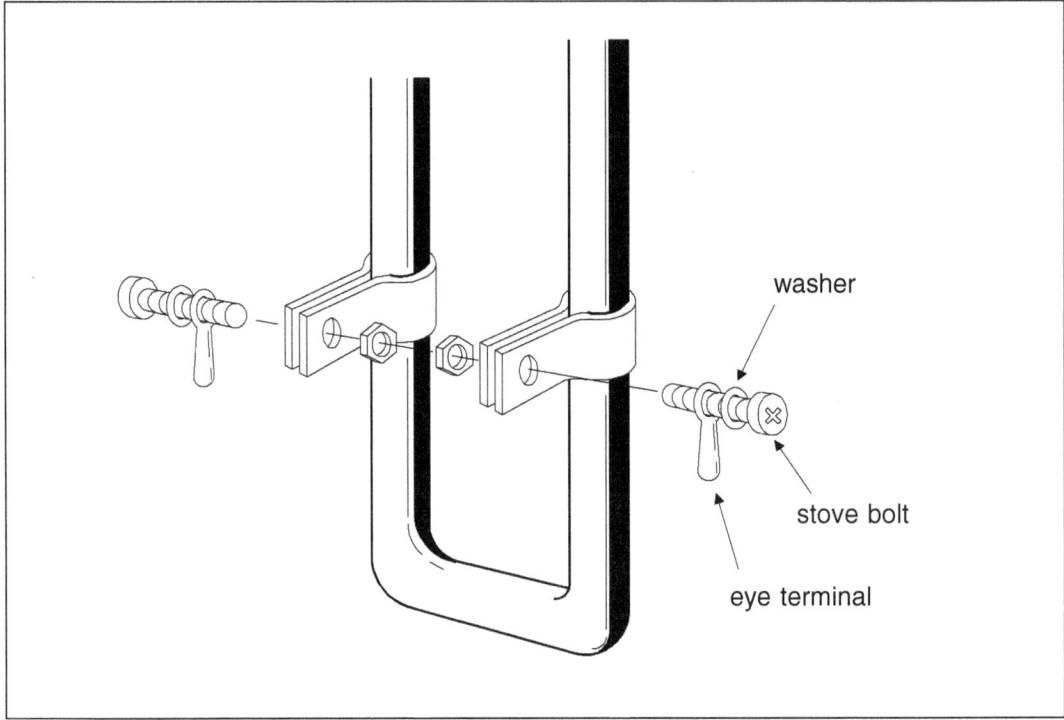

Figure 4.12   Assembling the feed point terminals.

The next step is to connect the coaxial cable to the feed point terminals. Prepare one end of the coaxial cable by separating the braid/shield from the inner conductor (see Figure 4.13).

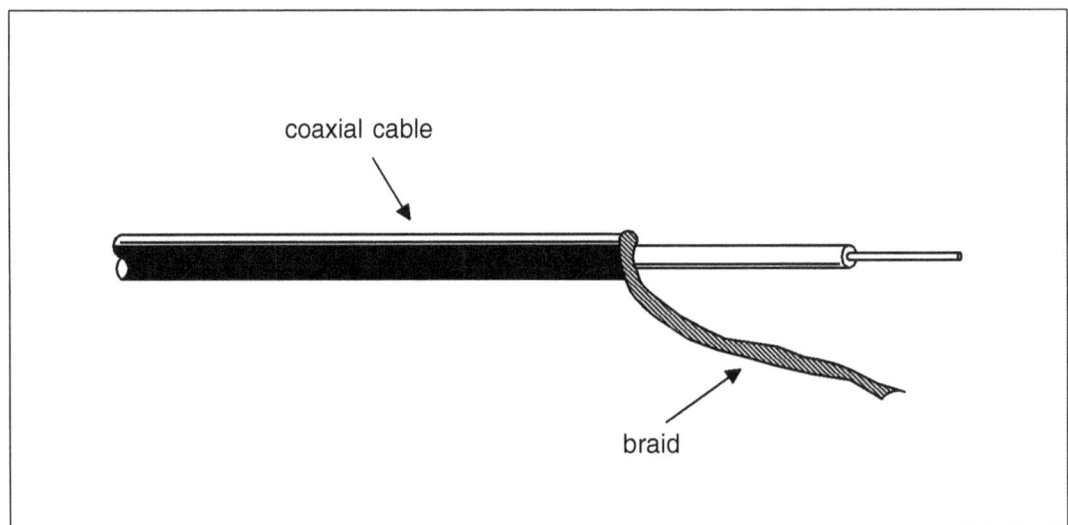

Figure 4.13   Making a pig tail.

Insert and solder the two conductors (braid and inner conductor) to the eye terminals attached in the feed point clamps. The braid must be connected to the shorter arm of the tube and the inner conductor must be connected to the longer arm (see Figure 4.14).

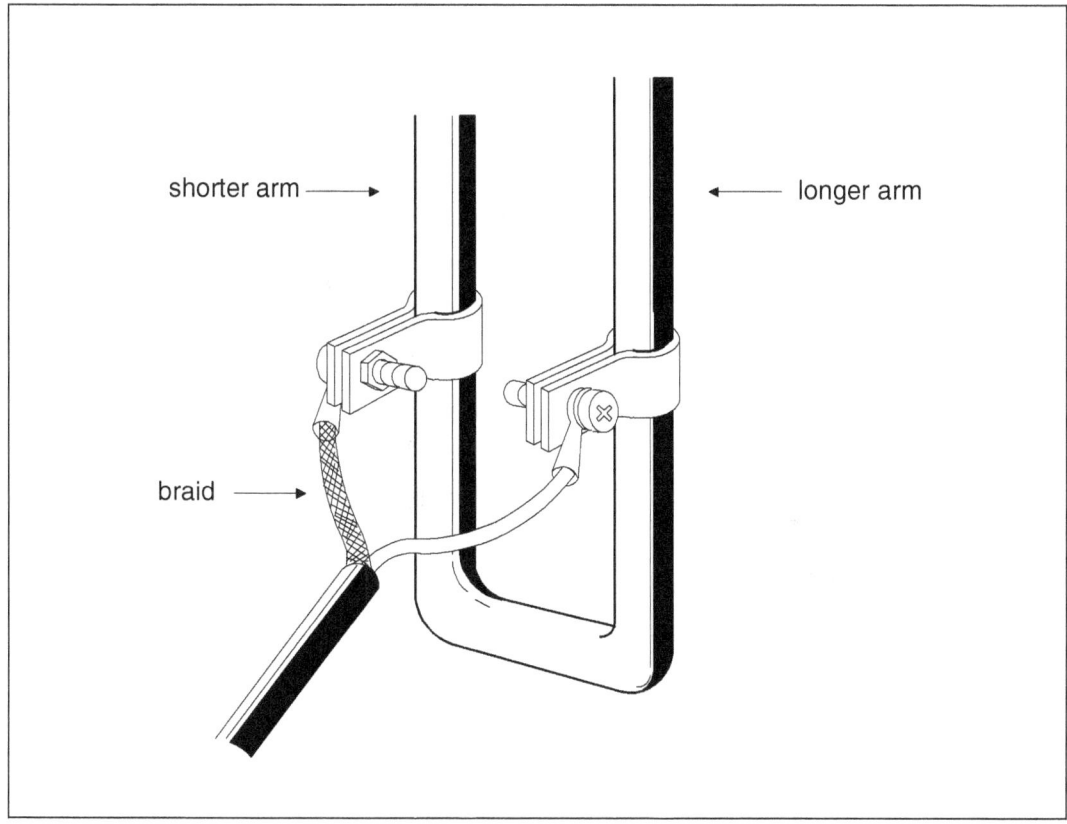

Figure 4.14   Connecting the coaxial cable to the feed point clamps.

Mount the antenna to the mast you intend to use. It is best to tune the antenna to resonance right at the mast where it will be installed permanently. Connect the coax cable to an SWR meter. The coaxial cable must be furnished with the right connectors for the particular type of SWR meter that you intend to use. Connect a transceiver to the input connector of the SWR meter (usually marked 'transmitter'). Set the transceiver to 145.00 MHz, and key the PTT to transmit. Note the SWR reading on the meter. While the transceiver is on standby , move both feed point clamps higher or lower than the initial setting until you get a low SWR response over the entire frequency range (140.00 MHz to 150.00 MHz specifically). Move the clamps about 1/4" at a time (see Figure 4.15 on the next page)

IMPORTANT:

Do not move the clamps while the rig is transmitting, and do not touch any part of the antenna when reading the SWR response. The position of the clamps must be moved always at the same level at the same time.

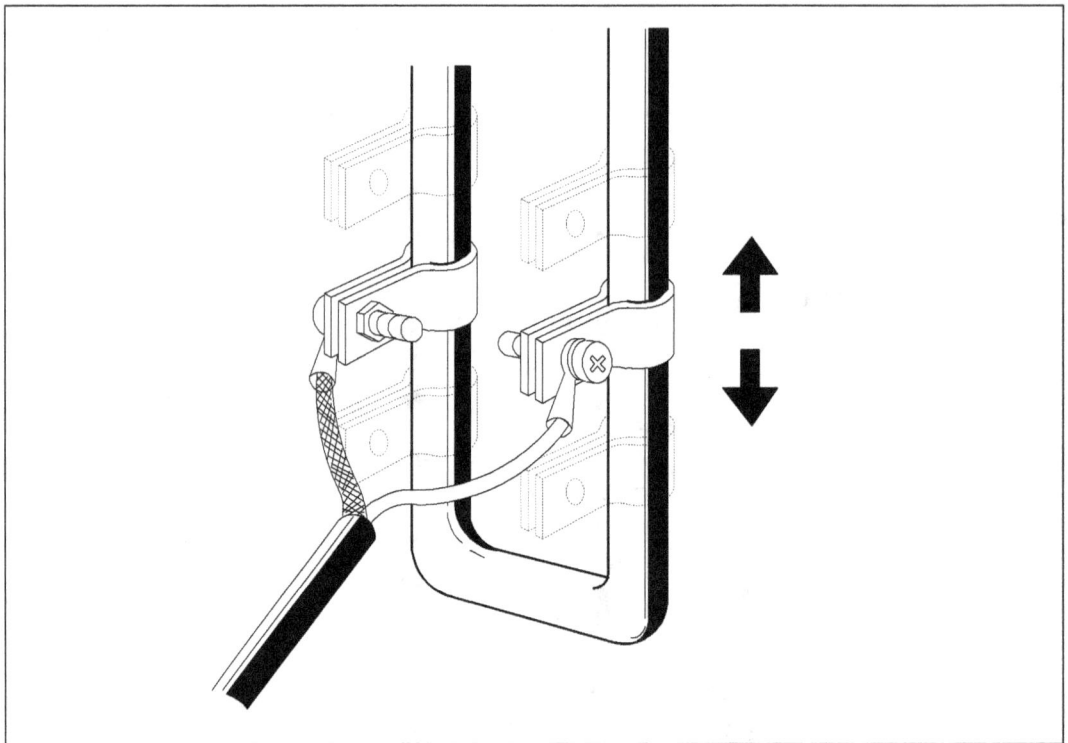

Figure 4.15 Adjusting the position of the clamps to tune the antenna.

After you have tuned the antenna to resonance, tighten the nuts holding the feed point clamps permanently, and fix the coaxial cable to the mounting plate with plastic clamps
(see Figure 4.16).

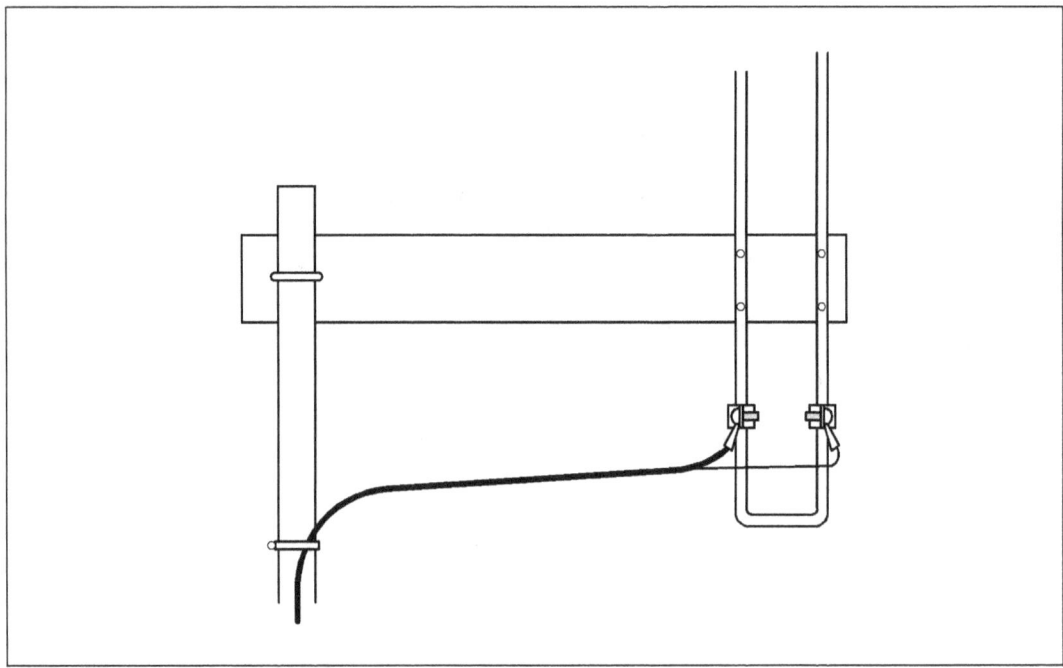

Figure 4.16 Final mounting of the J-fed antenna.

REVIEW QUESTIONS

1. How is a J-fed vertical antenna able to function without a grounding system?

2. What is the function of the shorter arm of the tube?

3. How is the antenna tuned to resonance?

4. What does 'tuning to resonance' mean?

5. By studying the design of this antenna, is it good to use a metallic plate to mount it? Why?

6. What does feed point mean?

Table 4.1 Conversion Table: fraction and decimal of an inch to millimeter

in inch	in mm	in inch	in mm	in inch	in mm
1/64 = 0.015	0.396	23/64 = 0.359	9.127	45/64 = 0.703	17.858
1/32 = 0.031	0.793	3/8 = 0.375	9.525	23/32 = 0.719	18.255
3/64 = 0.047	1.190	25/64 = 0.391	9.921	47/64 = 0.734	18.652
1/16 = 0.063	1.587	13/32 = 0.406	10.318	3/4 = 0.750	19.050
5/64 = 0.078	1.984	27/64 = 0.422	10.715	49/64 = 0.766	19.446
3/32 = 0.094	2.381	7/16 = 0.438	11.112	25/32 = 0.781	19.842
7/64 = 0.109	2.778	29/64 = 0.453	11.508	51/64 = 0.797	20.239
1/8 = 0.125	3.175	15/32 = 0.469	11.905	13/16 = 0.813	20.637
9/64 = 0.141	3.571	31/64 = 0.484	12.302	53/64 = 0.828	21.033
5/32 = 0.156	3.968	1/2 = 0.500	12.700	27/32 = 0.844	21.429
11/64 = 0.172	4.365	33/64 = 0.516	13.096	55/64 = 0.859	21.827
3/16 = 0.188	4.762	17/32 = 0.531	13.492	7/8 = 0.875	22.225
13/64 = 0.203	5.159	35/64 = 0.547	13.890	57/64 = 0.891	22.621
7/32 = 0.219	5.556	9/16 = 0.563	14.287	29/32 = 0.906	23.017
15/64 = 0.234	5.952	37/64 = 0.578	14.683	59/64 = 0.922	23.414
1/4 = 0.250	6.350	19/32 = 0.594	15.080	15/16 = 0.938	23.812
17/64 = 0.266	6.746	39/64 = 0.609	15.477	61/64 = 0.953	24.208
9/32 = 0.281	7.143	5/8 = 0.625	15.875	31/32 = 0.969	24.604
19/64 = 0.297	7.540	41/64 = 0.641	16.271	63/64 = 0.984	25.002
5/15 = 0.313	7.937	21/32 = 0.656	16.667	1 = 1.000	25.400
21/64 = 0.328	8.334	43/64 = 0.672	17.064		
11/32 = 0.344	8.730	11/16 = 0.688	17.462		

5   COAXIAL DIPOLE

Model CD-2

The coaxial dipole described here has the advantage of having lower resistance to wind compared to the ground plane designs. It also has a narrow form that some radio operators find beautiful. The following illustrations in Figure 5.1 show how the coaxial dipole was developed from a basic dipole antenna.

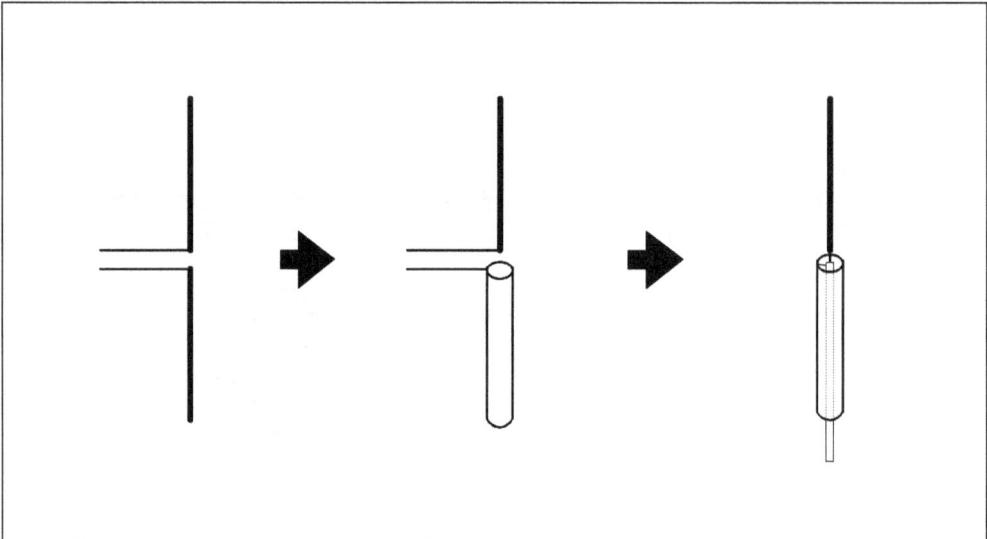

Figure 5.1   Development of the coaxial dipole from a basic dipole antenna.

As shown in the illustration, one of the elements is enlarged to form a tube. The coaxial transmission cable is then inserted through this tube, with the inner conductor of the coaxial cable connected to the radiating element, and the shield connected to the tube. The tube functions as a ground plane.

The CD-2 coaxial dipole has an operational bandwidth of 140-150 MHz. It exhibits an SWR response of less than 1.5:1 over the entire band. It has a power gain of 1 dB (unity gain) compared to a standard dipole reference. The RF signal radiates from the antenna in an omnidirectional pattern. Likewise, it receives signal equally well from all directions.

This unit is designed to be installed primarily in base stations, but it could be used for mobile applications, too. The radiating element must be detached when transporting the antenna.

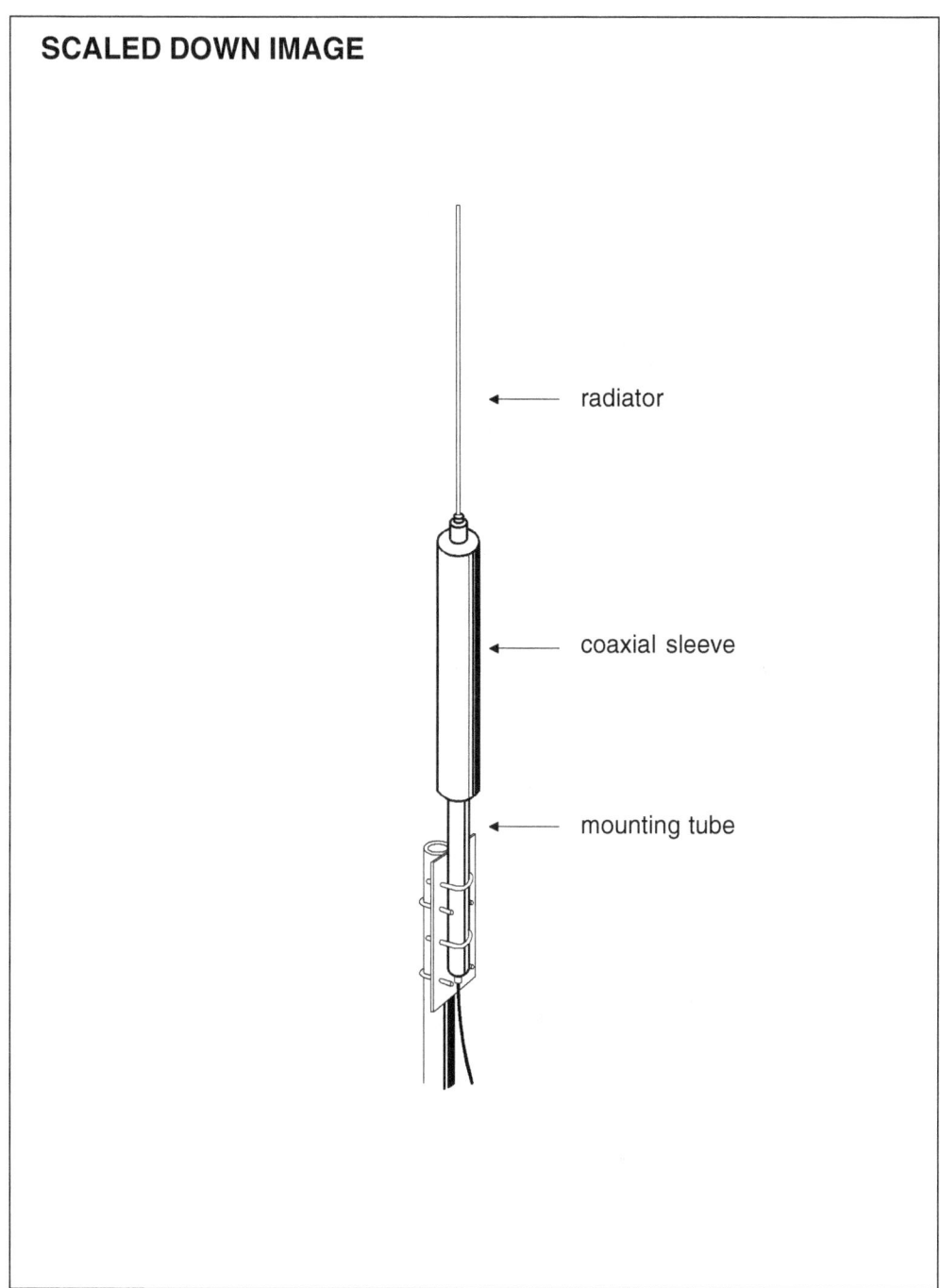

SCALED DOWN IMAGE

radiator

coaxial sleeve

mounting tube

Figure 5.2   Coaxial Dipole Model CD-2

Materials List

Quantity	Specification/Description	Dimensions
1	Aluminum tube	3" x 18"
1	Aluminum tube	1" x 36"
1	PL-259 VHF male connector	
2	SO-239 VHF female connector	
1	Brass rod 1/8" od* - the brass rod for acetylene welding is recommended	
2	Aluminum bushing - see main text for exact dimensions	
4	U-bolts - with accompanying hex nuts and lock washers	
1	Aluminum Plate or GI	3" x 6"
9	Self-tapping metal screws	1/8" x 1/2"
1	Short length of coaxial cable RG-58/U	37" long

*od- outside diameter

Construction

First prepare the two aluminum tubes of different diameters. Cut the tubes to their proper lengths as shown in Figure 5.3

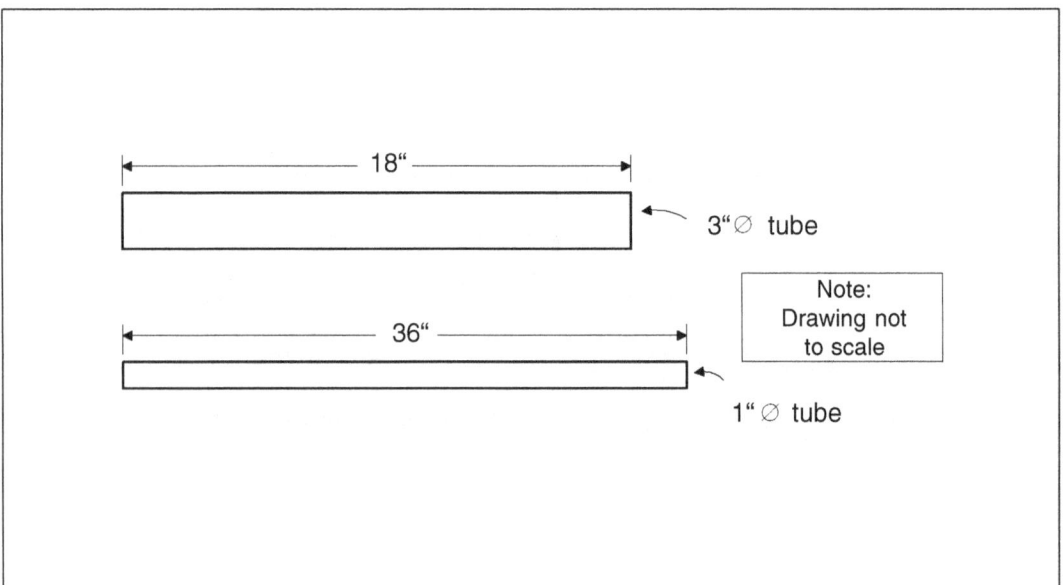

Figure 5.3 Cutting the tubes to their proper lengths.

Next, drill three holes (1/8" diameter) at *both ends* of the longer tube. The holes must be equally spaced from each other (see Figure 5.4).

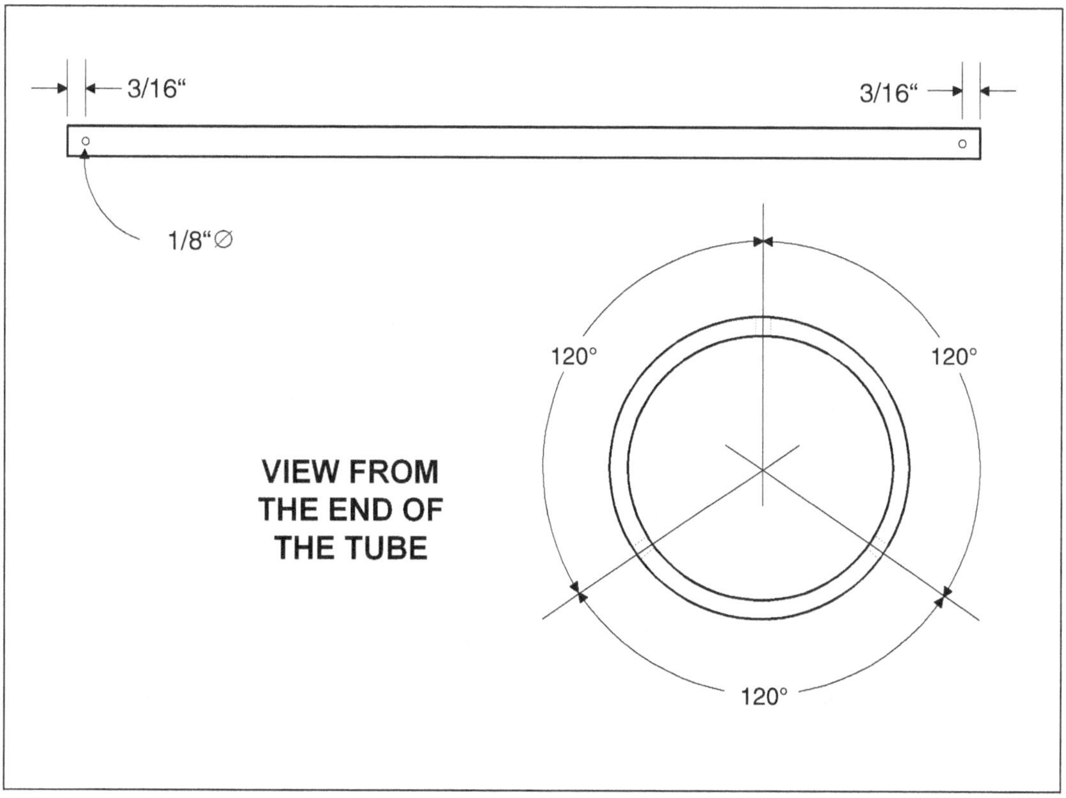

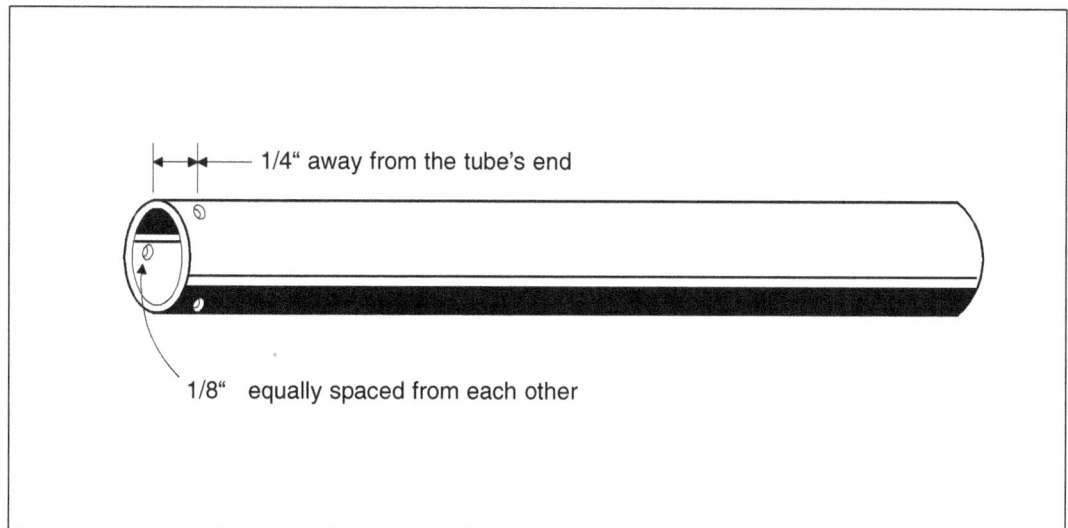

Figure 5.4   Drilling three holes at both ends of the long tube.

Drill three holes (1/8" diameter) at *one end* of the shorter tube. The holes must be 1/4" away from the edge, and equally spaced from each other (Figure 5.5).

Figure 5.5   Drilling a hole at one end of the short tube.

Machine the smaller bushing from a thick aluminum slab or rod to its proper size.
Follow the dimensions in Figure 5.6.

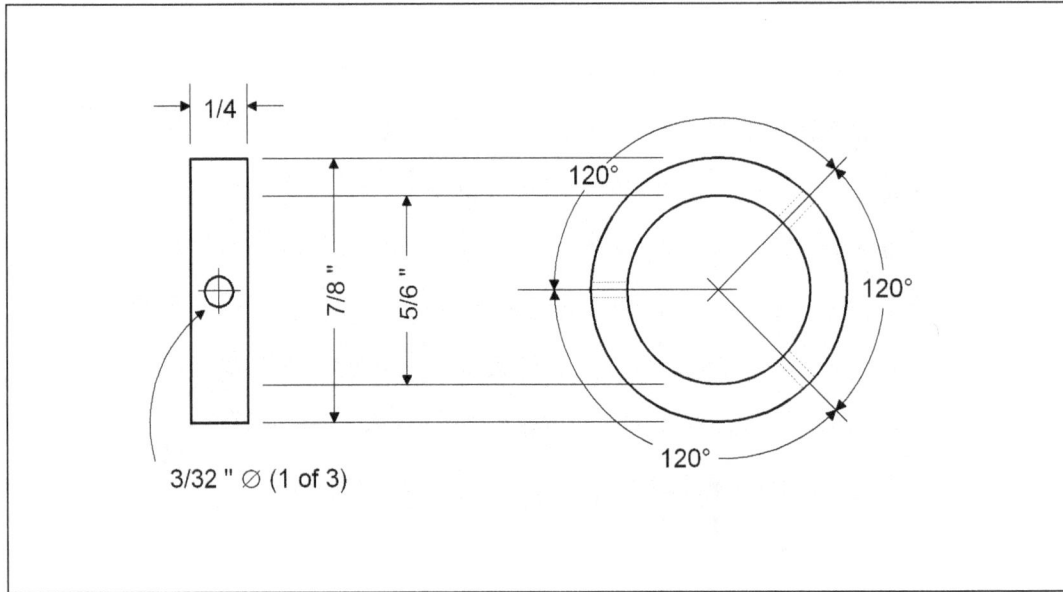

Figure 5.6   Smaller bushing dimensions.

Next, machine the larger bushing from similar material. Follow the dimensions shown in Figure 5.7.

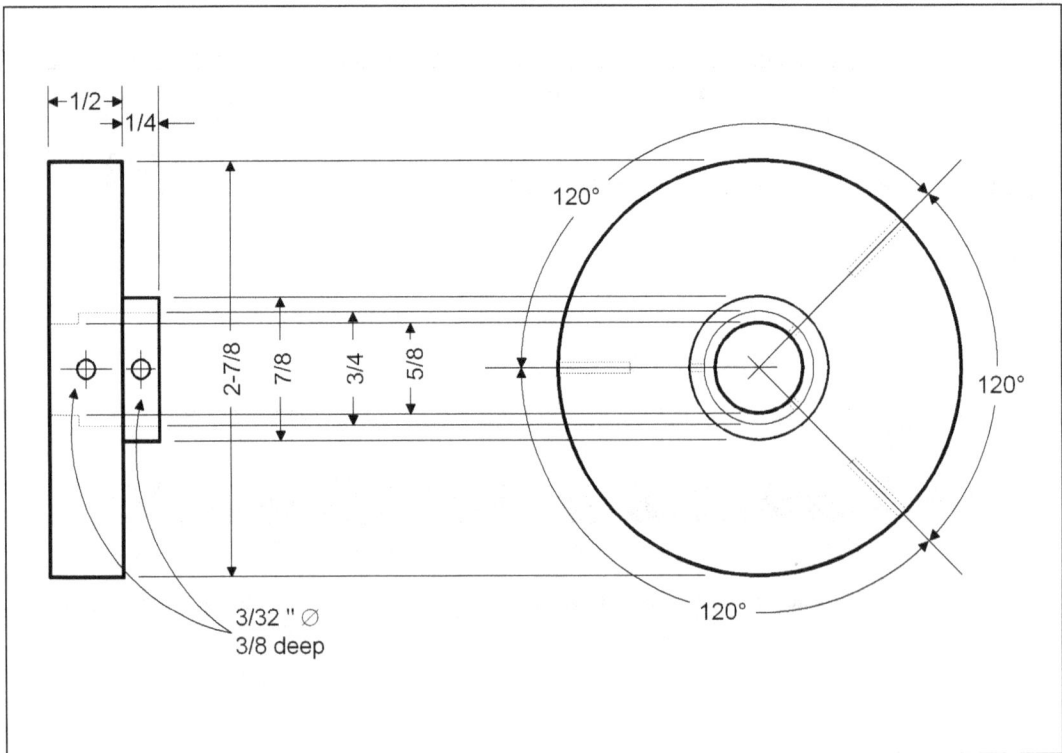

Figure 5.7   Dimensions of the larger bushing.

File away a small portion at one end of the brass rod, reducing it to smaller diameter enough to fit inside the center pin of the PL-259 connector. Solder the brass rod into the center pin of the PL-259 connector (see Figure 5.8).

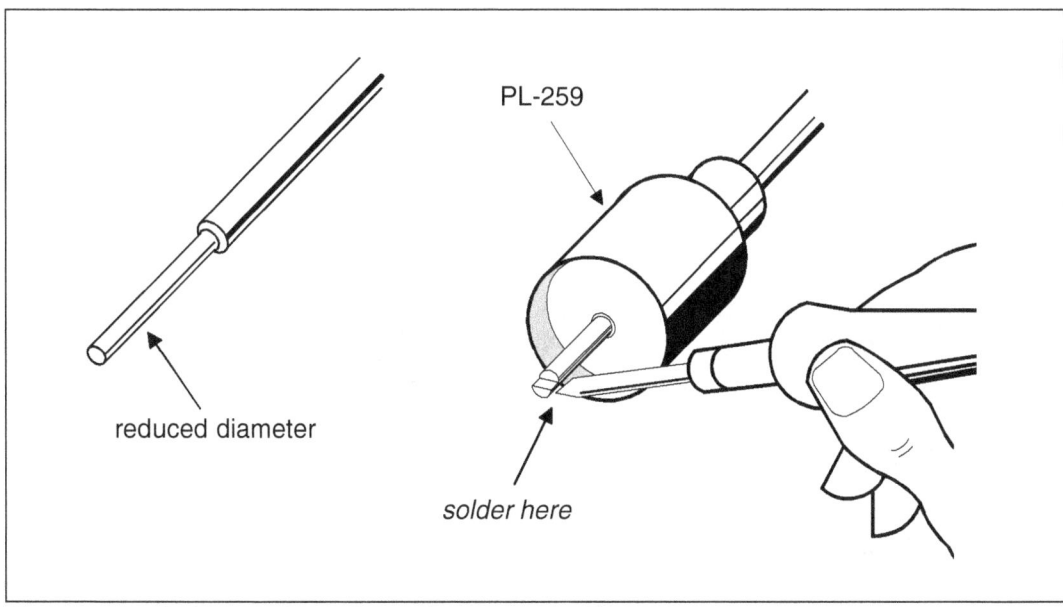

Figure 5.8   Soldering the radiator element to the PL-259.

Cut a small length of coaxial cable (about 2 inches) and remove its inner conductor and braid/shield. You need only the vinyl outer jacket. Insert it into the brass rod all the way inside the PL-259 connector. Cut away any protruding vinyl portion. The jacket serves as an insulator between the brass rod and the body of PL-259 (see Figure 5.9).

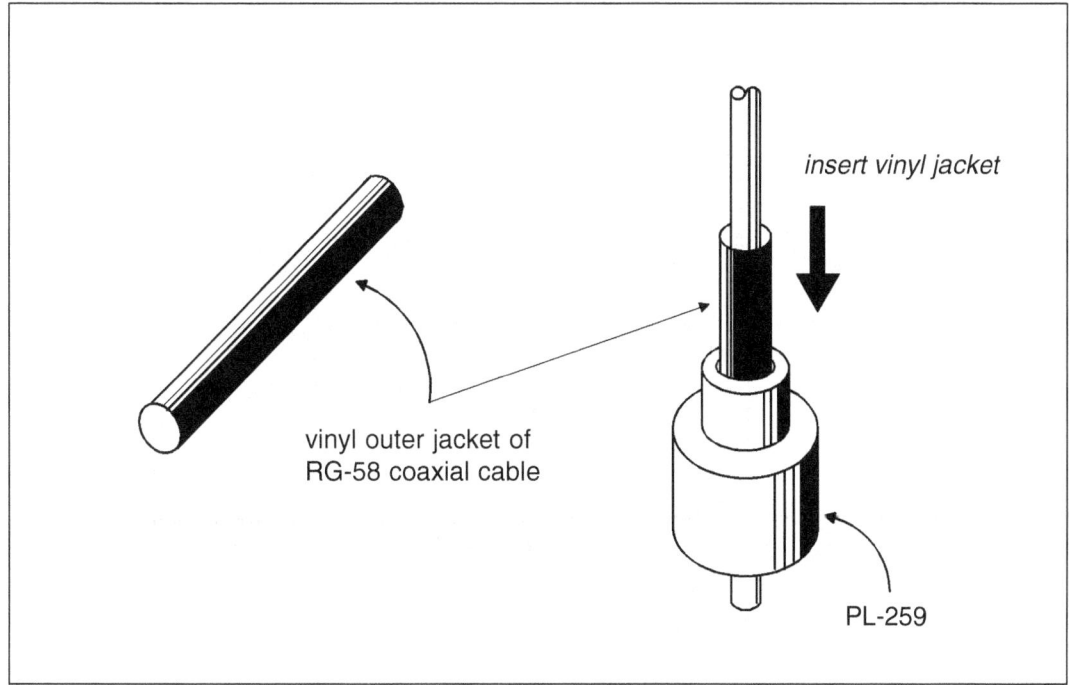

Figure 5.9   Inserting the insulating jacket into the PL-259.

Mix equal amount of epoxy glue, and place it over and around the protruding part of the vinyl jacket (see Figure 5.10). The epoxy serves as a sealant to avoid the seepage of rainwater and other moisture inside the PL-259 connector. Let the epoxy set and dry.

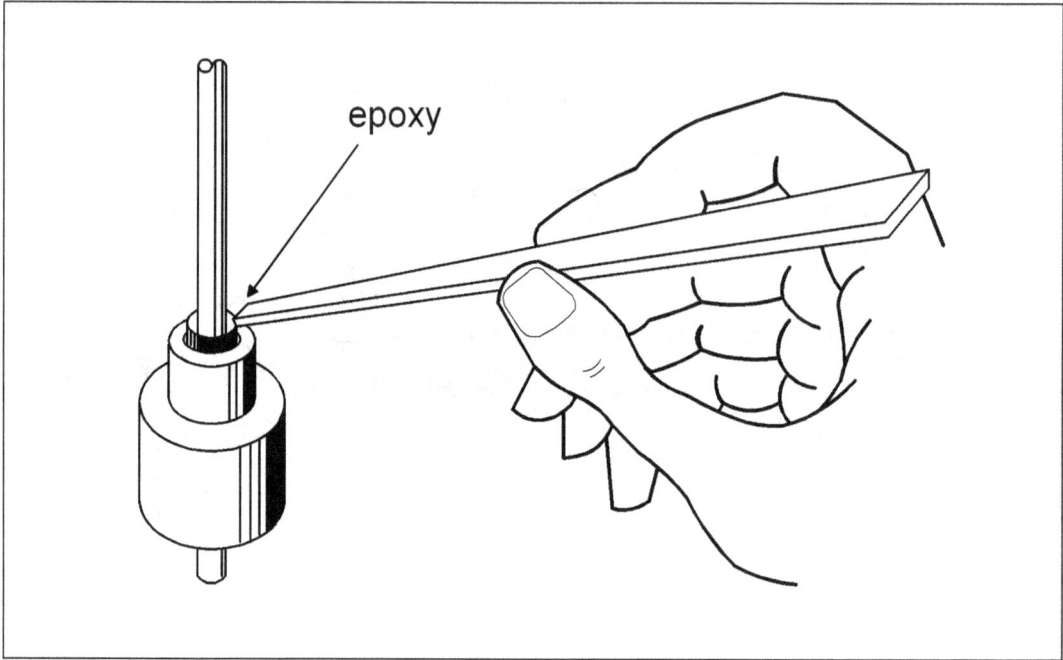

Figure 5.10 Sealing the PL-259 with epoxy glue.

Assembly

First attach the two SO-239 connectors into the two aluminum bushings as shown in Figure 5.11. Don't forget to include its grounding ring or solder lug.

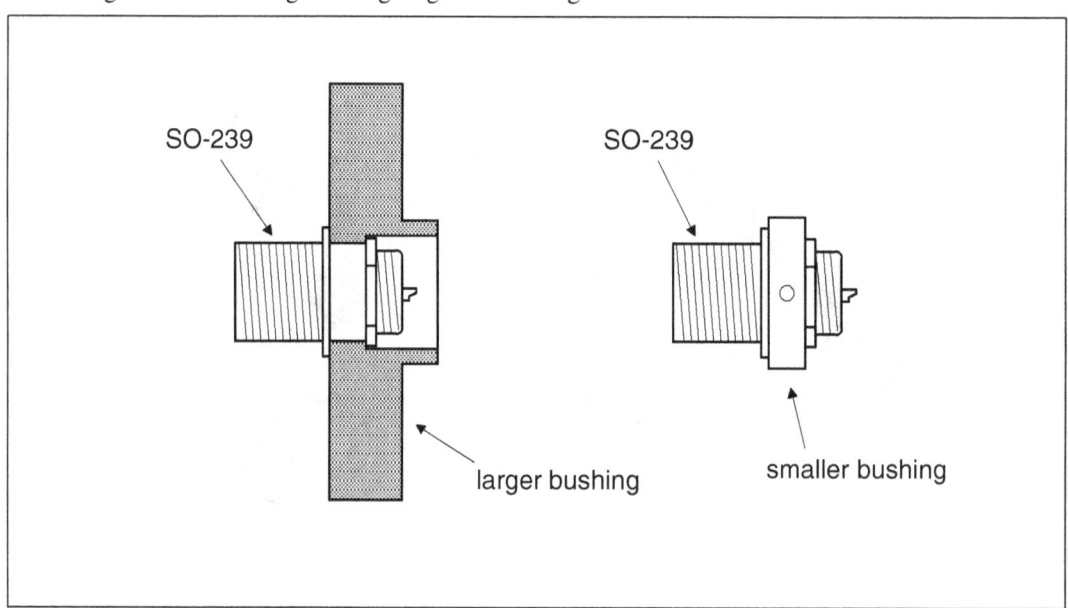

Figure 5.11 Mounting the SO-239 into the larger bushing.

Cut a 36 inch long coaxial cable (RG-58/U), and solder its conductors at one end to the SO-239 connector attached to the larger bushing (see Figure 5.12).

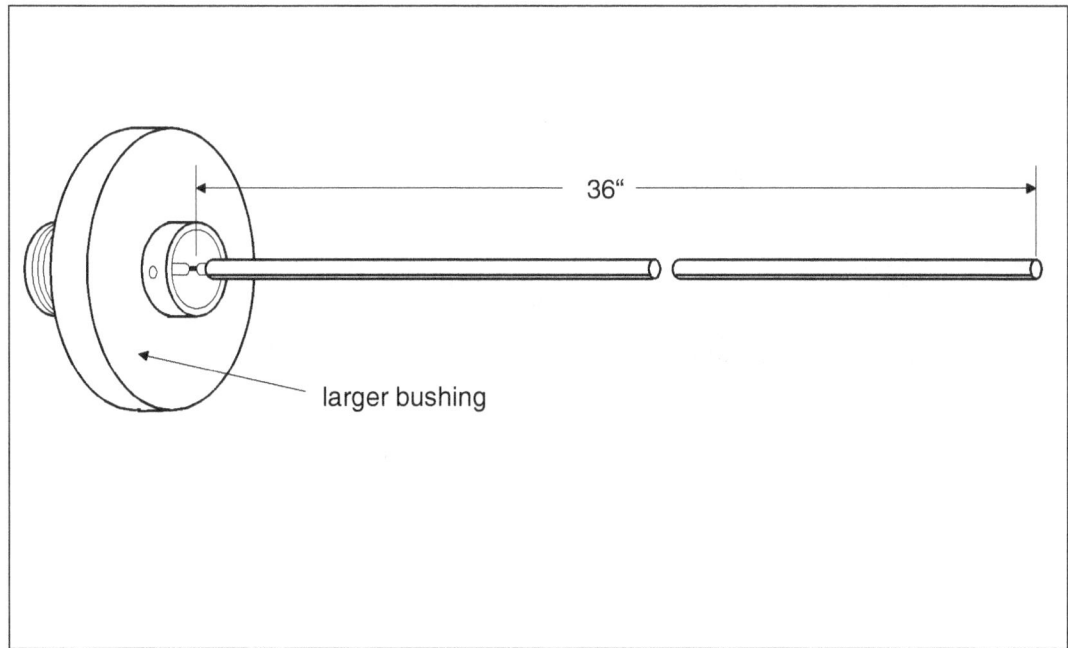

Figure 5.12   Soldering the coaxial cable to the SO-239.

Lay the coaxial cable and the longer tube side by side as they would be when they are finally assembled together. Trim the free end of the coaxial cable at the point 3/8" away from the end of the longer tube (see Figure 5.13).

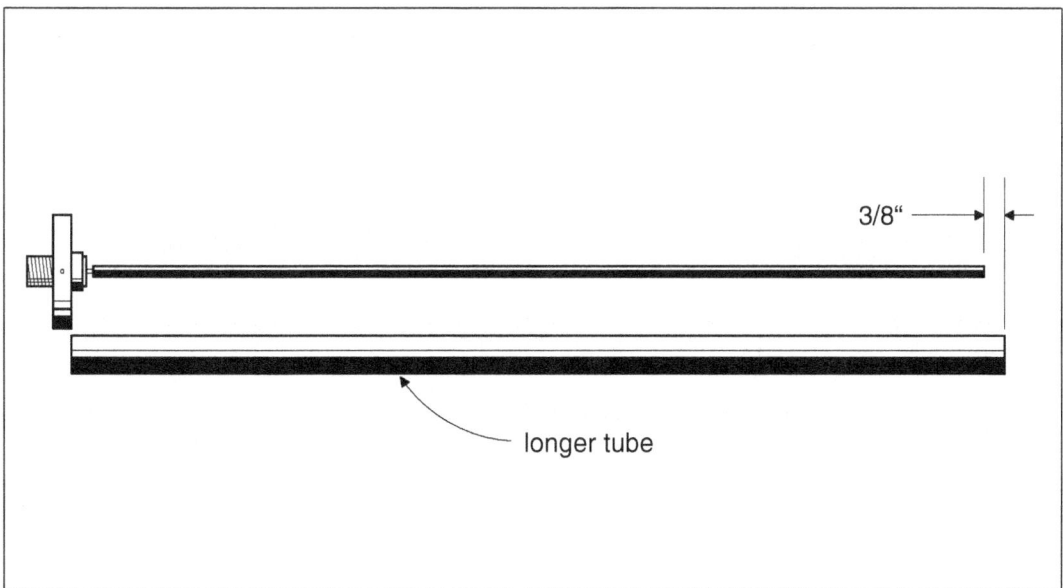

Figure 5.13   Trimming the coaxial cable to the proper length.

Solder the free end of the coax cable into the remaining SO-239 connector attached to the smaller bushing (see Figure 5.14).

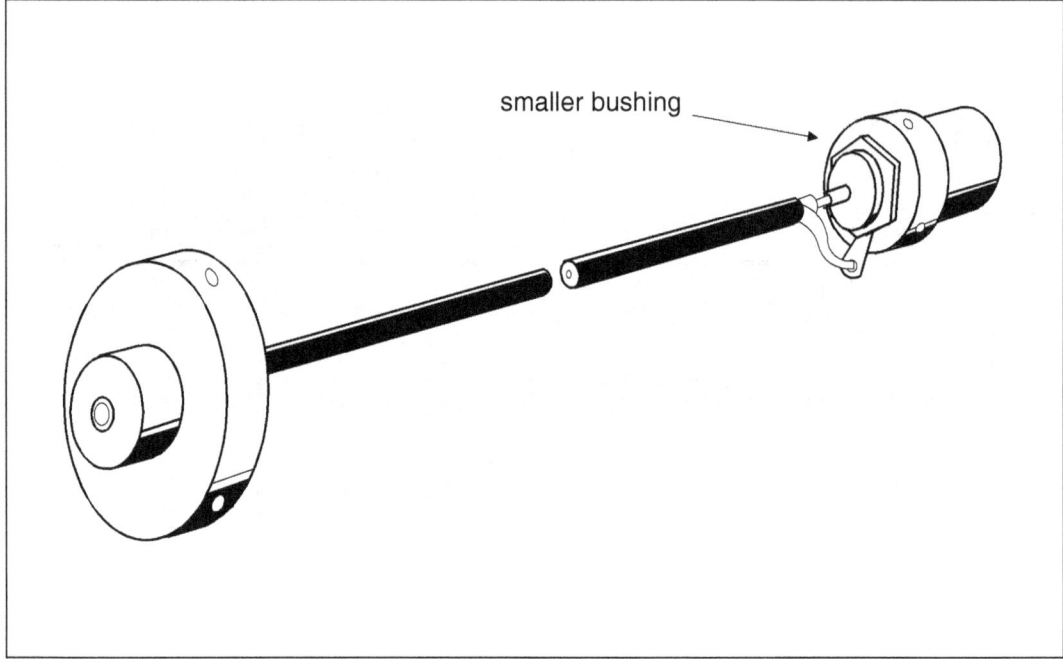

Figure 5.14   Soldering the other end of coaxial cable to the other SO-239.

Insert the smaller bushing, coaxial cable, and the large bushing all the way inside the longer tube until the holes in the two bushings are aligned to the holes in the tube itself. If in your first try you do not manage to align the holes, then maybe a slight re-trimming of the coax cable is needed, or the SO-239 connector must be repositioned or re-soldered. After a few trials you should have it right (see Figure 5.15).

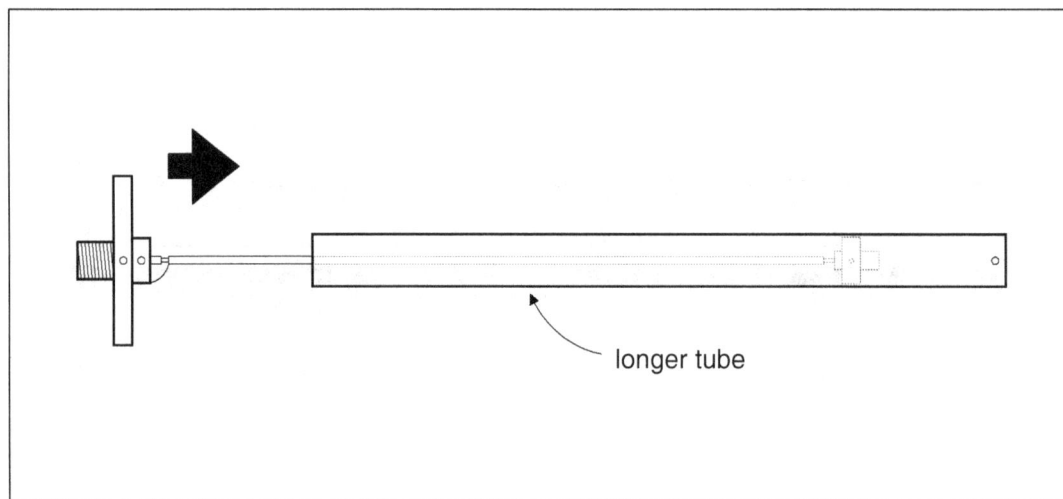

Figure 5.15   Inserting the feeder coaxial cable into the long tube.

Secure the two bushings permanently into the tube using self-tapping metal screws (see Figure 5.16).

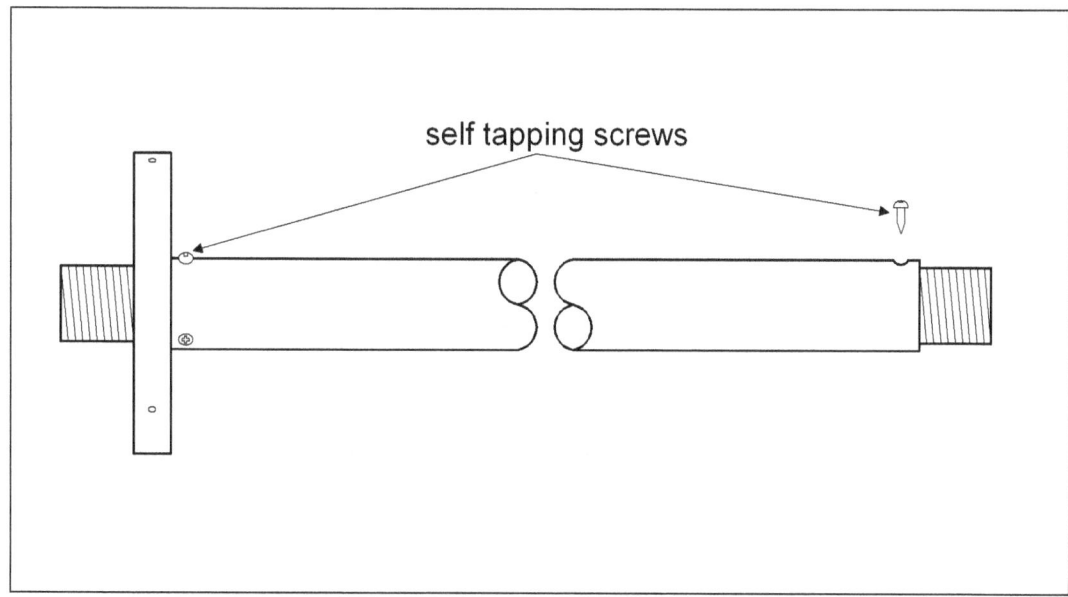

Figure 5.16   Securing the bushings and the tube together.

Insert the longer tube and the large bushing inside the shorter tube (see Figure 5.17). Align the holes in the large bushing to the holes in the shorter tube and place self-tapping screws through the holes to fix the bushing firmly inside the short tube.

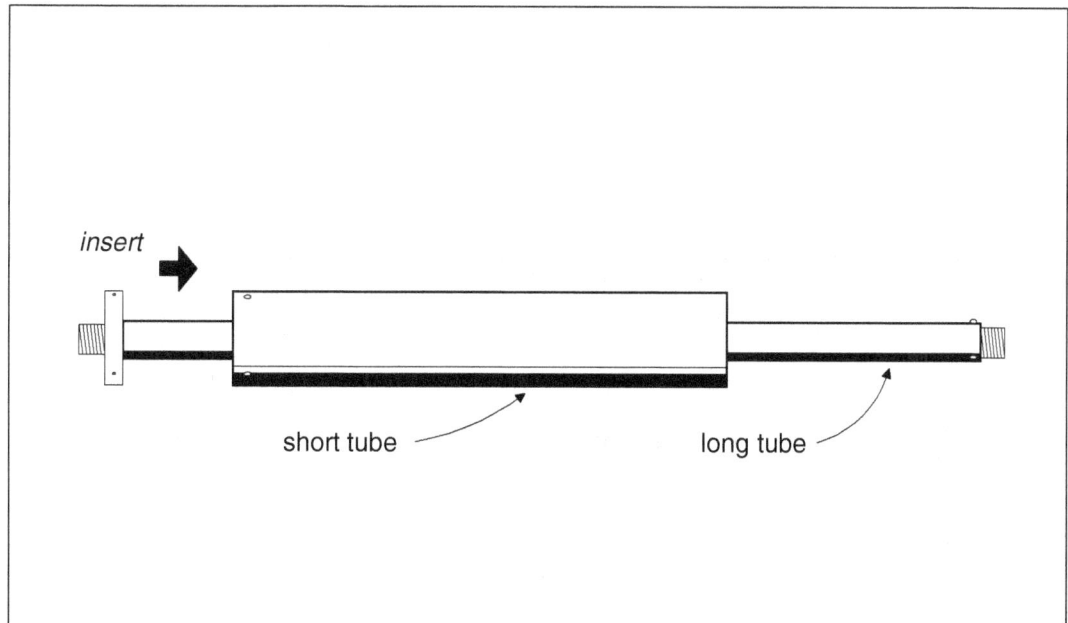

Figure 5.17   Assembling the two tubes together.

Attach the radiator element to the SO-239 connector in the upper larger bushing
(see Figure 5.18).

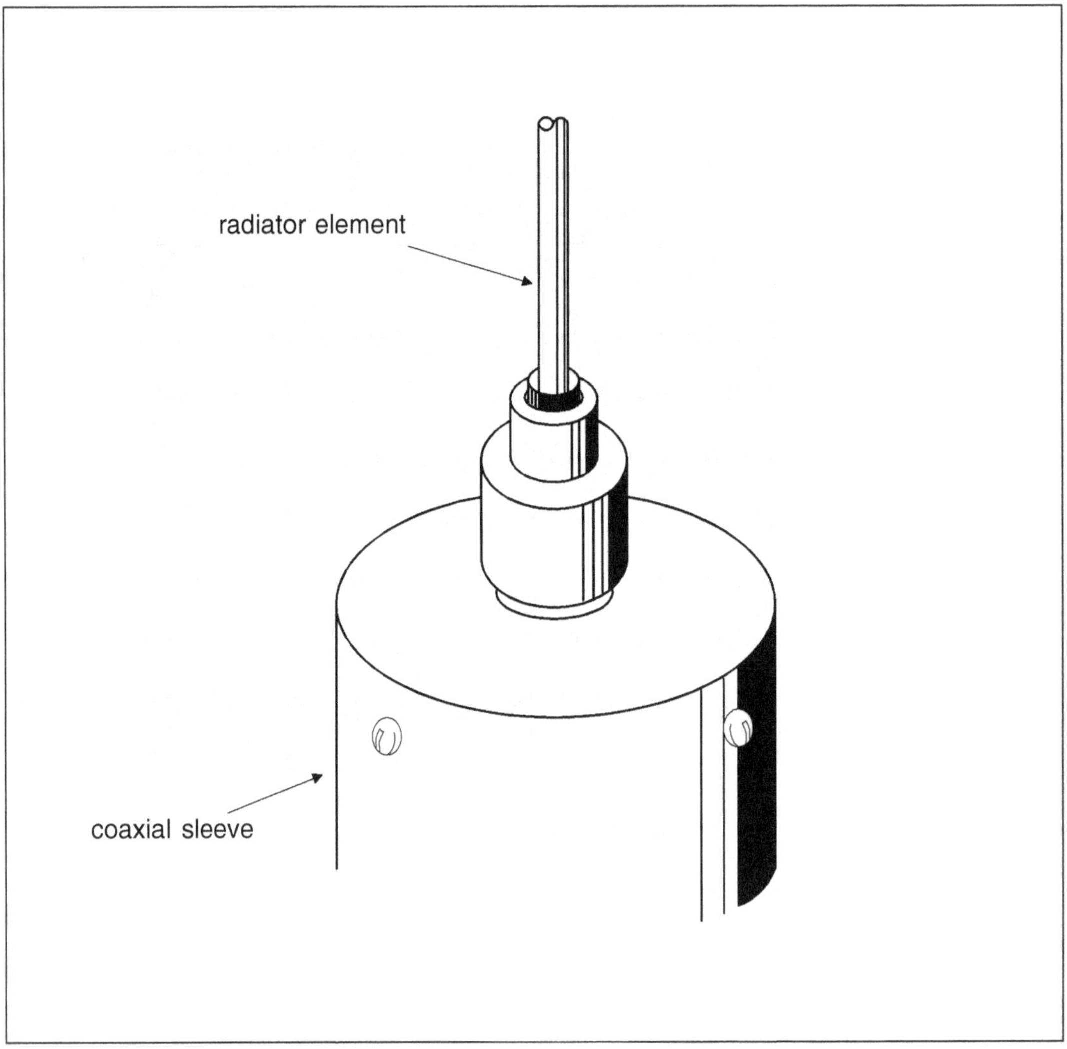

radiator element

coaxial sleeve

Figure 5.18   Installing the radiator element into the antenna base.

REVIEW QUESTIONS

1. What is the advantage of using a coaxial dipole?

2. What is the function of the outer tube?

3. What is the function of the inner tube?

Installation of the coaxial dipole Model CD-2

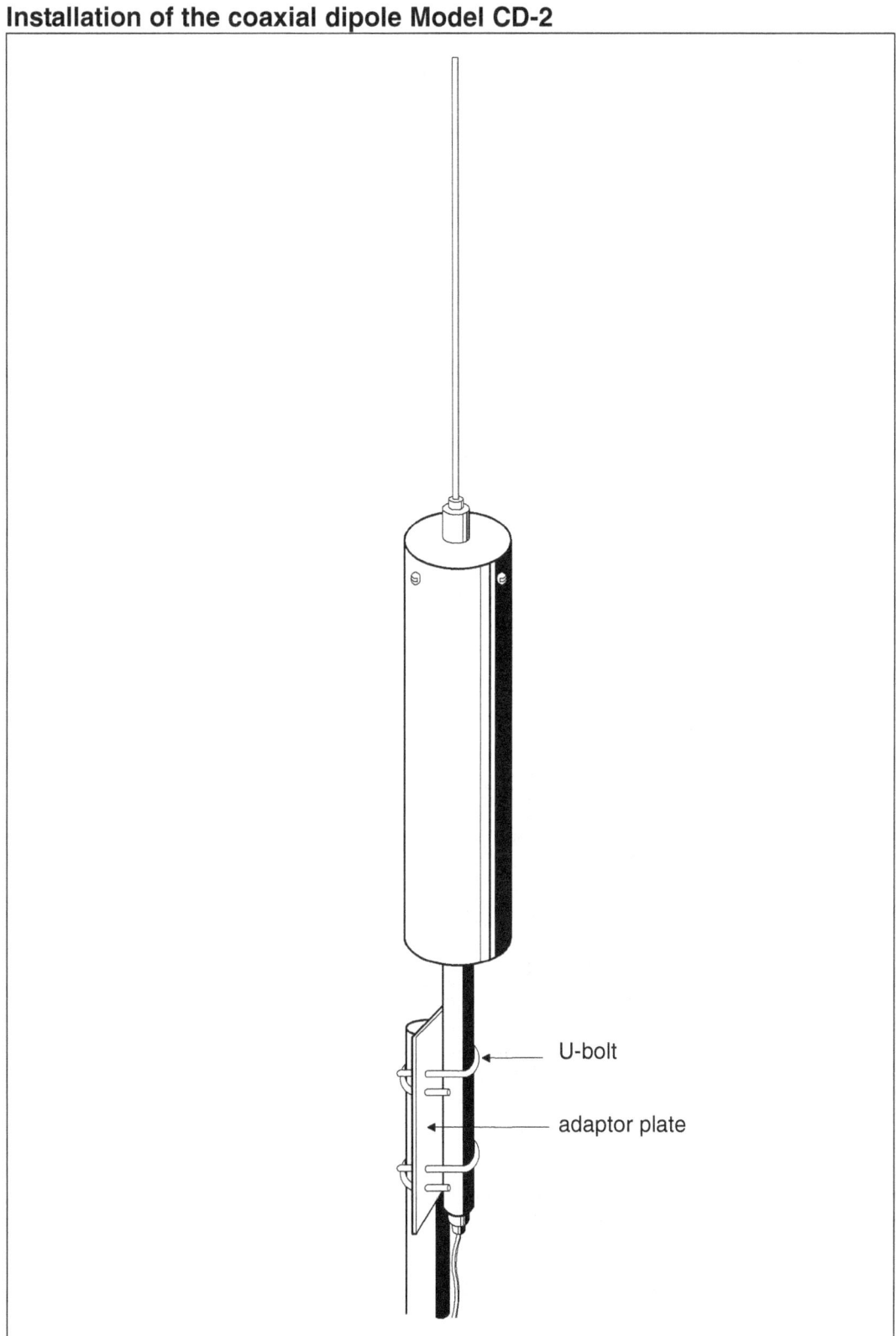

U-bolt

adaptor plate

Figure 5.19   Installation of Model CD-2

6 DIPOLE ANTENNA
(gamma fed)

Model DP-2F

This particular design of a dipole antenna is very popular in VHF applications because of its capability to be fine-tuned during tuning procedures. Tuning is accomplished by a so-called gamma matching system connected near the center of the dipole element. Gamma matching is based on the principle of the delta match system, whereby the transmission line can be directly connected near the center of a continuous half wave conductor, and fanned out and tapped at the point of most efficient power transfer.

The middle of a half wave dipole is electrically neutral, which means that there is no RF voltage present. Thus, the outer conductor of the coax cable can be connected directly to the element at this point. The inner conductor of the coaxial cable carries an RF current, so it is tapped into the dipole element at the matching point.

Careful observation of this design reveals that the center conductor of the coax cable is not directly connected to the dipole element, but is instead coupled via a short tube called the "gamma tube". The combination of the short tube and the coaxial cable inside it provides the capacitance needed to cancel the inductance of the dipole element to attain an electrical balance. The gamma match therefore achieves two functions at the same time: to match the impedance of the transmission line to the impedance of the antenna; and to couple the unbalanced coaxial cable to the symmetrical dipole element. This method makes it unnecessary to use a separate balancing transformer. Fine tuning of the antenna can be done by adjusting the shorting bar that connects the gamma tube to the dipole element until the lowest SWR response is achieved.

The Model DP-2F is also used as a basic driven element for high gain Yagi and collinear antenna designs. Because the middle of the dipole element is electrically inactive, it is not required to be insulated from its mounting boom, thereby simplifying the mechanical construction. Lightning protection for this antenna system is also improved, because all the metallic parts of the antenna are grounded via its mast or tower.

The dipole design described in this chapter is designed to operate in the frequency band of 140-150 MHz. If properly tuned, it exhibits an SWR of less than 1.4:1 over the entire band. It radiates its signal in an omnidirectional pattern. It has a gain of 1 dB (unity gain) compared to a standard dipole reference.

This antenna is intended primarily for fixed installations. However, some radio operators are able to use it successfully in mobile operations by modifying its mechanical construction.

Some antenna constructors choose to build this antenna because it offers them deeper understanding of the electrical principles of antennas, compared to other simpler designs like ground planes or coaxial dipoles. If you are the experimenter type of radio operator, then this design is for you.

SCALED DOWN IMAGE

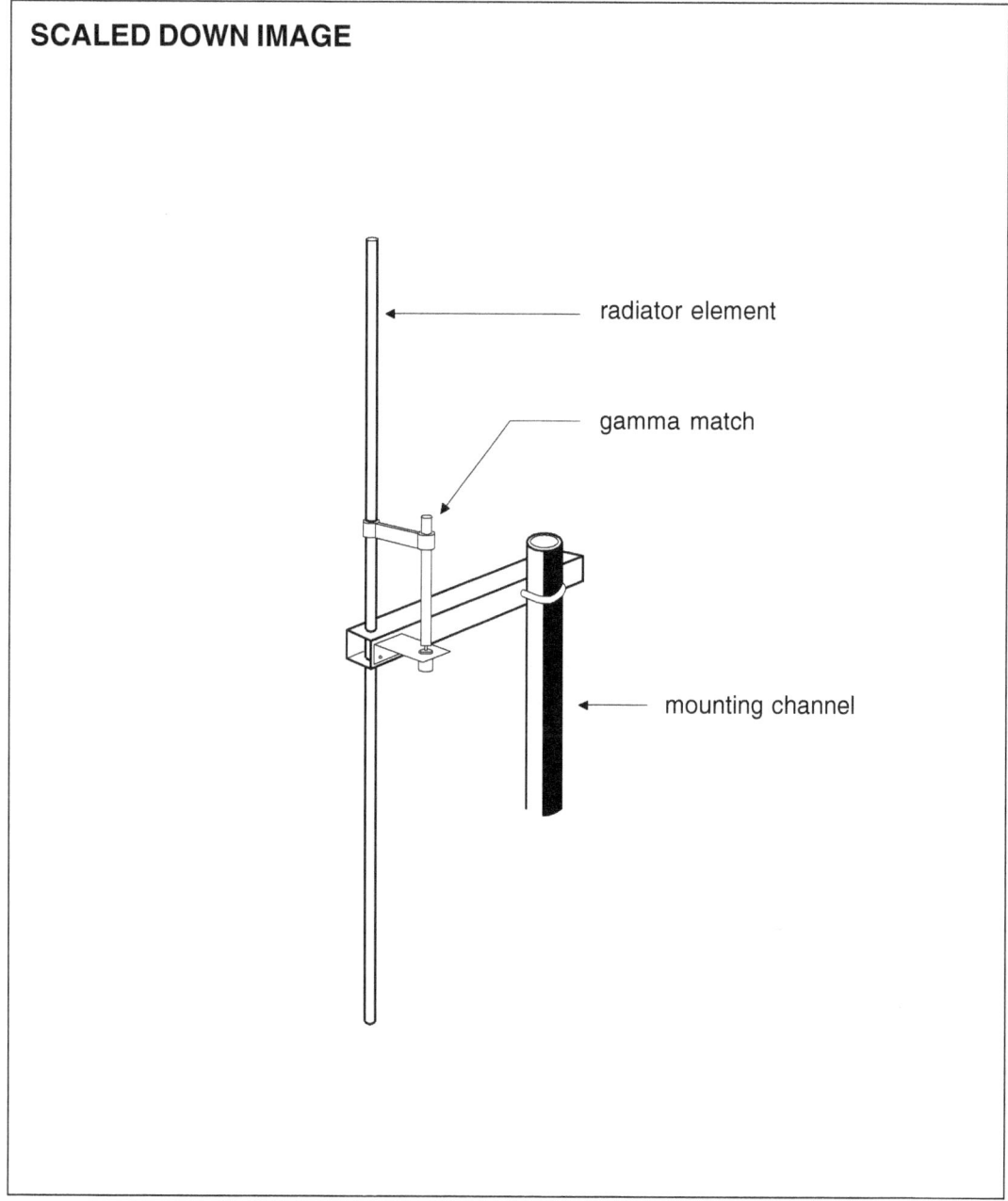

radiator element

gamma match

mounting channel

Figure 6.1   Gamma fed dipole antenna Model DP-2F

Materials List

Quantity	Specification/Description	Dimensions
1	Aluminum tube	3/8" id* 38" long
1	Aluminum tube	3/8" id* 6" long
1	Aluminum square channel	1" x 1" x 12"
1	Aluminum strip - see text for fabrication	1/2" x 4"
1	Coax cable RG-58/U	6"
1	BNC female connector	
2	Stove bolts - brass or GI	1/8" x 3/8"
2	Hex nuts - brass or GI	1/8" id
1	U-bolt with accompanying hex nuts and washers	
3	Self-tapping metal screws	1/8" x 1/2"

Miscellaneous: Epoxy glue

*id- inside diameter

Construction

The radiator element is made from 3/8 od* aluminum tube cut to a length of 38 inches. Drill a hole (1/16" diameter) through and through at the middle of its length (see Figure 6.2).

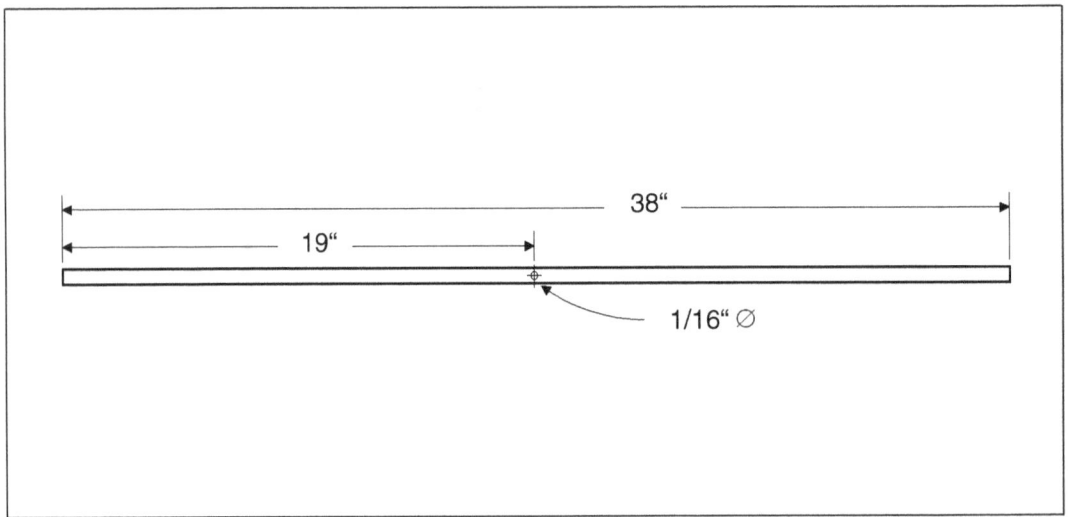

Figure 6.2 Drilling a hole through the middle point of the radiator element.

Next prepare the mounting channel by drilling a hole at one end (see Figure 6.3). The diameter of the hole must accommodate the aluminum tube that will be inserted into it. The hole is 3/8" and slightly oversized, so that the tube will not be scratched upon insertion, but not too loose as to sacrifice rigidity.

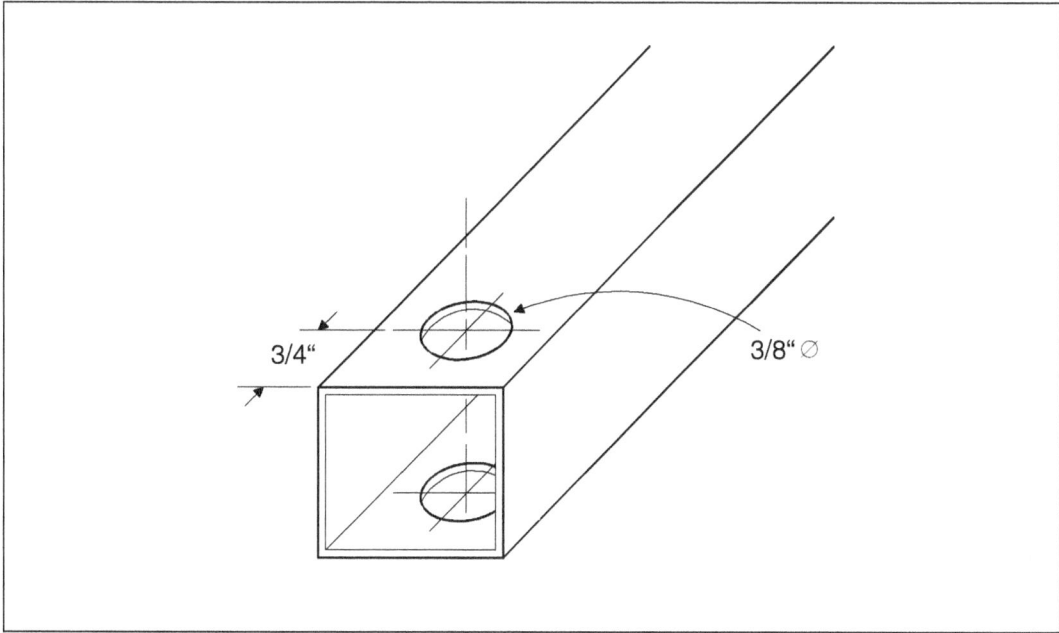

Figure 6.3   Preparing the mounting channel.

Next, drill two small holes (1/16" diameter) at one side of the channel perpendicular to the axis of the bigger hole (see Figure 6.4).

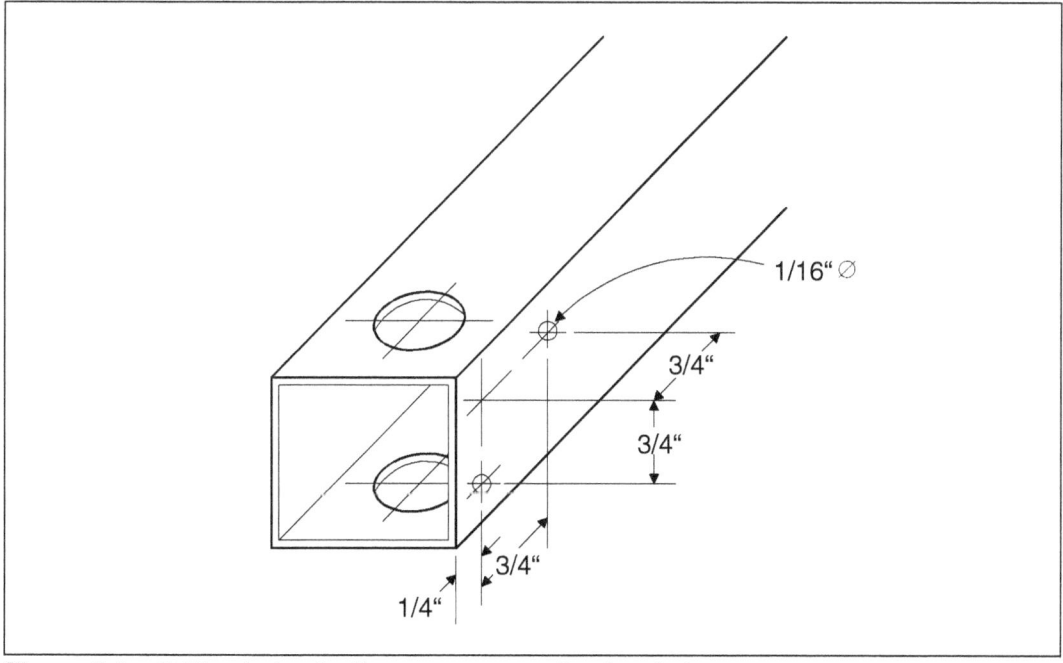

Figure 6.4   Drilling holes for the gamma mounting bracket.

Drill another pair of holes (3/16" diameter) at the same side, but at the opposite end of the channel (see Figure 6.5). Drill the hole through and through. The size of the holes and the distance between them must conform to the dimensions of the U-bolt used.

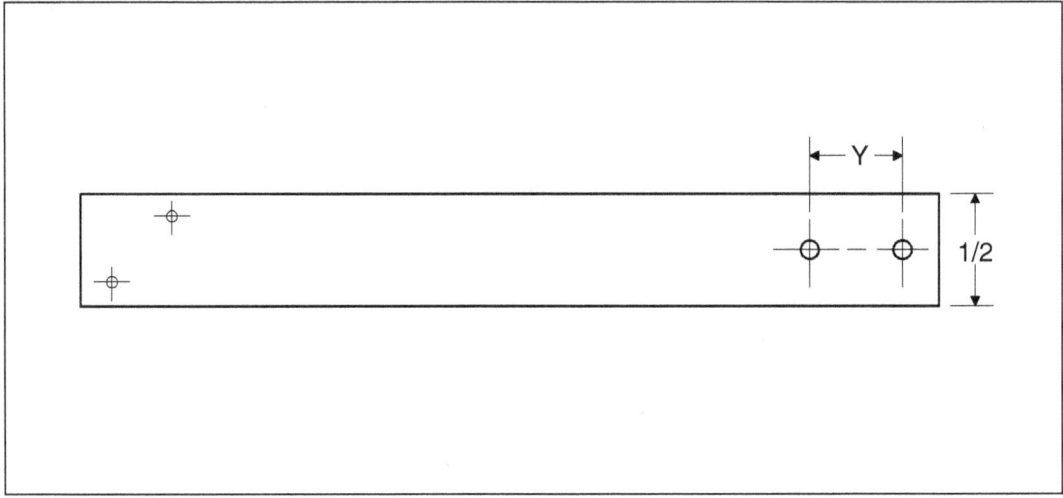

Figure 6.5   Drilling holes for the U-bolt at the opposite end .

Drill a single 1/8" diameter hole at the other side opposite to the two small holes (1/16" diameter). See the dimensions in the following illustration.

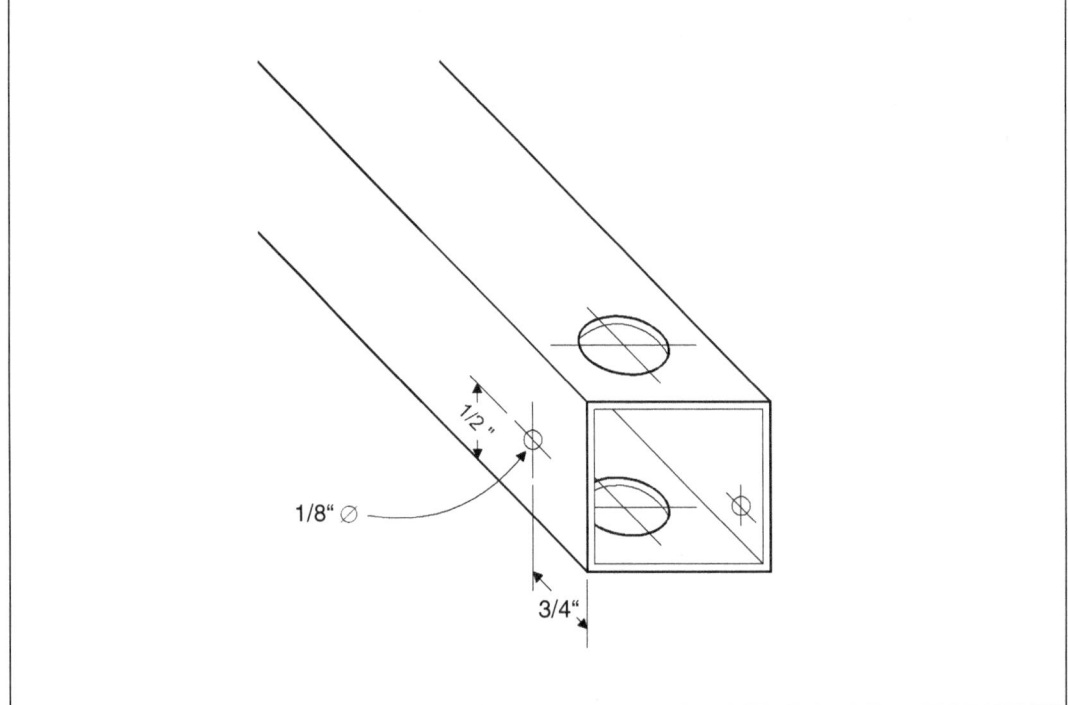

Figure 6.6   Drilling a single hole.

Insert the aluminum tube through the large hole, and align the hole at its middle part to the 1/8" diameter hole at the side of the channel (see Figure 6.7 on the next page).

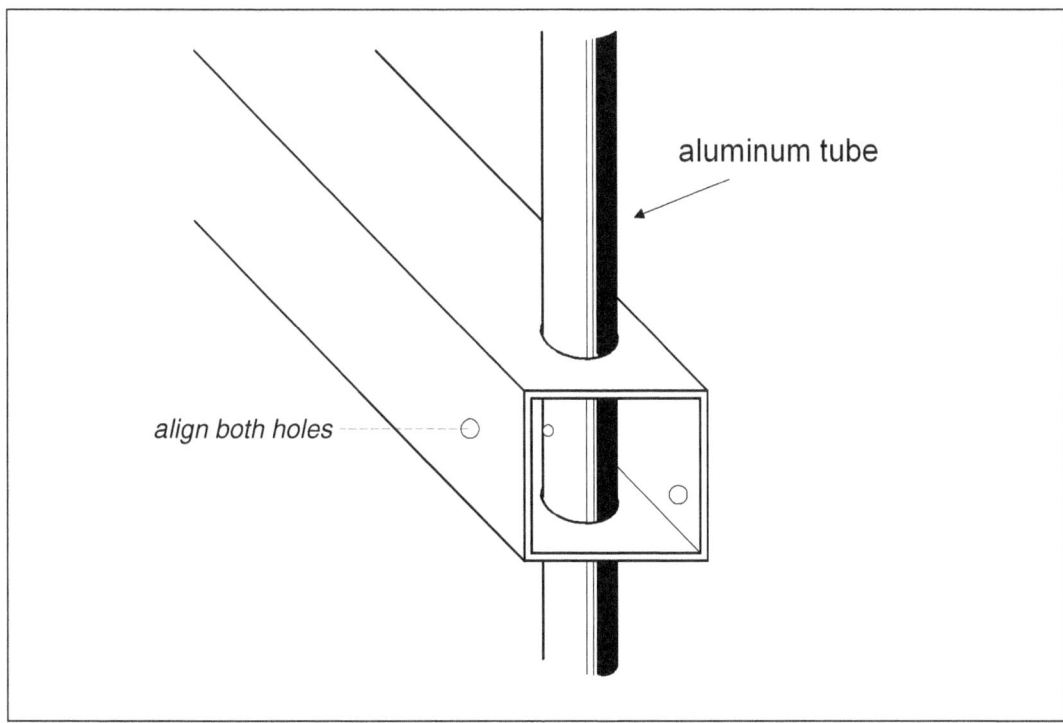

Figure 6.7   Inserting the aluminum tube into the mounting channel.

Insert a self-tapping screw through the side hole, and forcibly screw it into the smaller hole of the tube inside (see Figure 6.8). Tighten the screw until the aluminum tube is rigidly held in the aluminum channel.

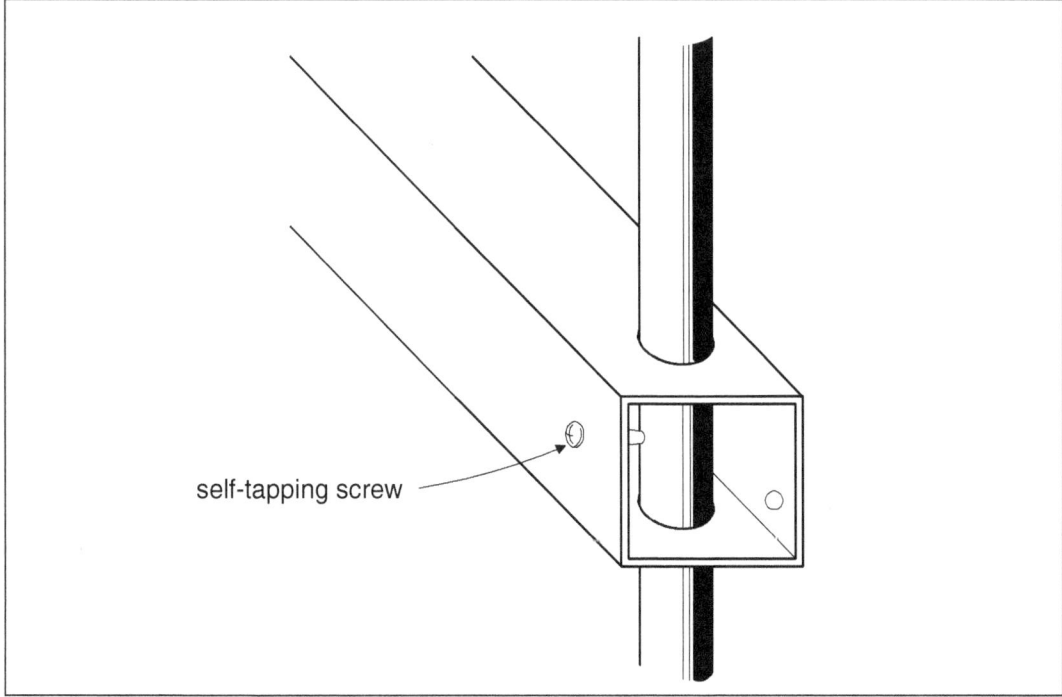

Figure 6.8   Locking the tube with a self-tapping screw.

Prepare the feed point angle bracket. The bracket is cut from a small strip of aluminum and bent into a right angle. An alternative method is to saw off a portion of a 1" x 3" rectangular aluminum channel. This will give you a more durable bracket with a near perfect angle.

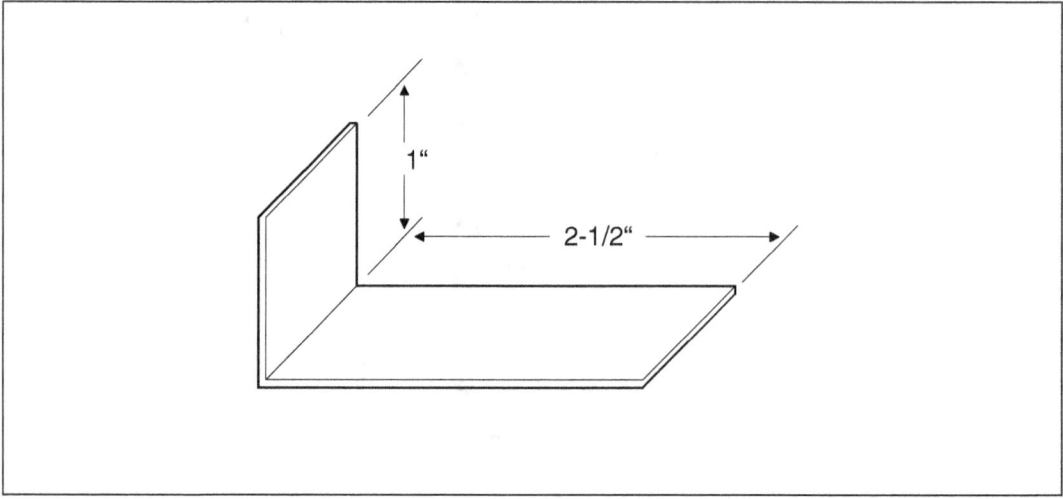

Figure 6.9   Preparing the feed point bracket (gamma mounting bracket).

Drill two small holes (1/8" diameter) at one side of the angle bracket. Drill another hole at the other side of the bracket. This lone hole must be large enough to accomodate the BNC female connector (see Figure 6.10).

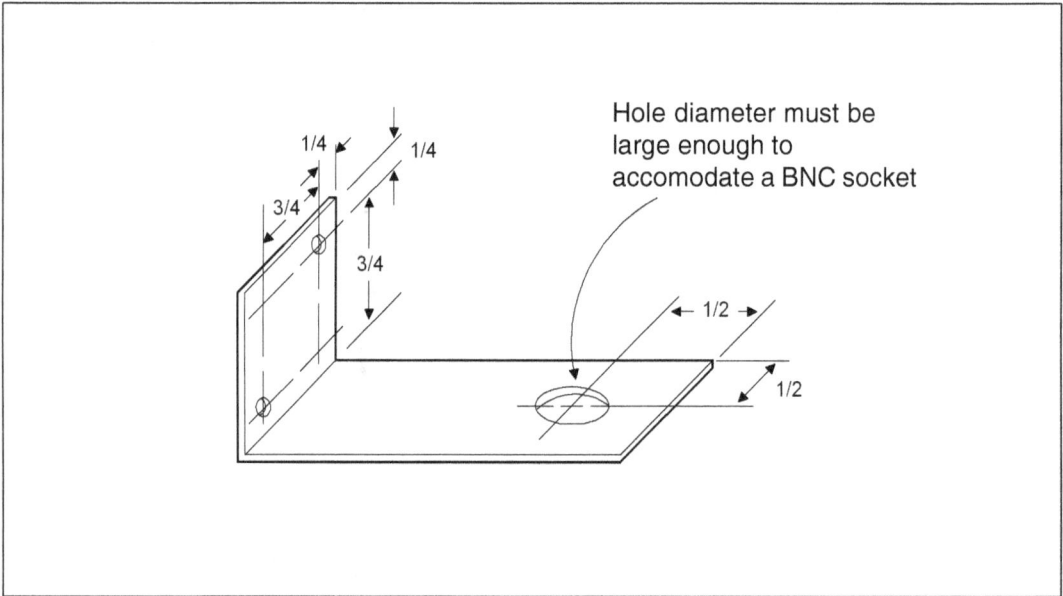

Figure 6.10   Drilling mounting holes in the bracket.

Attach the bracket into the mounting channel by screwing it with small self-tapping screws (see Figure 6.11 on the next page).

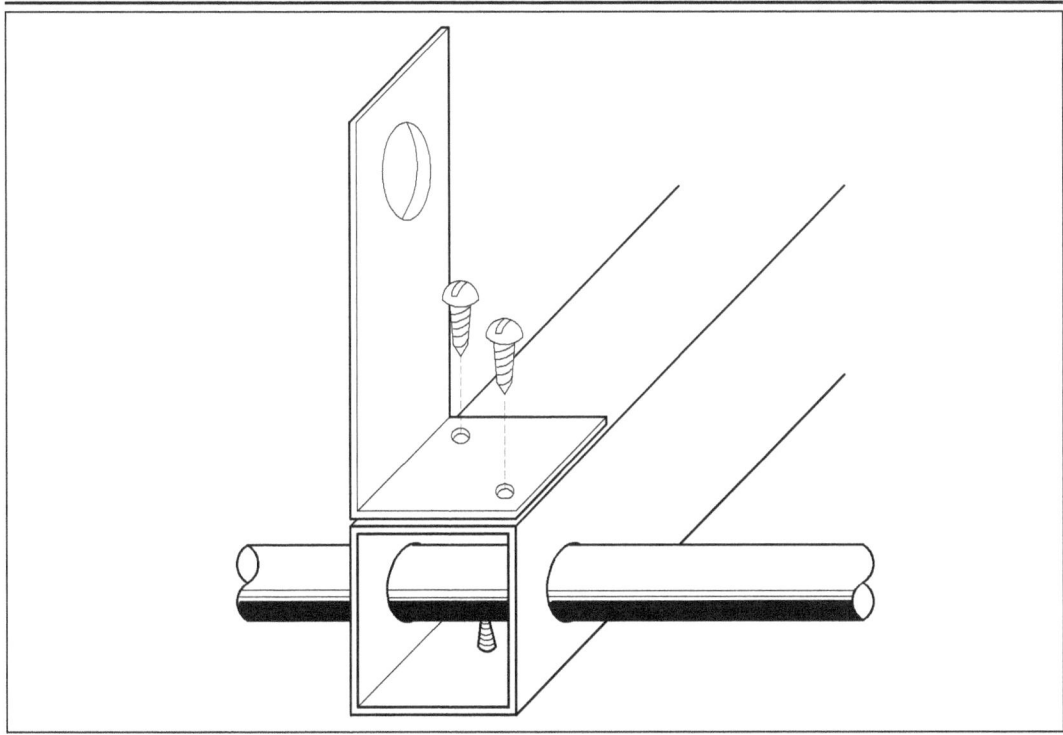

Figure 6.11 Fixing the feed point bracket on the mounting channel.

Insert the BNC female connector in an upside down position into the large hole of the feed point bracket, and secure it with its nut (see Figure 6.12).

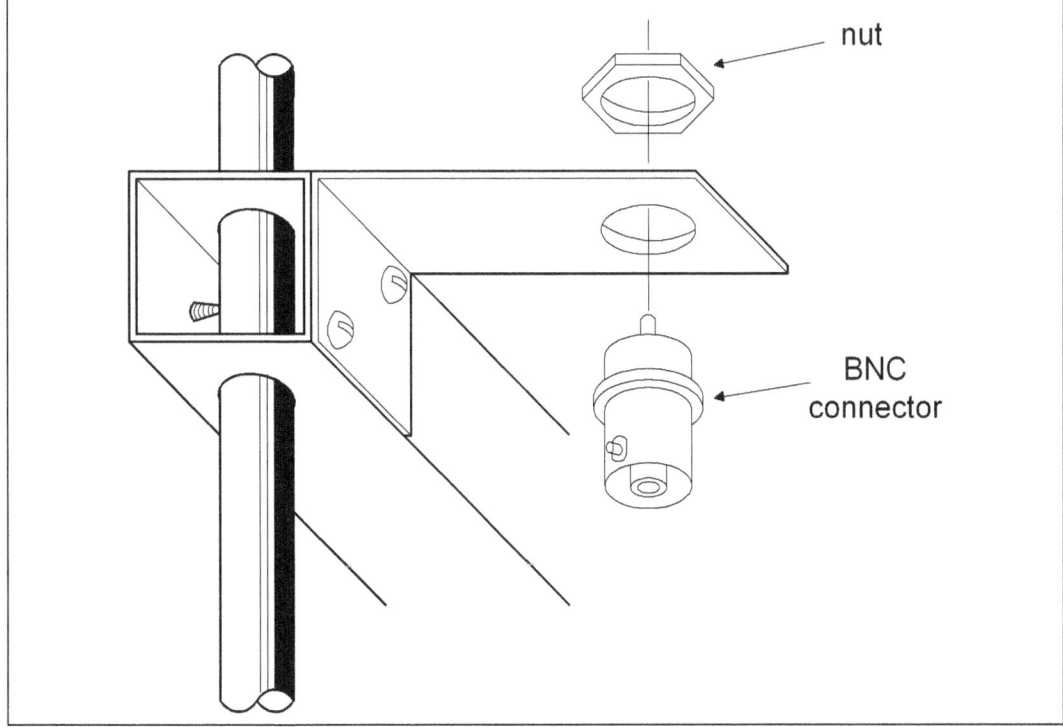

Figure 6.12 Installing the BNC connector into the feed point bracket.

The next step is to prepare the tuning clamp and the gamma matching tube. First, fabricate a flat strip from a scrap tube (about 4 inches long) by pressing it in a table vise until it is completely flattened. Cut about 4 inches of the flat strip, and form both ends into a ring clamp by bending it around an aluminum tube. The ends must be formed to fit around the tube (see Figure 6.13).

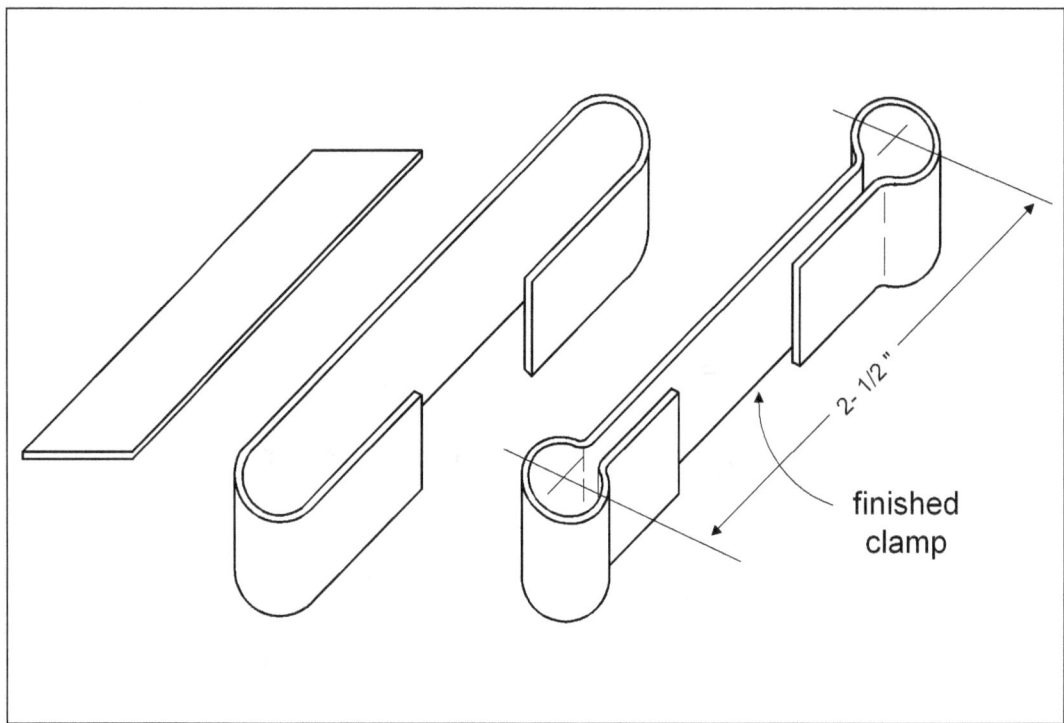

Figure 6.13 Preparing the tuning clamp.

Drill two holes (1/8" diameter) in the tuning clamp (see Figure 6.14).

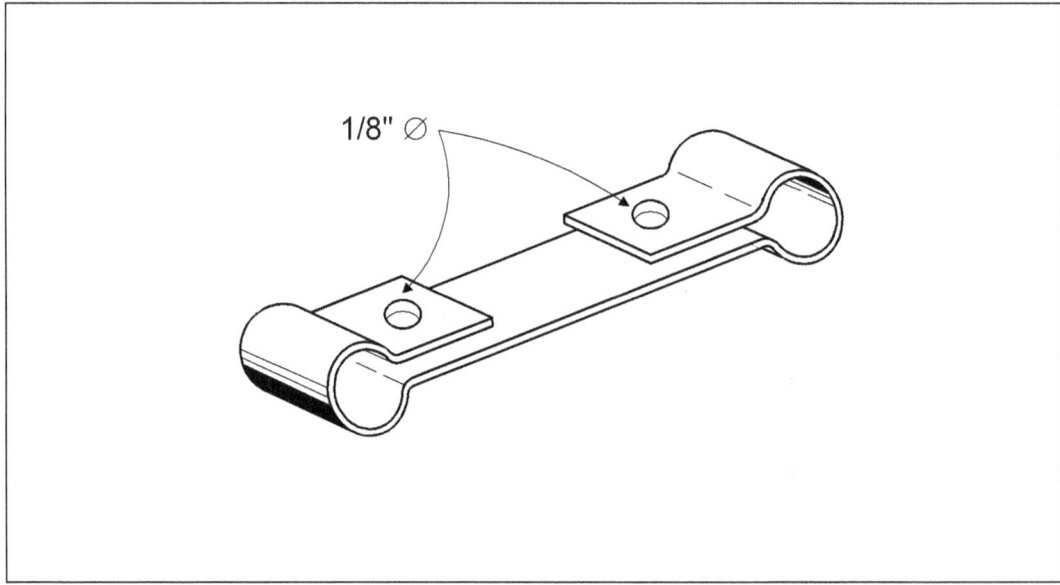

Figure 6.14 Drilling holes in the tuning clamp.

Insert the 6-inch long tube into one loop of the clamp, and secure the clamp with a 1/8" x 3/8" stove bolt. Attach a nut to the bolt and tighten it lightly (see Figure 6.15).
Don't tighten this bolt too much at this time!

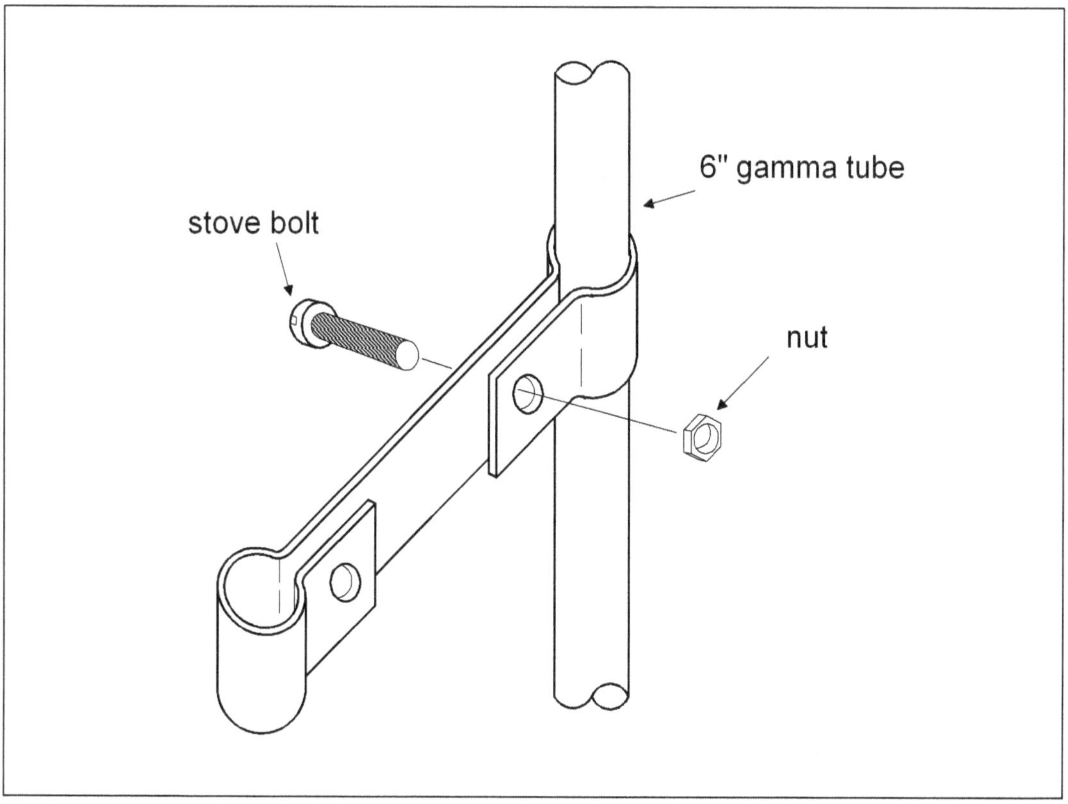

stove bolt

6" gamma tube

nut

Figure 6.15   Inserting the 6-inch long gamma tube into the tuning clamp.

Waterproof the top end of the 6-inch gamma tube by inserting a substantial volume of cotton wad inside the open end. The cotton wad serves as a stopper for the epoxy. Place epoxy glue over the cotton wad inside. Let the epoxy set and dry (see Figure 6.16).

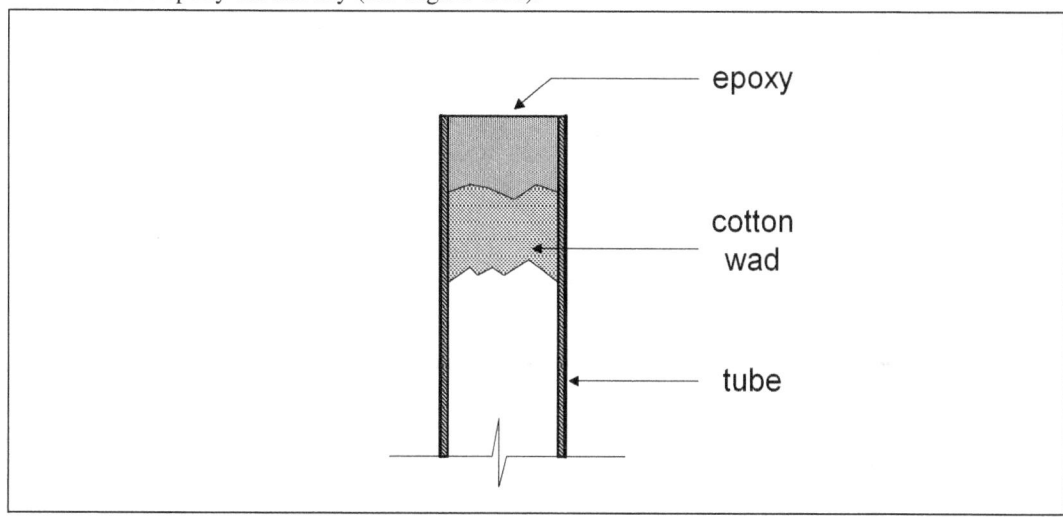

epoxy

cotton wad

tube

Figure 6.16   Sealing the top end of the gamma tube.

While the epoxy is drying, prepare the gamma match from a short length of RG-8/U coaxial cable. Cut a piece 6 inches long, and remove its vinyl outer jacket and its braid. Cut away a small portion of the PE inner insulator, exposing the copper conductor inside (see Figure 6.17).

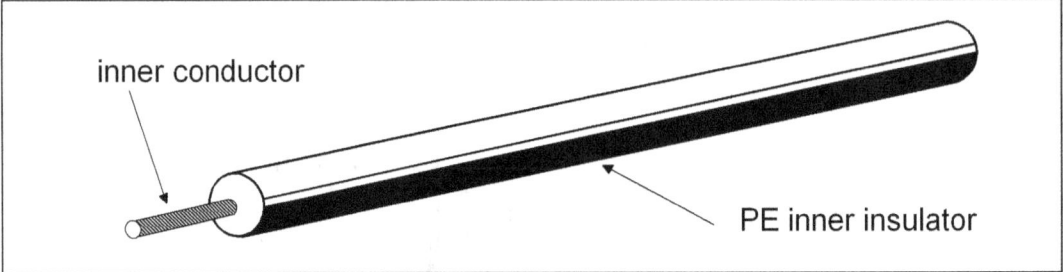

Figure 6.17   Preparing the gamma match from a short length of RG-8/U.

Solder the exposed copper conductor of the gamma match directly into the center pin of the BNC connector attached to the mounting channel.

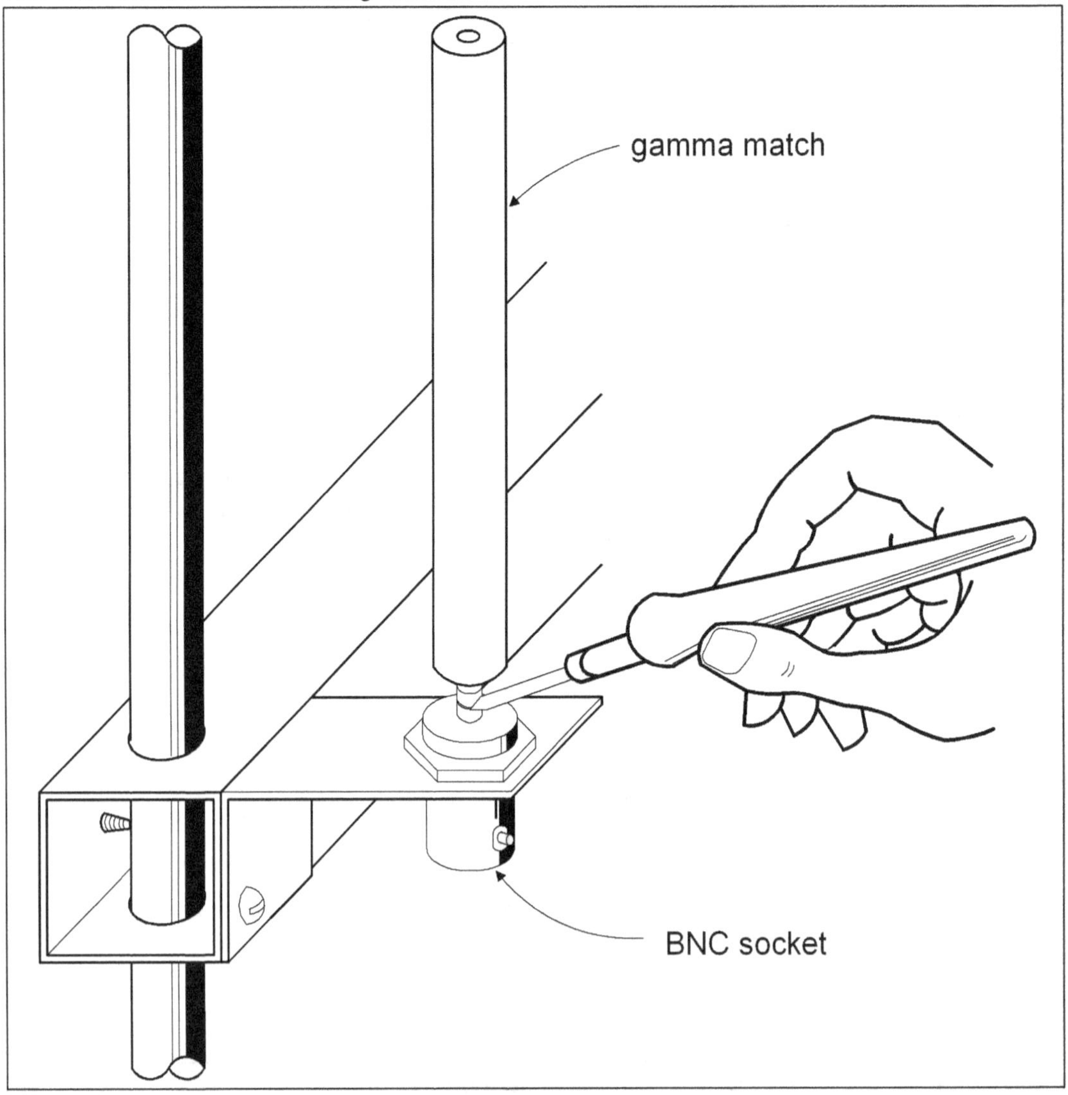

Figure 6.18   Soldering the gamma match into the BNC connector.

Next, carefully insert the free end of the tuning clamp into the radiator element (long tube), starting at the top end. As the tuning clamp is lowered down along with the gamma tube attached to it, insert the gamma match into the gamma tube (see Figure 6.19).

Stop the gamma tube just about 1/2 inches above the plane of the angle bracket, or just enough for the gamma tube to cover the whole PE insulator. Insert a bolt into the tuning clamp attached around the radiator element, and tighten it lightly enough to hold the gamma matching assembly in place. The antenna is already mechanically ready at this point; it only needs to be tuned to resonance for proper operation.

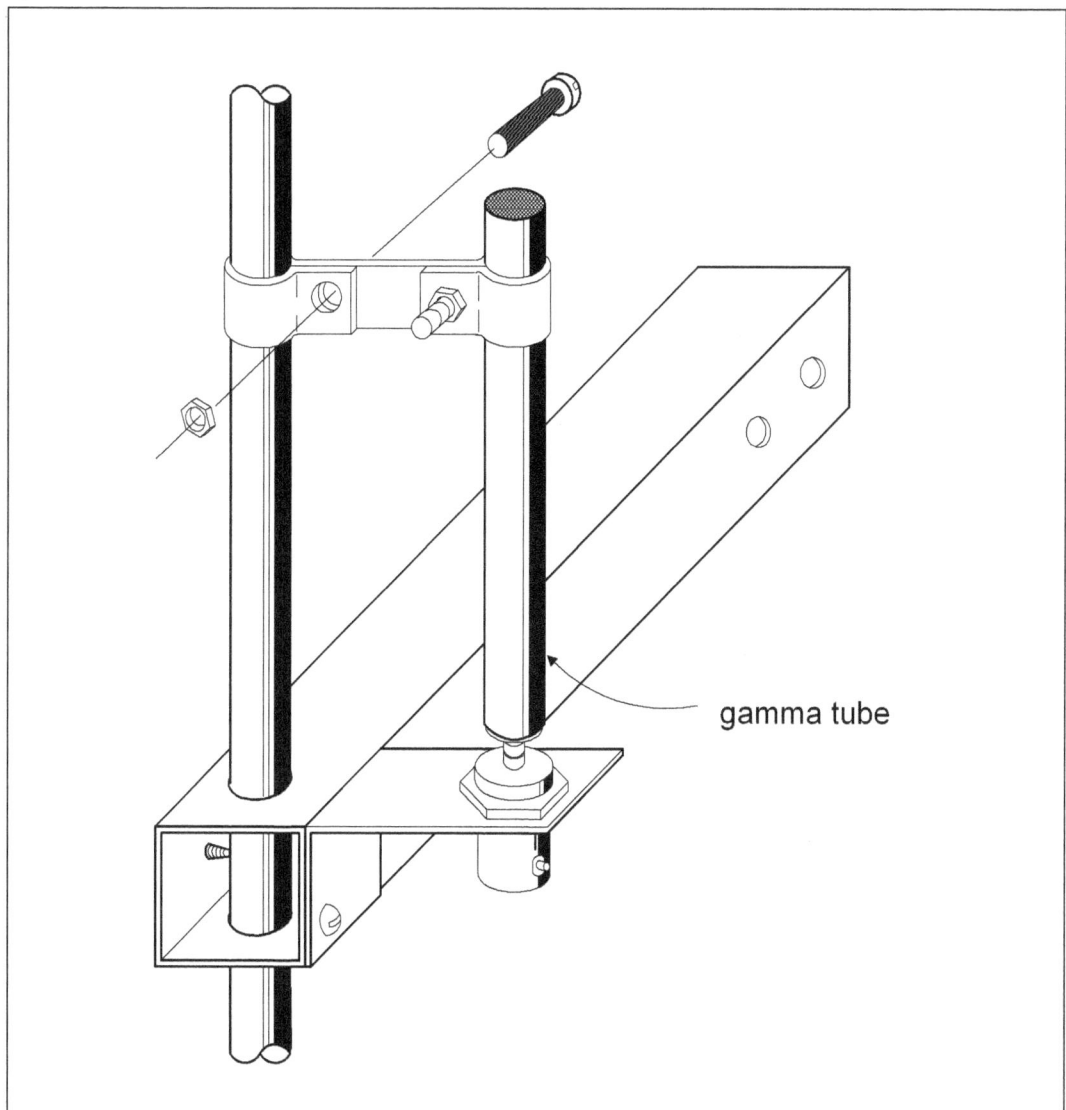

gamma tube

Figure 6.19 Assembling the gamma tube into the gamma match.

In tuning the antenna to resonance, install it to the mast or tower as in Figure 6.1. Connect the coaxial cable to the BNC connector in the angle bracket and connect the other end to a suitable SWR meter. Connect a VHF transceiver to the SWR meter (usually marked 'transmitter') and set its frequency to either the center or extreme frequencies of the band. Key the PTT to transmit and read the SWR response.

To tune the antenna: move the tuning clamp and find the position where you can get the lowest SWR response for the center frequency and a relatively flat response curve over the entire band. Move the tuning clamp either lower or higher about 1/4 inch at a time (see Figure 6.20).

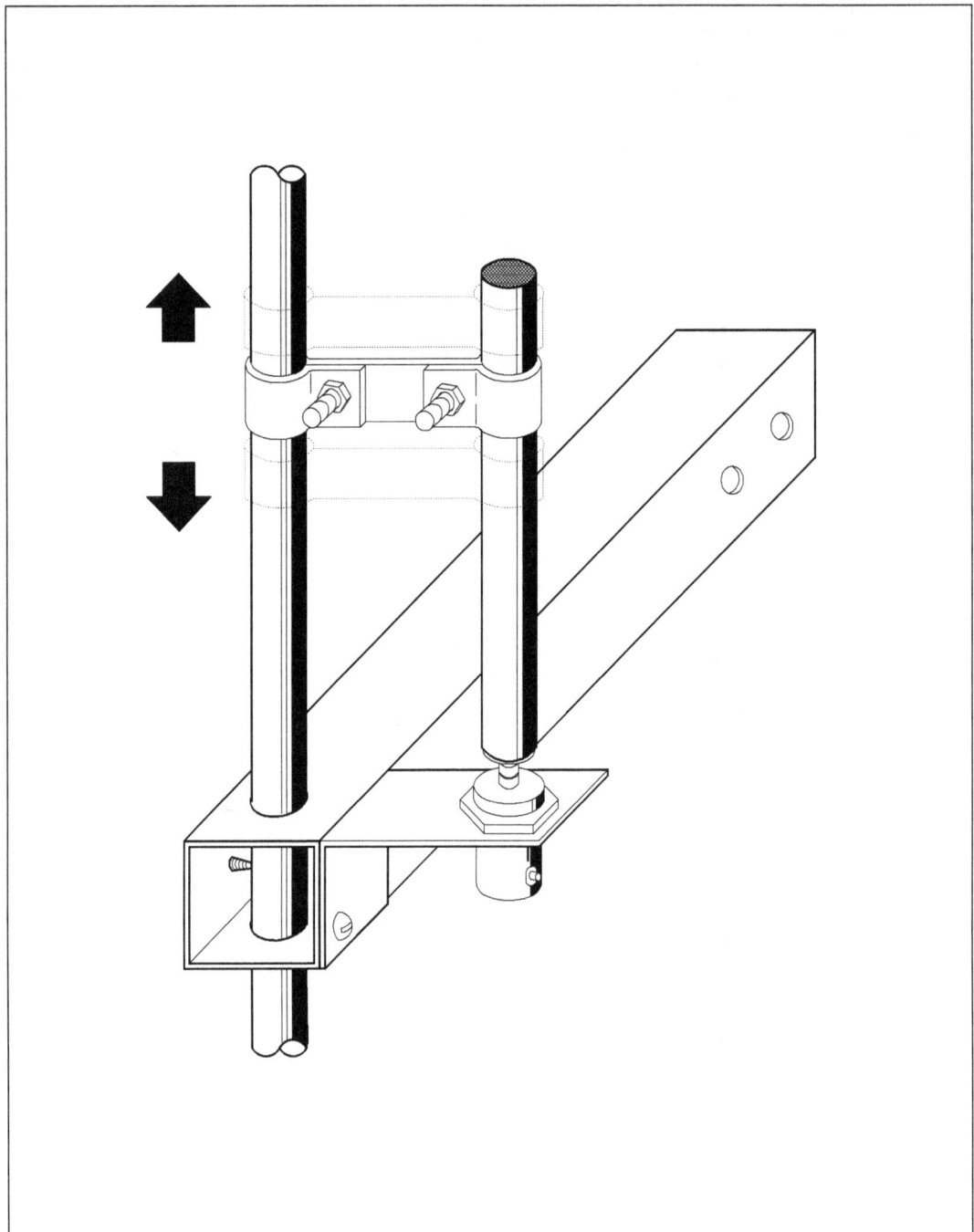

Figure 6.20   Raising or lowering the tuning clamp to find the best match.

If you have finally found the right position (after several trials) then you have succesfully tuned the antenna to resonance. Tighten the nuts at the tuning clamp permanently and place a moderate amout of silicone sealant (RTV compound) around the open lower end of the gamma matching tube to seal it off from moisture and rainwater (see Figure 6.21).

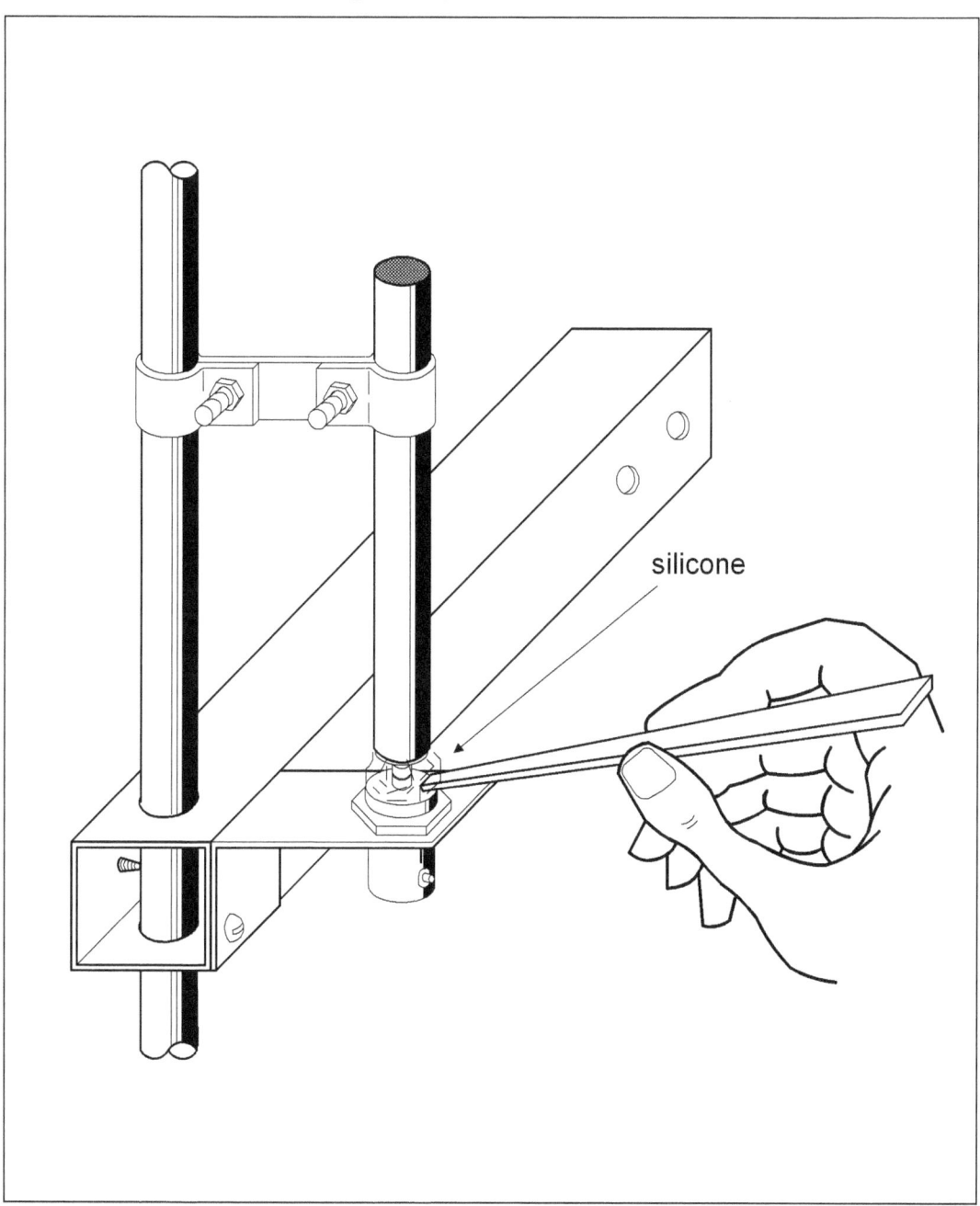

silicone

Figure 6.21   Sealing off the lower end of gamma tube with silicone.

IMPORTANT:

Do not touch any part of the antenna while keying the rig or reading the SWR response.

Do not move the tuning clamp while the transceiver is still transmitting.

REVIEW QUESTIONS

1. What is the advantage of using a gamma-fed dipole antenna?

2. What is a gamma match?

3. What is the function of the coaxial cable inside the short tube?

4. Why is it not necessary to use a balancing tranformer for this particular antenna design?

5. In a mechanical point of view, what are the advantages of this antenna?

6. What is the function of the shorting bar?

7. How is this dipole tuned to resonance?

7 QUAD LOOP ANTENNA

Model QA-2F

In preceding chapters all of the various antenna designs presented are assemblies of linear half wave (or approximately half wave) dipole elements. On the other hand other element forms may also be used to effectively function as an antenna. One example is the quad antenna described in this chapter. This is the type of antenna with a radiating element made of a loop having a perimeter of one wavelength and used in much the same way as a dipole.

The quad antenna was originally designed in the late 1940's. Since then it has been the subject of controversy whether it performs better than a dipole. The debate continues but after some years several facts have become apparent. It was found out that the quad has a slight gain of approximately 2 dB over a dipole. It is also said to cover a wider area in the vertical plane and exhibits broadband characteristics.

The quad antenna model QA-2F is specifically designed to operate in the frequency band of 140-150 MHz. It displays a bi-directional radiation pattern with maximum radiation in the direction perpendicular to the plane of the loop. By carefully following the instruction for constructing this antenna you should be able to get an SWR response of less than 1.5:1 over the entire band.

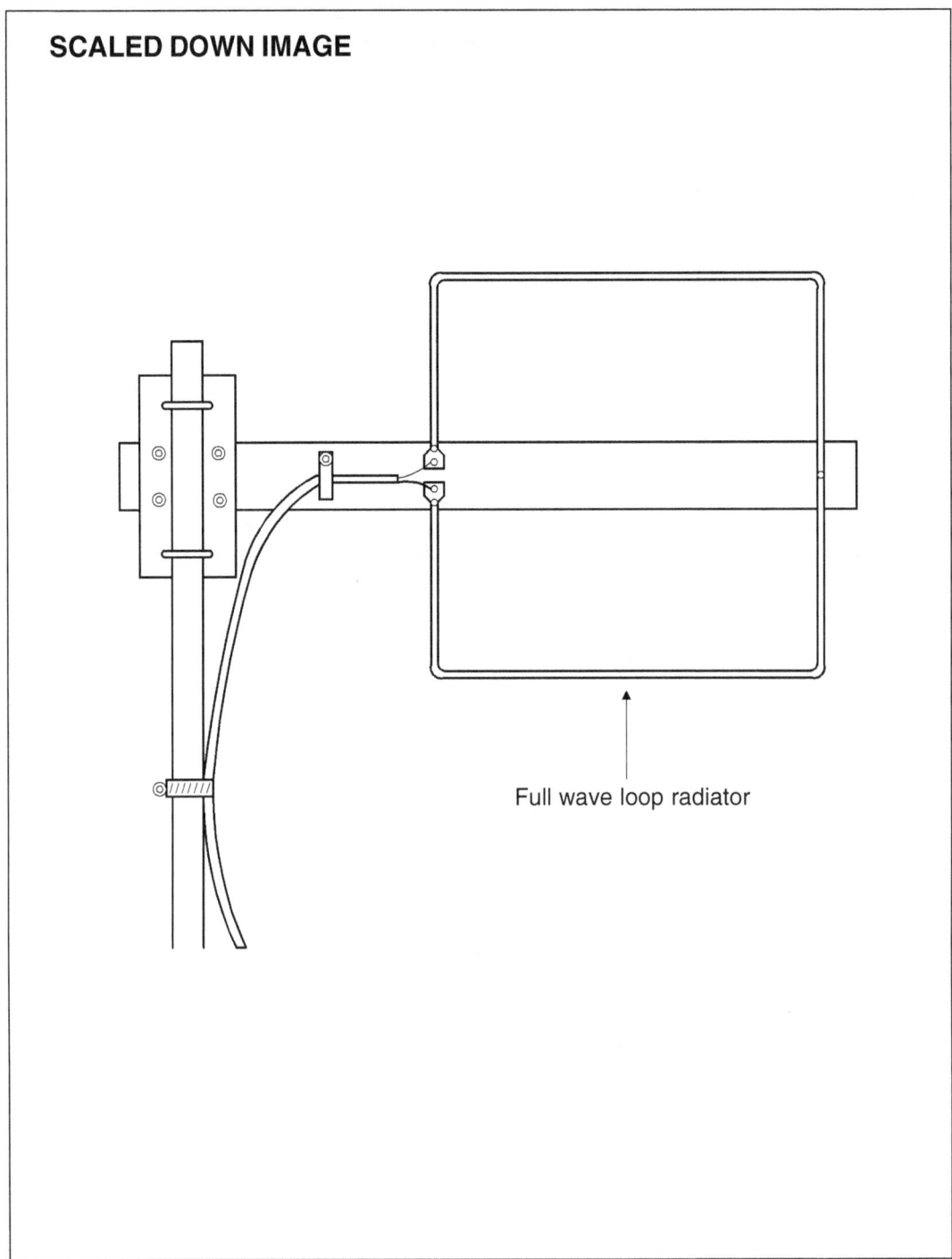

SCALED DOWN IMAGE

Full wave loop radiator

Figure 7.1   A quad loop antenna model QA-2F.

Materials List

Quantity	Specification/Description	Dimensions
1	Aluminum tube	3/8" id* 82 „ long
1	Plastic plate	1/2 „ thick 32" long
2	Stove bolts - brass or GI	1/8" x 3/8"
3	Stove bolts - brass or GI	1/8" x 1"
4	Stove bolts - brass or GI	3/16" x 1"
2	U-bolts with accompanying hex nuts and lockwashers	
1	Plastic C-clamp - enough to hold a 3/8" cable	
1	Self tapping metal screw	1/8" x 3/8"
2	Eye terminals - vinyl insulated	
4	Plain washers - 1/8" id*	
1	Hose clamp - enough to hold a 1-1/2" tube	

*id- inside diameter

Construction

First, prepare the plastic mount with dimensions shown in Figure 7.2 on the next page.

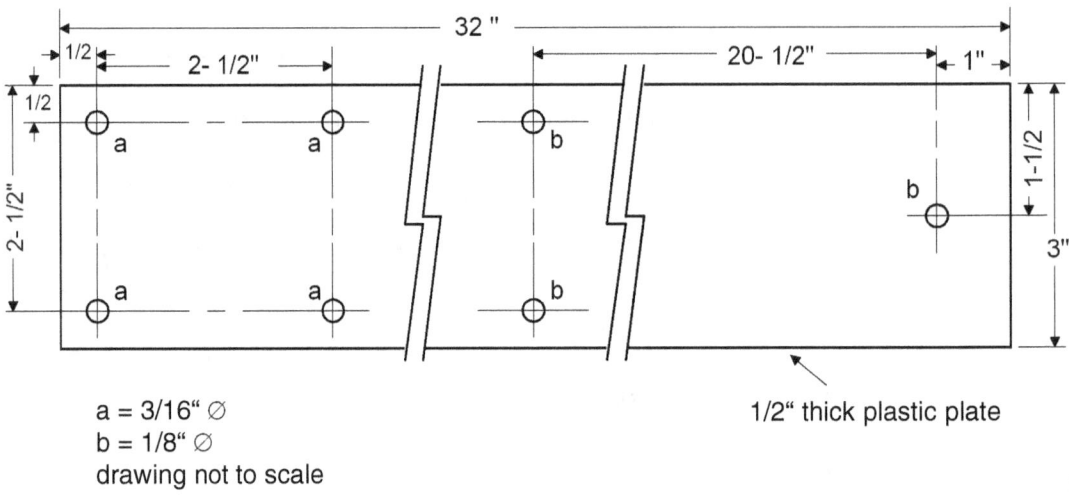

a = 3/16" ∅
b = 1/8" ∅
drawing not to scale

1/2" thick plastic plate

Figure 7.2 Plastic mount dimensions.

Next, prepare the metallic mast adaptor. As shown in the following illustration, the distance between one pair of 3/16" holes at the extreme ends is equal to the distance between the threaded ends of the U-bolt used (see Figure 7.3).

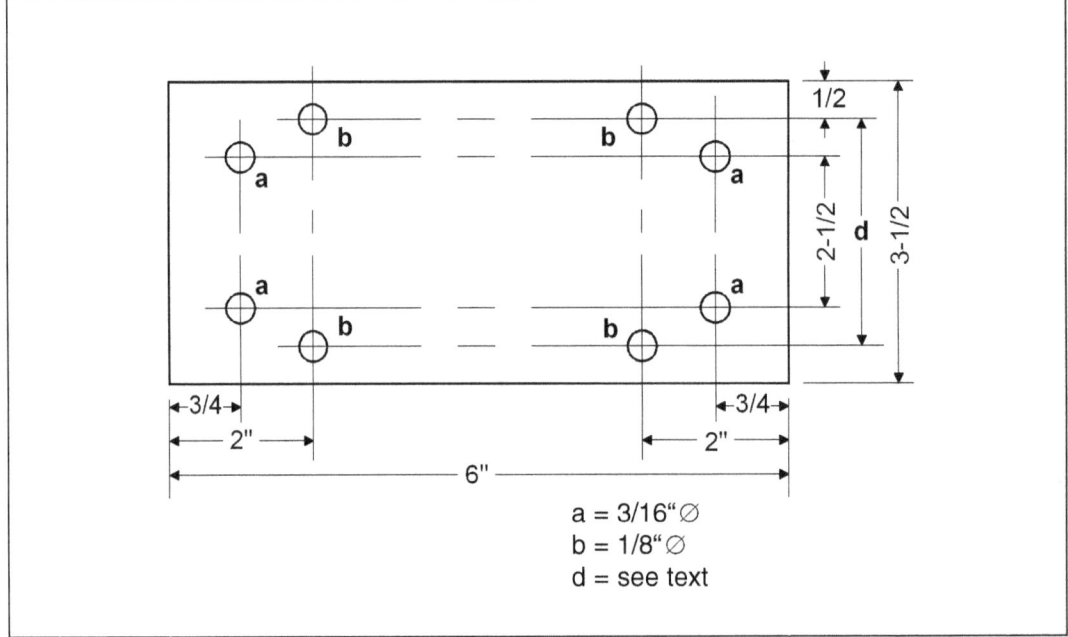

a = 3/16" ∅
b = 1/8" ∅
d = see text

Figure 7.3 Mast adaptor dimensions.

Join the two plates together using four 3/16" x 1" stove bolts made of rust resistant materials (e.g. brass or stainless steel). Do not forget to include a lock washer in each bolt (see Figure 7.4).

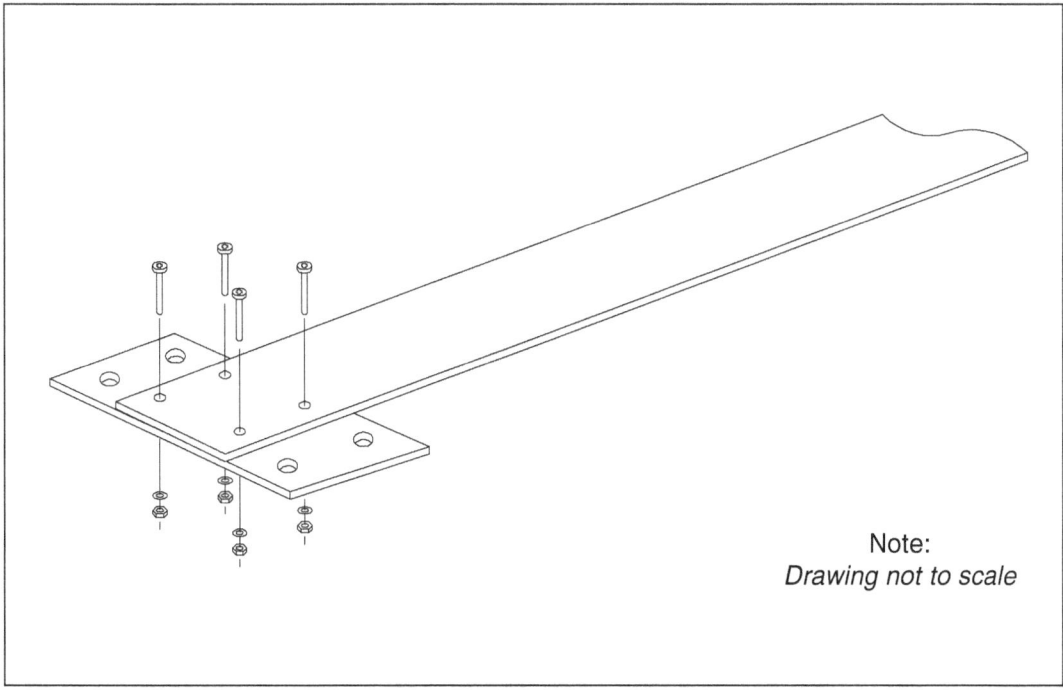

Figure 7.4 Joining the two plates together with stove bolts.

Bend the aluminum tube into a square loop with equal sides using a suitable tube bender (see Figure 7.5). Cut away the excess tube.

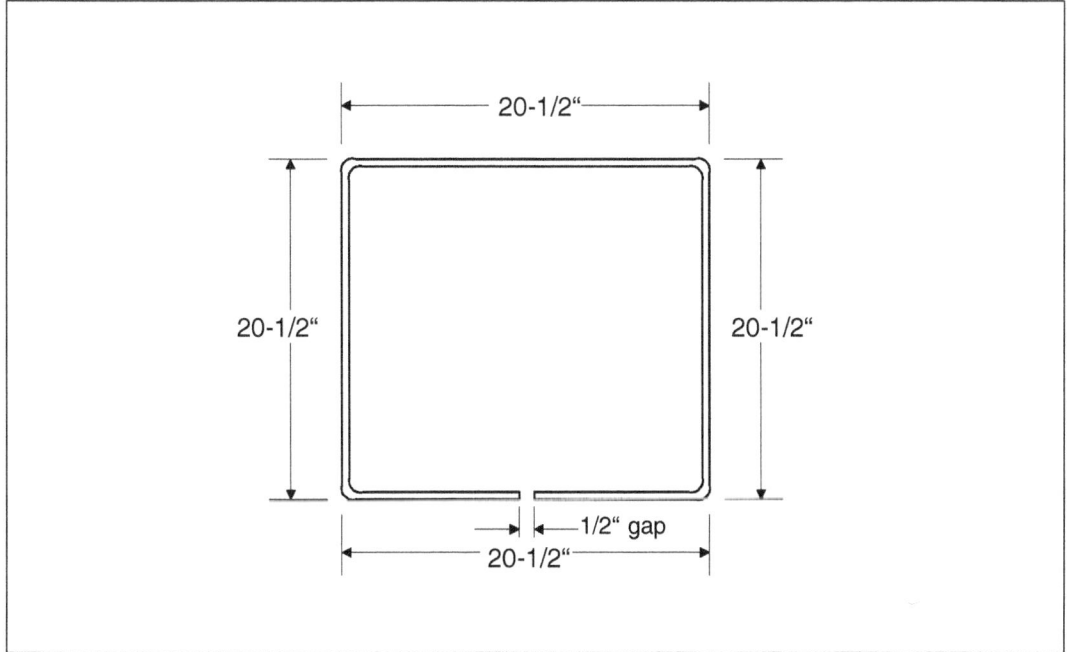

Figure 7.5 Forming the tube into a square loop.

Flatten a small portion at both ends of the tube and drill a hole (1/8" diameter) in each flattened end (see Figure 7.6).

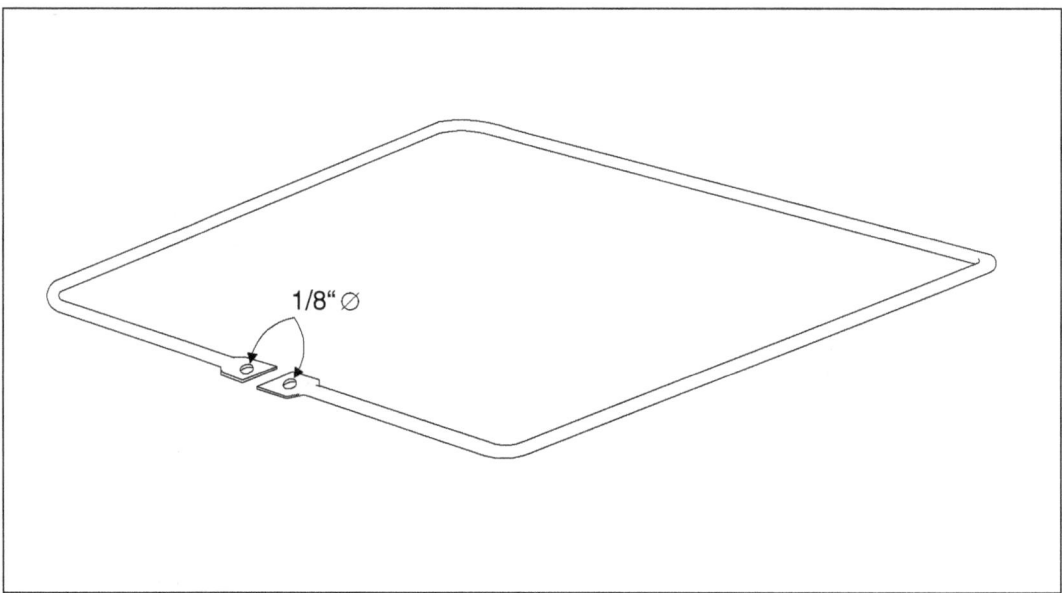

Figure 7.6   Flattening and drilling the ends of the tube.

Drill additional holes (1/8" diameter) in the tube as shown in the following illustration (Figure 7.7). The holes must be drilled through and through. Be careful in drilling the holes to avoid deforming the tube.

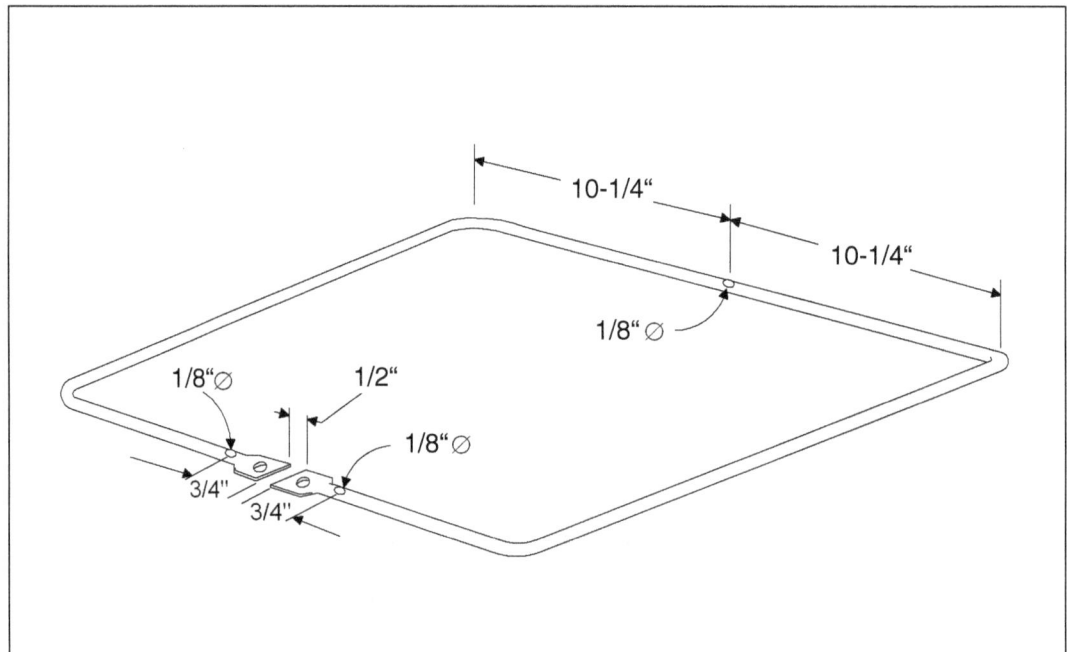

Figure 7.7   Drilling additional holes in the tube.

Insert two stove bolts (1/8" x 3/8") through the holes at both ends of the loop and attach the necessary hardware as shown in Figure 7.8.

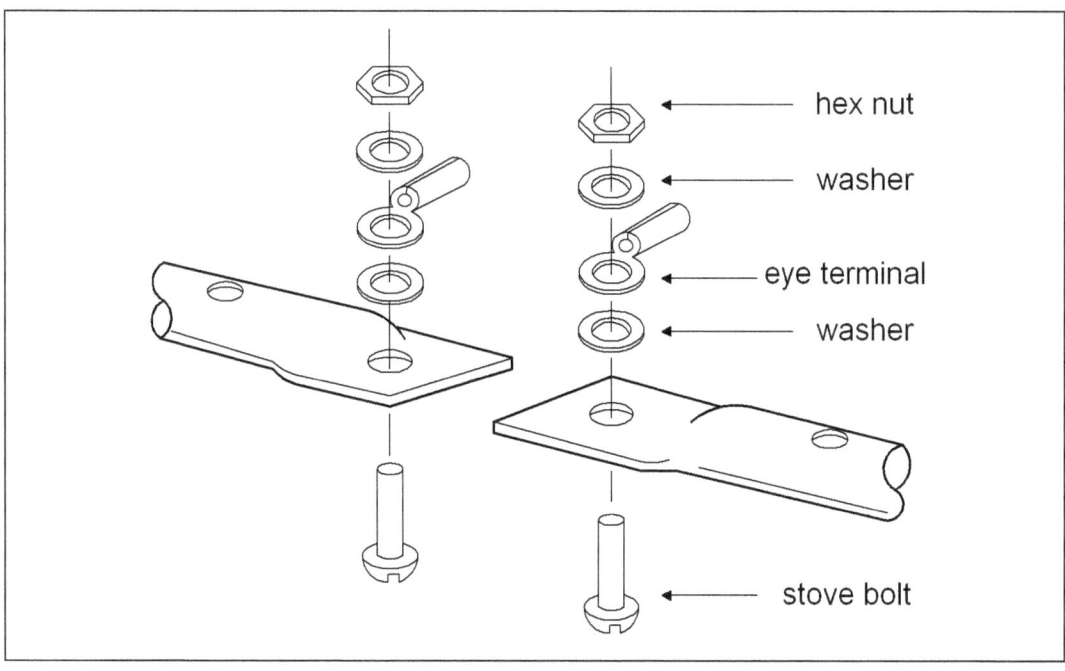

Figure 7.8   Installing the necessary hardware at both ends of the tube.

Attach the prepared loop into the plastic mounting plate by bolting it through the 1/8" diameter holes as shown in the following illustration. Use 1/8" x 1" stove bolts (brass or stainless steel).
See Figure 7.9.

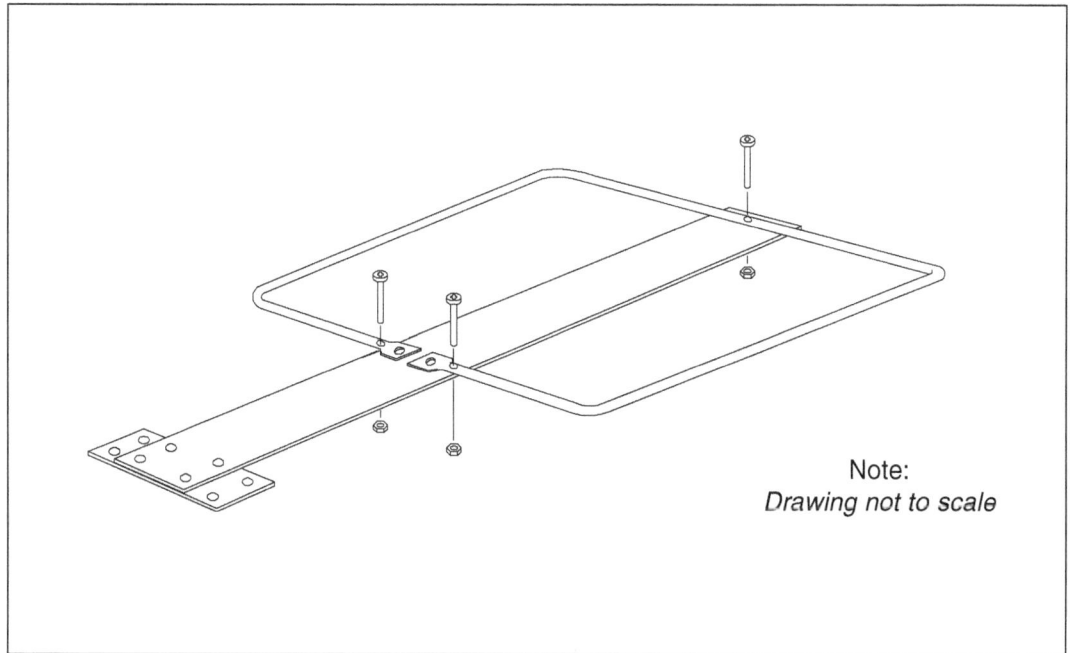

Figure 7.9   Fixing the loop on the plastic mount.

Prepare one end of the coax cable by separating the inner conductor from the copper braid. Solder the two conductors to the two eye terminals in the loop. The braid is costumarily connected to the lower terminal (see Figure 7.10).

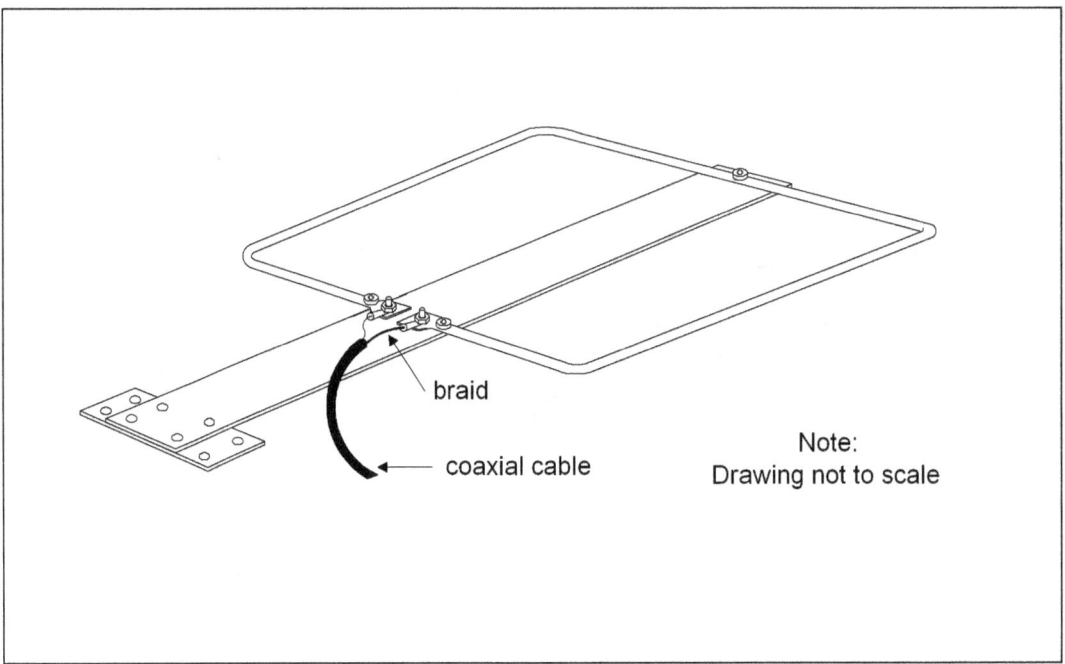

Figure 7.10   Connecting the coaxial cable to the loop element.

Clamp the coaxial cable to the plastic mounting plate (see Figure 7.11).

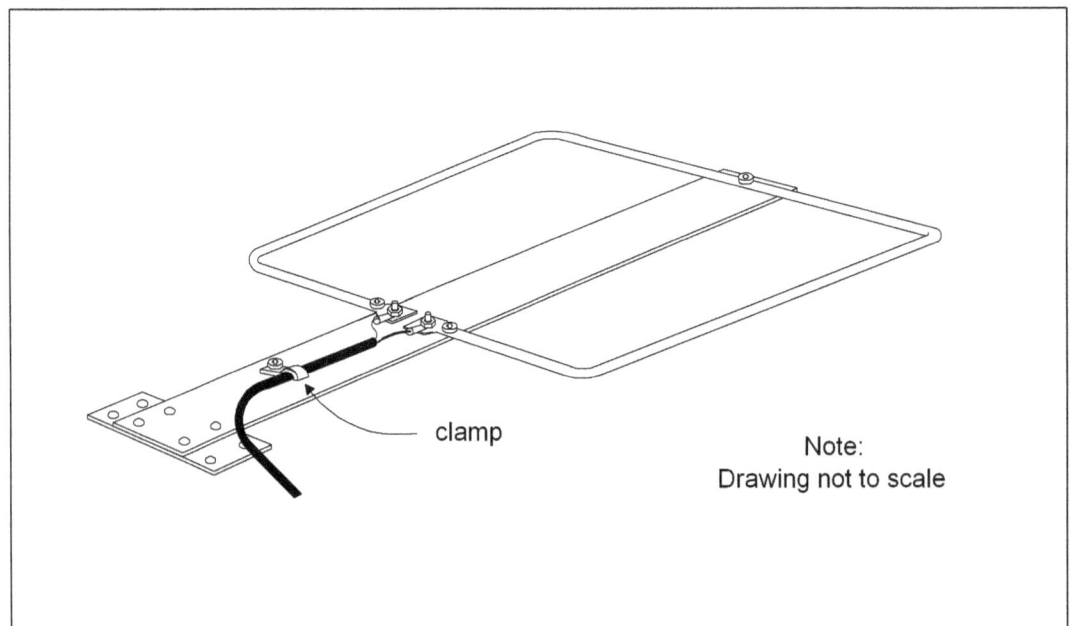

Figure 7.11   Clamping the coaxial cable to the plastic mount.

Installation of QA-2F

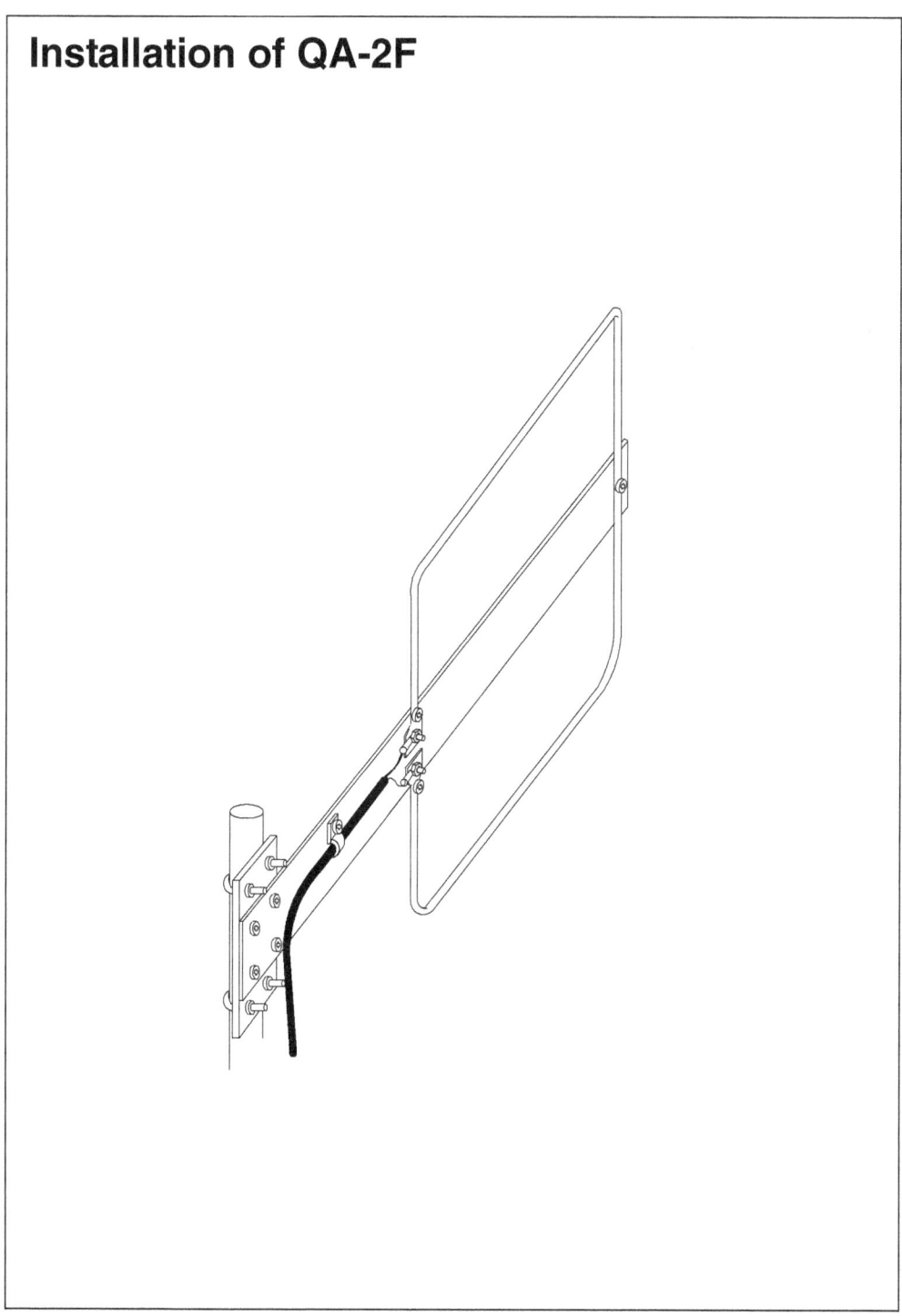

Figure 7.12   Installing the QA-2F to the mast.

REVIEW QUESTIONS

1. What is the advantage of using a quad antenna?

2. What is the radiation pattern of a quad?

3. By studying the design presented here, is it possible to replace the mounting plane with a metallic one?

4. Which end of the loop is the shield of the coax cable attached to?

5. What is the direction of the antenna's maximum radiation?

Table 7.1 American and English Wire gauges, diameter in inches and millimeter

The american standard wire gauge is based on the standards of the *Brown & Sharpe* company which uses numbers in identifying the wire size. In general, the abbreviation AWG (= *American Wire Gauge*) is used. In Great Britain, there are two standard wire gauges: *BWG* (= *Birmingham Wire Gauge*) and *ISWG* (= *Imperial Standard Wire Gauge*) or *SWG* (= *Standard Wire Gauge*). Both these standards also use numbers to identify the size of the wire.

Wire gauge Nr.	AWG diameter in inches	in mm	BWG diameter in inches	ISWG(SWG) diameter in mm.	in inches	in mm.
0000(4/0)	0.460	11.68	0.454	11.53	0.40	10.16
000(3/0)	0.409	10.41	0.425	10.80	0.372	9.45
00 *(210)*	0.365	9.27	0.380	9.65	0.348	8.84
0(110)	0.325	8.25	0.340	8.64	0.324	8.23
1	0.289	7.35	0.300	7.62	0.300	7.62
2	0.258	6.54	0.283	7.21	0.276	7.01
3	0.229	5.83	0.259	6.58	0.252	6.40
4	0.204	5.19	0.238	6.05	0.232	5.89
5	0.182	4.62	0.220	5.59	0.212	5.38
6	0.162	4.11	0.203	5.16	0.192	4.88
7	0.144	3.66	0.179	4.57	0.176	4.47
8	0.128	3.26	0.164	4.19	0.160	4.06
9	0.114	2.90	0.147	3.76	0.144	3.66
10	0.102	2.59	0.134	3.40	0.128	3.25
11	0.091	2.30	0.120	3.05	0.116	2.95
12	0.081	2.05	0.109	2.77	0.104	2.64
13	0.072	1.83	0.195	2.41	0.092	2.34
14	0.064	1.63	0.083	2.11	0.081	2.03
15	0.057	1.45	0.072	1.83	0.072	1.83
16	0.051	1.29	0.065	1.65	0.064	1.63
17	0.045	1.15	0.058	1.47	0.056	1.42
18	0.040	1.02	0.049	1.24	0.048	1.22
19	0.036	0.91	0.042	1.07	0.040	1.02
20	0.032	0.81	0.035	0.89	0.036	0.92

NOTE: Values in millimeter were rounded off. AWG 21 to 40 see Table 10.1 in page 119.

8 DISCONE ANTENNA

Model CD-2W

Most of the antenna designs described in the preceding chapters are all suitable for VHF work requiring omnidirectional pattern of radiation. Also in the mechanical viewpoint, these designs are simple and easy to construct which makes them very popular among radio operators. However all of them have a limited bandwidth of 140-150 MHz.

If one attempts to operate his transceiver outside these frequency limits (assuming he has a wideband transceiver) the signal response becomes weaker as the operating frequency of the transceiver is moved farther away from the operational bandwidth of the antenna. At the same time the SWR in the transmission line increases and can reach an intolerable point which may cause damage to the transceiver. Although this handicap can be avoided by using a different antenna tuned to a different frequency band, the process of changing antennas every time the operator changes his operating band becomes time-consuming and cumbersome. This problem can be solved by using a discone antenna described in this chapter.

The discone antenna is a broadband antenna. Meaning it can operate over a wide range of frequencies. Theoretically, a properly designed discone antenna can operate up to a frequency 10 times the value of its lowest operational frequency. Specifically speaking, if a discone antenna is designed to operate with a lowest operational frequency of 140 MHz, then it can be conveniently used up to 1.4 gigahertz! The lowest operational frequency is called cut-off frequency. Below this frequency the SWR will increase rapidly.

Amazing! Well, a discone antenna can achieve that because it functions more like a transformer than a conventional antenna. It couples the low impedance transmission line to the higher impedance of free space. Its signal pattern is similar to that of a quarter wave ground plane antenna. Radio waves from the transmission line emerge at the feed point (cone apex) and travel along the antenna surface to the edges of the cone and disc. In designing the discone, the dimensions of the antenna are carefully computed so as to make the impedance at its edges similar to that of free space. Naturally the discone radiates a signal because there is a maximum transfer of energy when the impedances are matched.

The discone antenna described in this chapter is made of wire screen mesh. This material is purposely used to minimize the effect of wind to the antenna. The thin metal strips used to clamp the two overlapping edges of the cone are for mechanical reasons only -- the RF waves travel down to the cone edge and not around it, so an electrical connection is not important.

This unit has an operational frequency bandwidth of 140 MHz up to 1.4 Gigahertz, although best results can be obtained if its use is limited up to 1 Gigahertz only. SWR is measured to be less than 2:1 over the entire bandwidth. Power gain is 1 dB (unity gain), and its installation is fixed. The discone antenna does not need tuning after construction. It is also popular for use in automatic scanning wideband monitors.

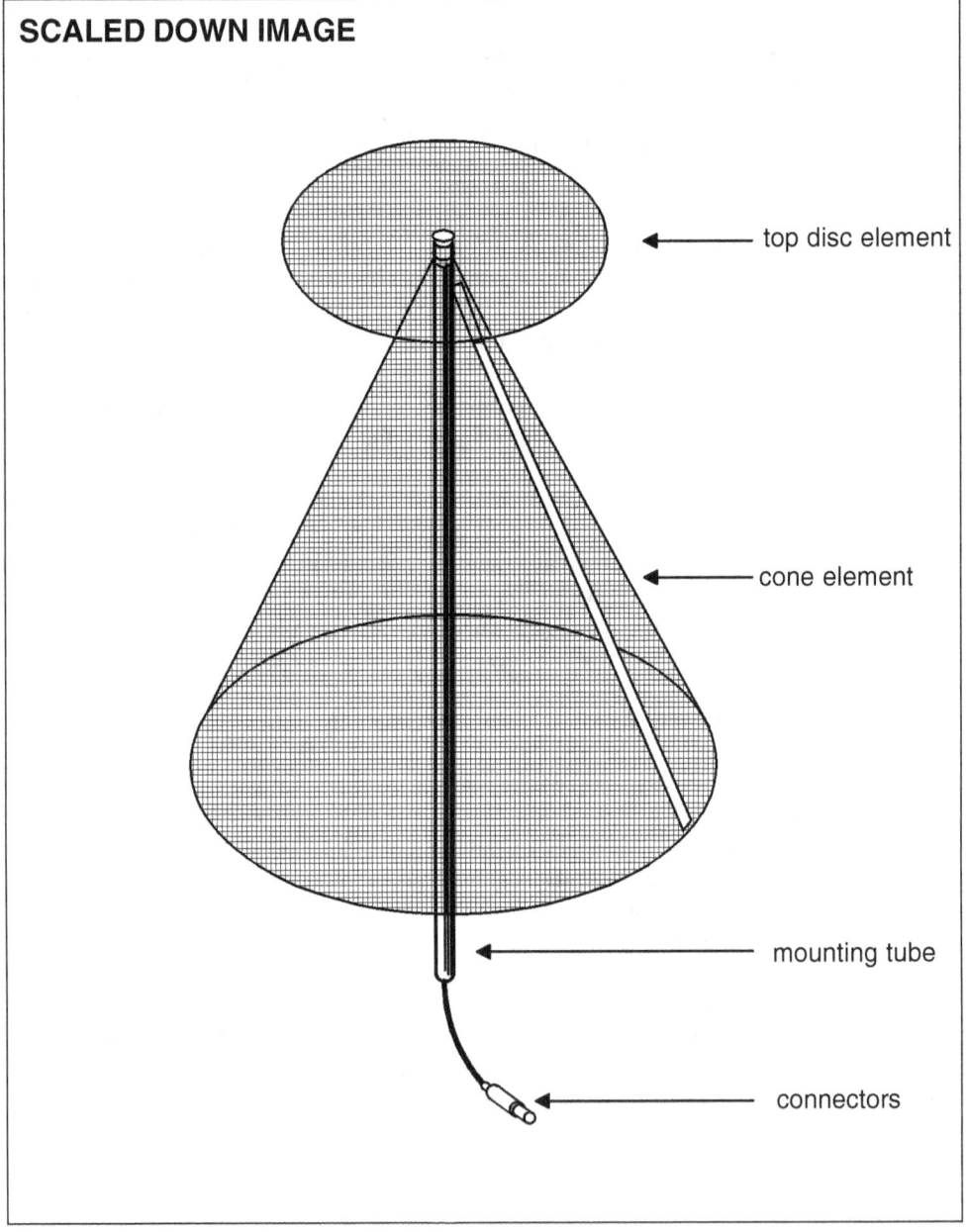

Figure 8.1   Discone Antenna CD-2W

Materials List

Quantity	Specification/Description	Dimensions
1	Aluminum or GI wire screenmesh medium gauge - enough to support itself without reinforcement	
1	Aluminum tube	1" id* 25" long
1	PL-259 VHF male connector	
1	PL-258 VHF straight connector	
1	Coaxial cable RG-58/U	30" long
2	Eye terminals	1/8"
1	Washer - aluminum (customized dimensions see text)	
1	Plastic bushing (customized dimensions see text)	
1	Stove bolt - brass or GI	1/8" x 1-3/4"
1	Hex nut	1/8" id*
1	Hose clamp - stainless steel	1-1/2" diameter
1	Metal plate 1/8" thick	3" x 6"
2	Aluminum strip gauge 14 or 16	1/2" x 22"
4	U-bolt with hex nuts and lock washers	

*id- inside diameter

Construction

First, prepare the customized aluminum washer to be used as a disc holder. Machine it from a thick aluminum plate or rod, following the dimensions shown in Figure 8.2.

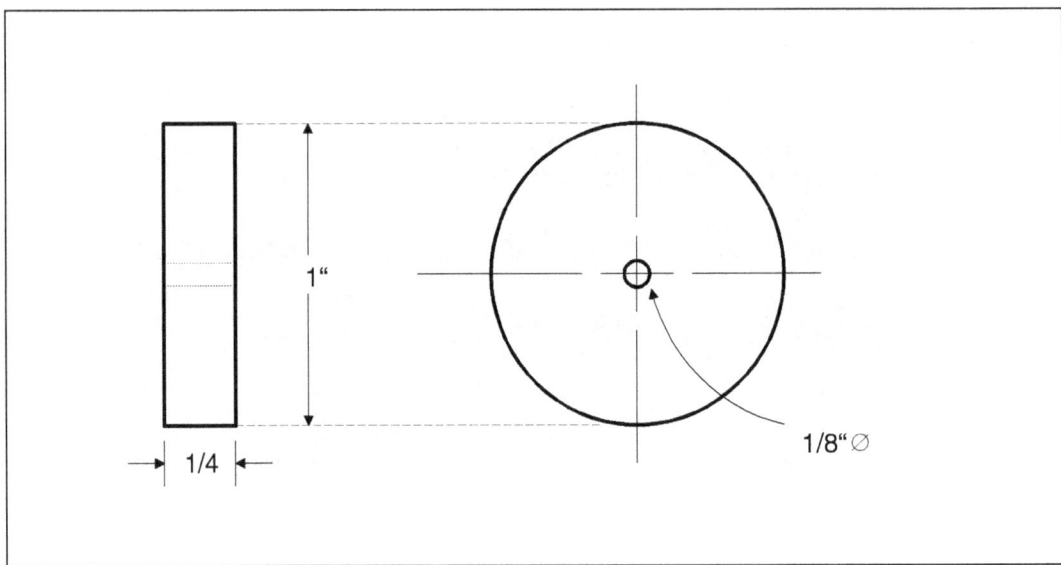

Figure 8.2   Customized aluminum washer dimensions.

Next, prepare the plastic bushing from a small piece of engineering plastic rod with the required diameter. Machine it according to the dimensions shown in Figure 8.3.

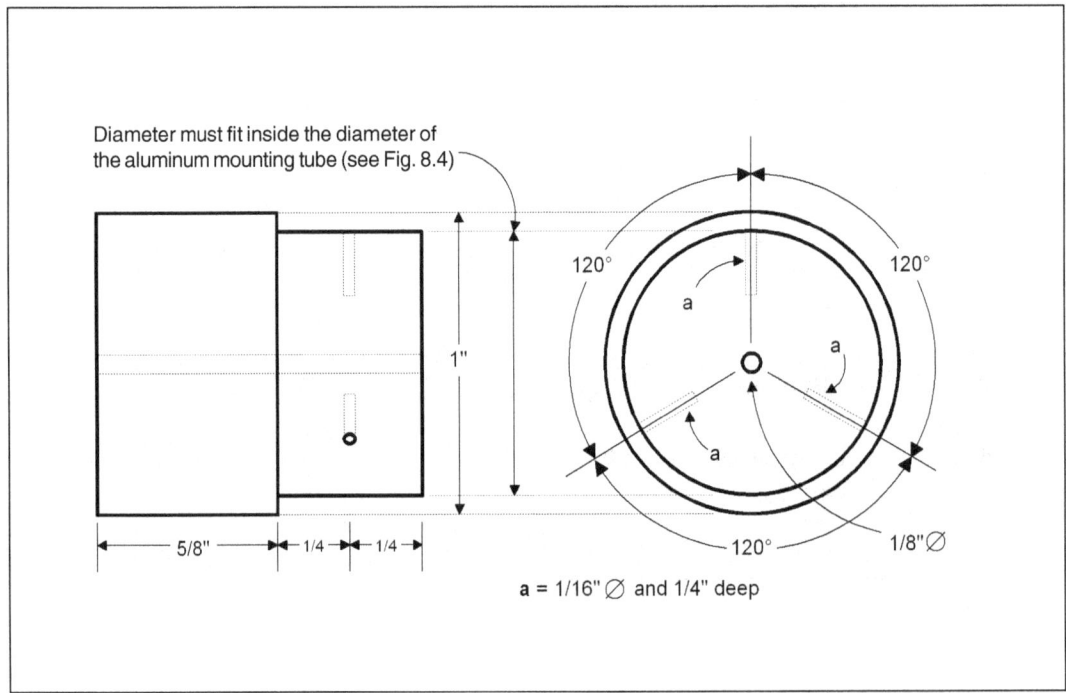

Figure 8.3   Plastic bushing dimensions.

Next, prepare the aluminum mounting tube. The tube must be 1" in diameter and 25 inches long. Drill three holes (1/8" diameter) around one end of the tube, with the holes equally spaced between each other (see Figure 8.4).

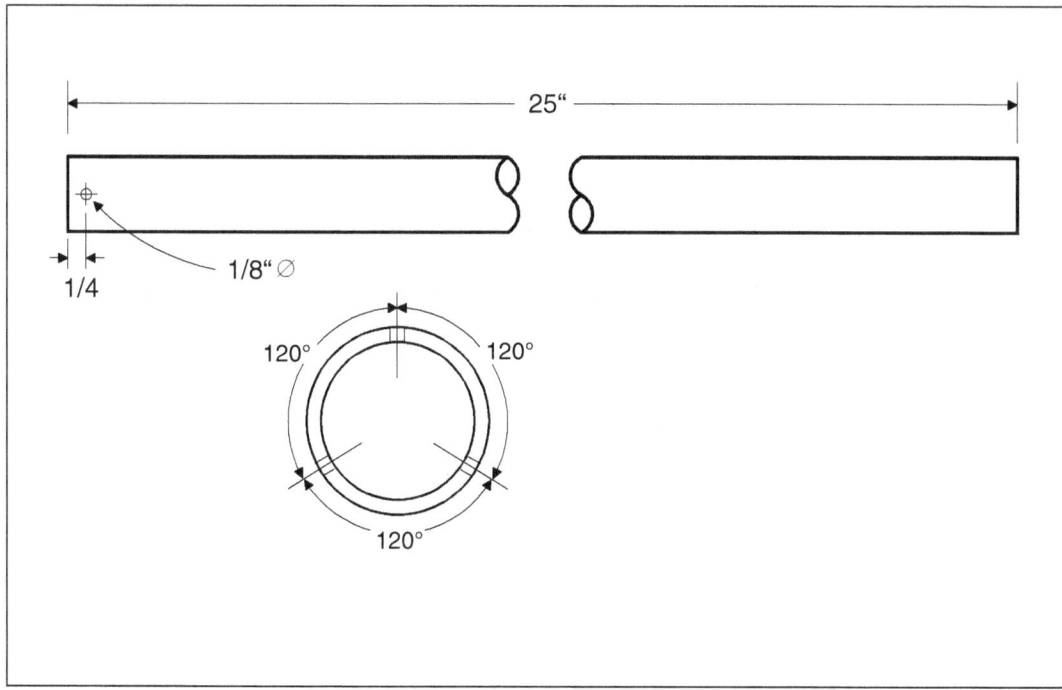

Figure 8.4   Drilling holes at one end of the mounting tube.

Drill a single hole (1/8" diameter) at the same end, but slightly lower than the first three holes (see Figure 8.5). This hole will accomodate the screw to hold the coaxial braid inside the tube as described later in the final steps.

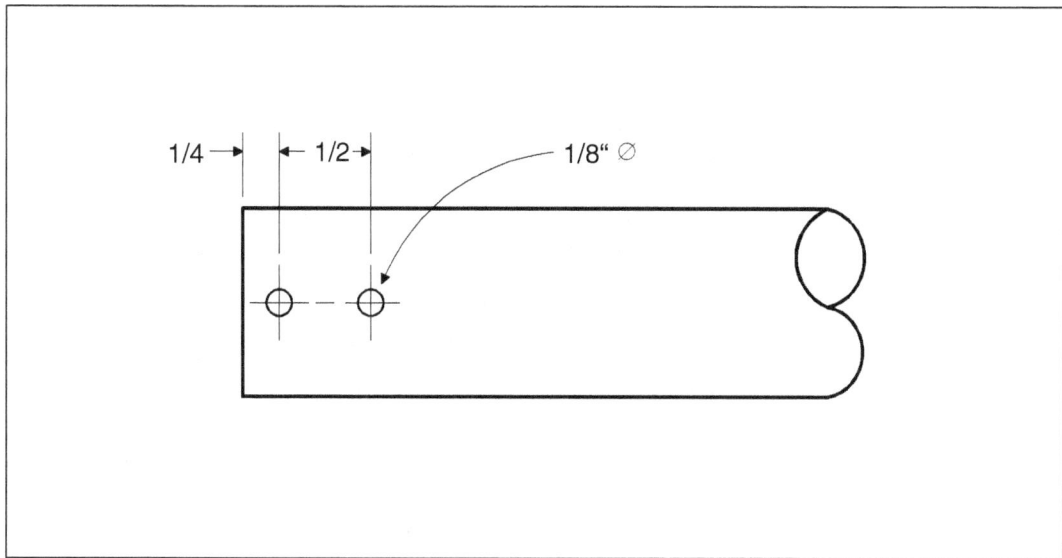

Figure 8.5   Drilling the hole which accomodates the coaxial braid lock screw.

Preparing the disc and cone

Cut the disc element from the aluminum screen mesh using a suitable tin snip (see Figure 8.6).

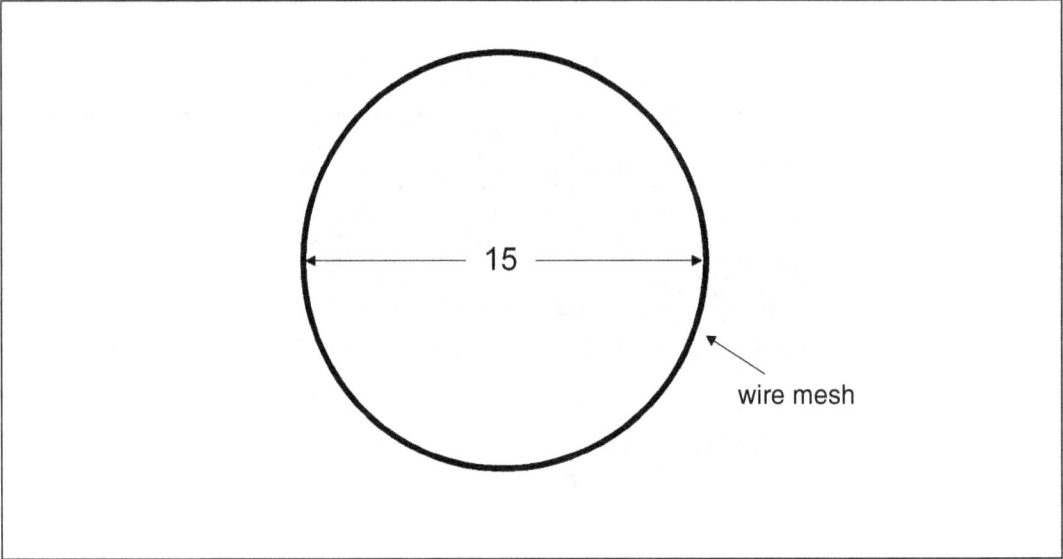

Figure 8.6   Disc element dimension.

Next, prepare the cone element from a similar material. Follow the dimensions shown in Figure 8.7.

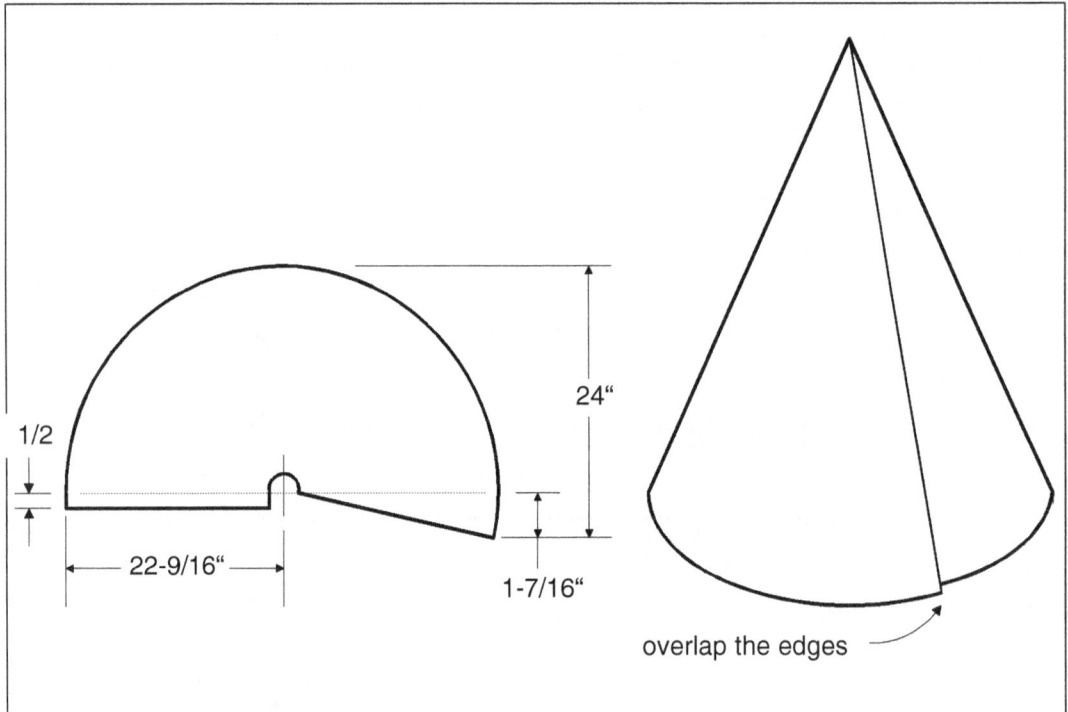

Figure 8.7   Cone element dimensions.

Prepare the aluminum strips according to the dimensions shown in Figure 8.8. These two strips will be used to clamp the two overlapping edges of the cone permanently.

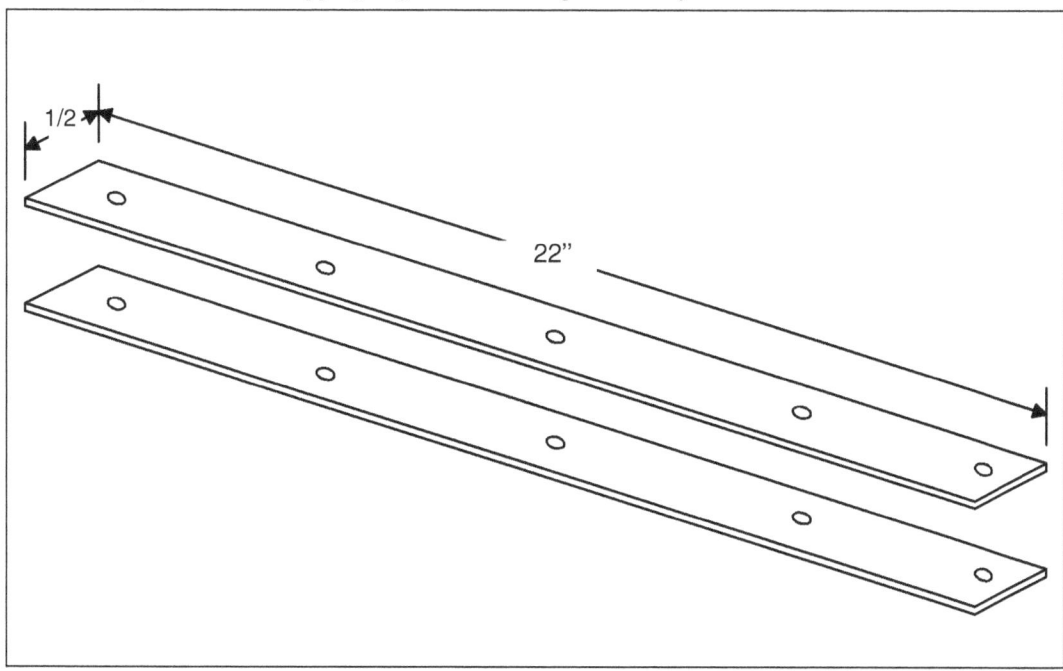

Figure 8.8   Preparing the clamping strips.

Place one strip along the overlapping edges under the cone, and place the other strip over the overlap outside the cone. Align the holes in both strips, and rivet the two pieces together. The rivet must pierce through the two overlapping edges (see Figure 8.9). The riveted strips must sandwich the screen mesh and hold the cone form rigidly.

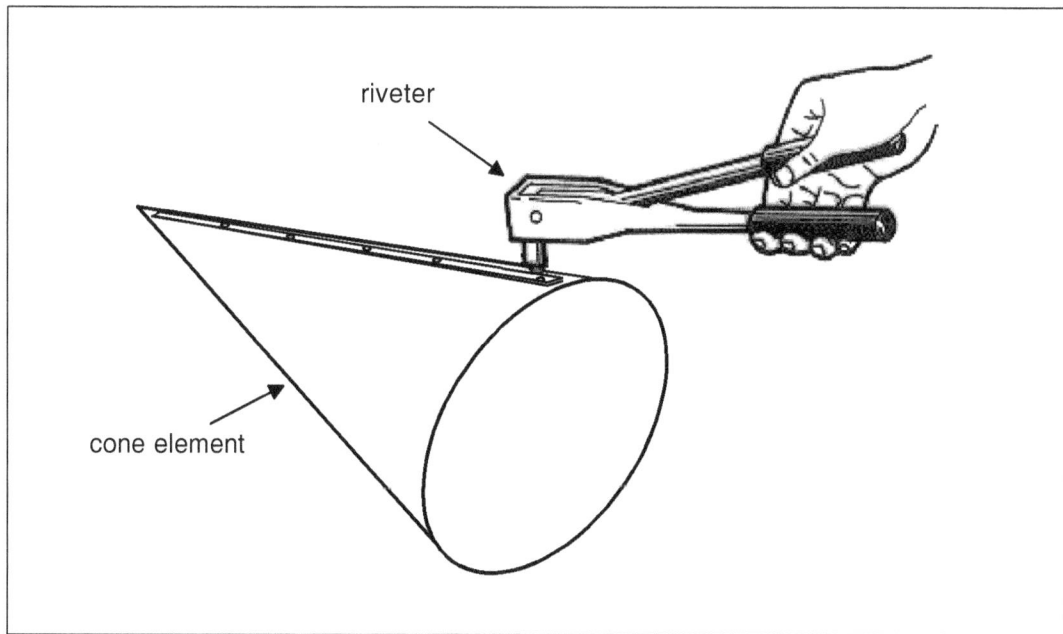

Figure 8.9   Riveting the clamping strips.

By using a tin snip, make crosscuts on the apex of the cone to make a hole large enough for the mounting tube to go through. Follow the illustration in Figure 8.10 carefully . The cuts must result in an opening equal to the diameter of the tube -- around 1 inch.

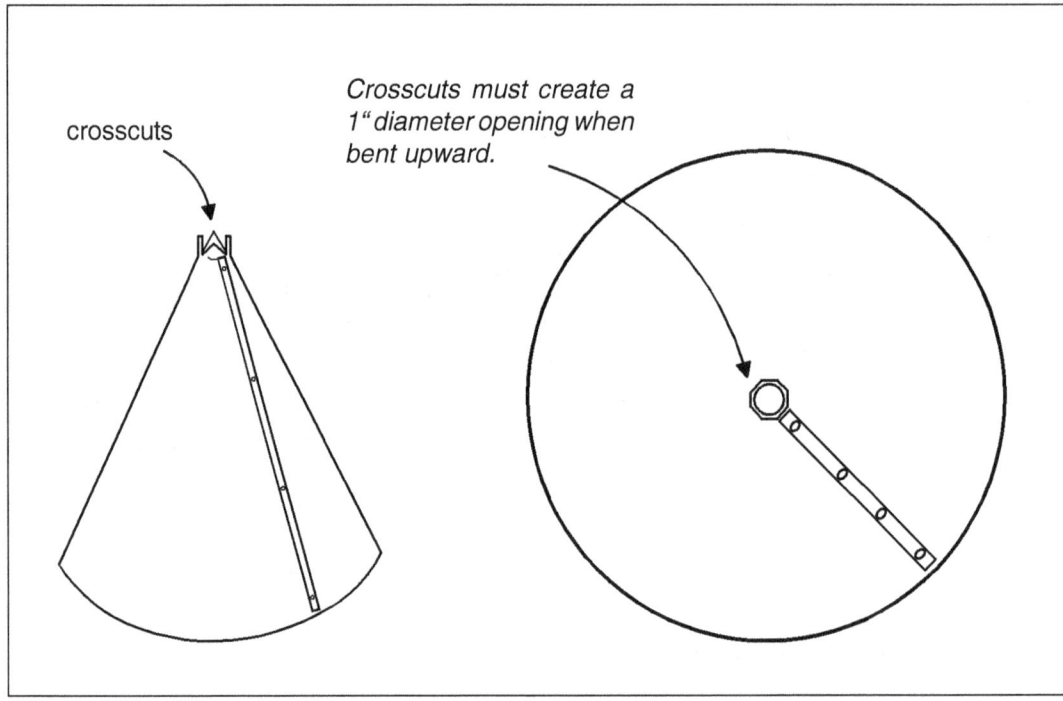

Figure 8.10   Making crosscuts at the apex of the cone.

Assemble the top disc elements following the illustrated steps in Figure 8.11.

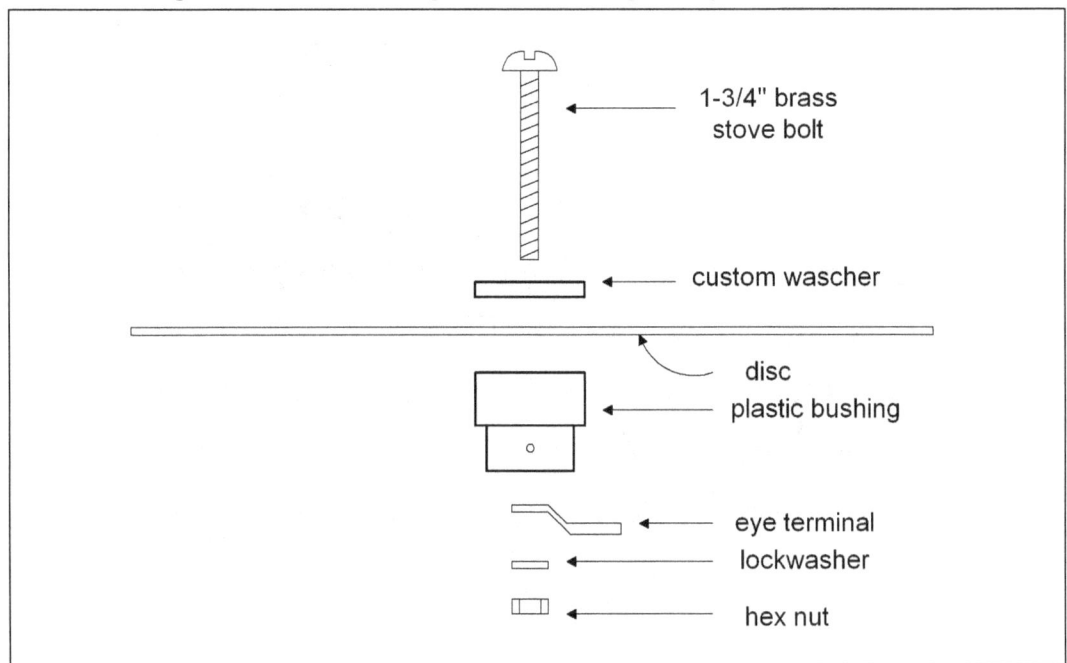

Figure 8.11   Assembling the top disc hardware.

Cut a 30-inch long RG-58/U coaxial cable, and solder the inner conductor at one of its ends into the eye terminal held by the bolt below the plastic spacer. Attach an eye terminal into its braid (see Figure 8.12).

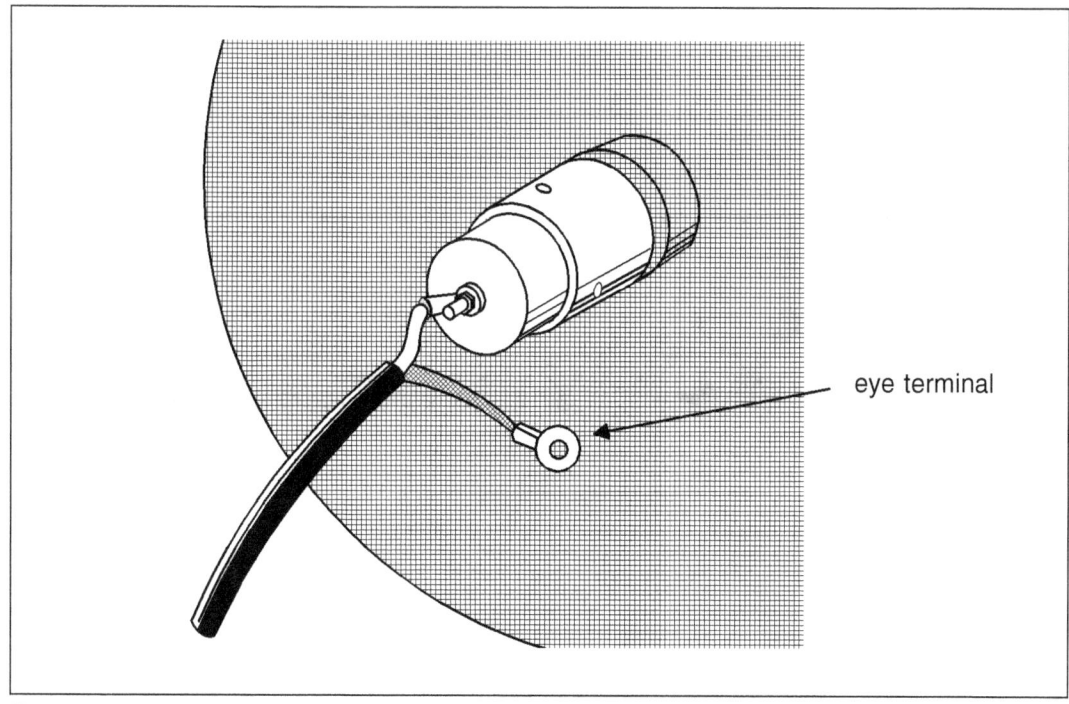

eye terminal

Figure 8.12   Connecting the coaxial cable to the top disc element.

Insert the free end of the coaxial cable into the mounting tube starting from the tube's end with side holes (see Figure 8.13).

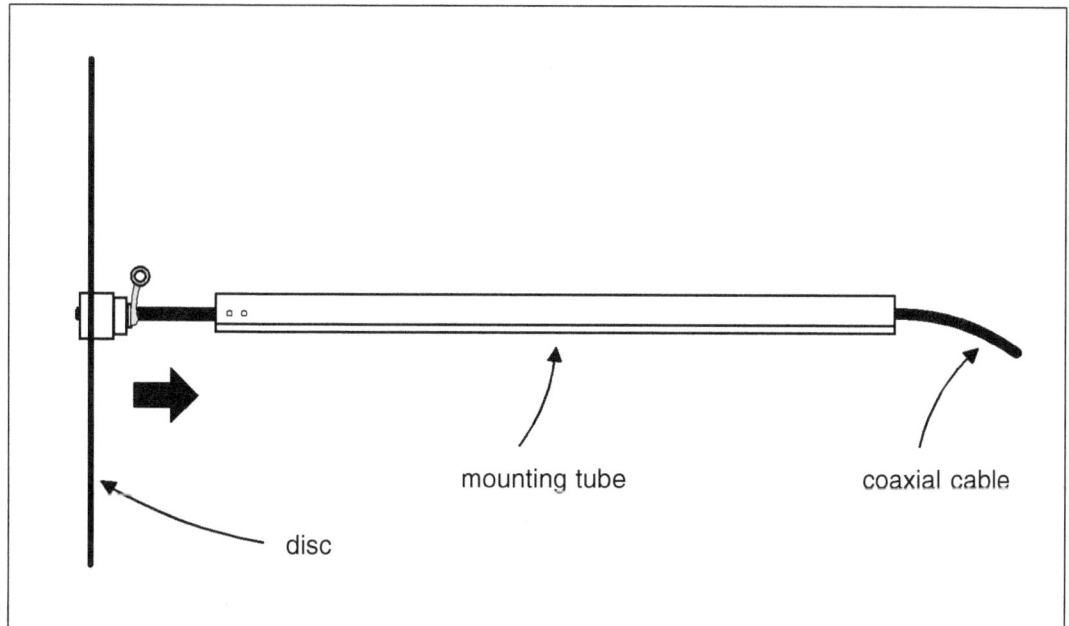

mounting tube   coaxial cable

disc

Figure 8.13   Inserting the coaxial cable into the mounting tube.

Insert the plastic spacer/bushing holding the top disc element inside the tube and align its hole to the tube's side holes. Fix the bushing to the tube permanently using metal screws (see Figure 8.14).

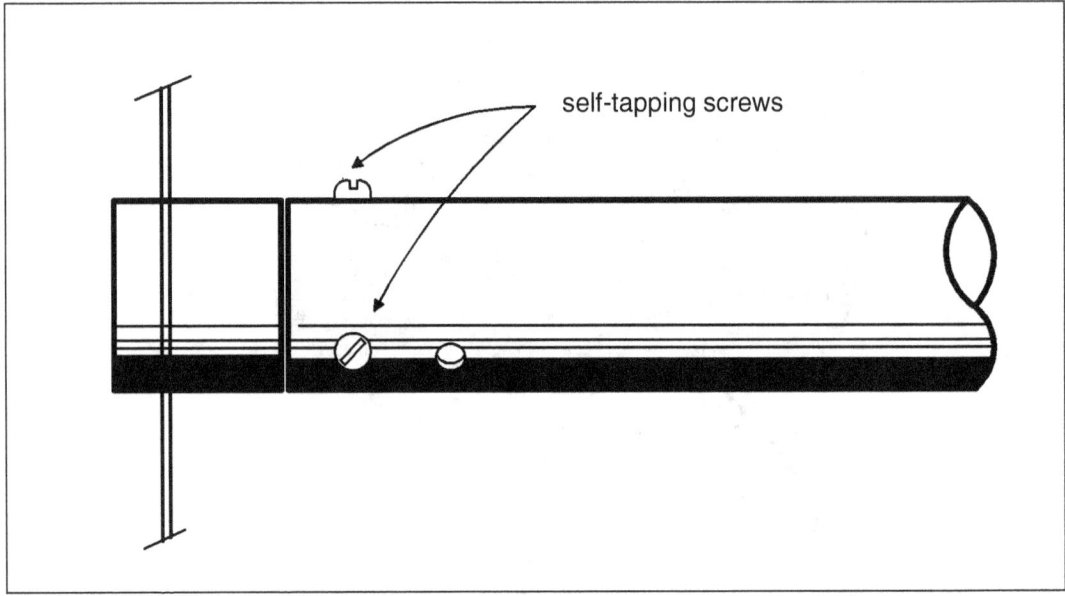

Figure 8.14   Securing the top disc element to the tube.

Align the lower hole to the eye terminal of the braid inside the tube. If it is not aligned yet, insert a slender stick inside the tube, and remotely move the eye terminal until you can see it through the hole outside. Insert a metal screw into the hole, and turn it until it catches the eye terminal inside. Tighten the screw to hold the terminal firmly (see Figure 8.15). This procedure is the main reason why you should use an eye terminal with a 1/16" diameter eye.

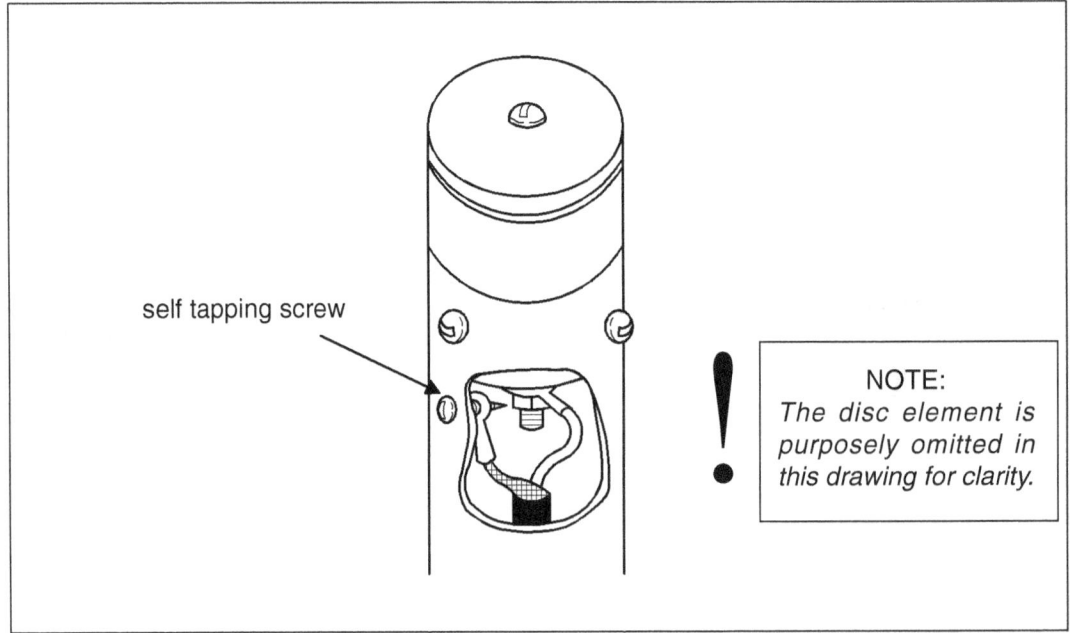

Figure 8.15   Securing the braid inside the mounting tube.

Solder a PL-259 to the free end of the coax cable, and connect a straight connector (PL-258) into it prior to the final installation of the antenna (see Figure 8.16).

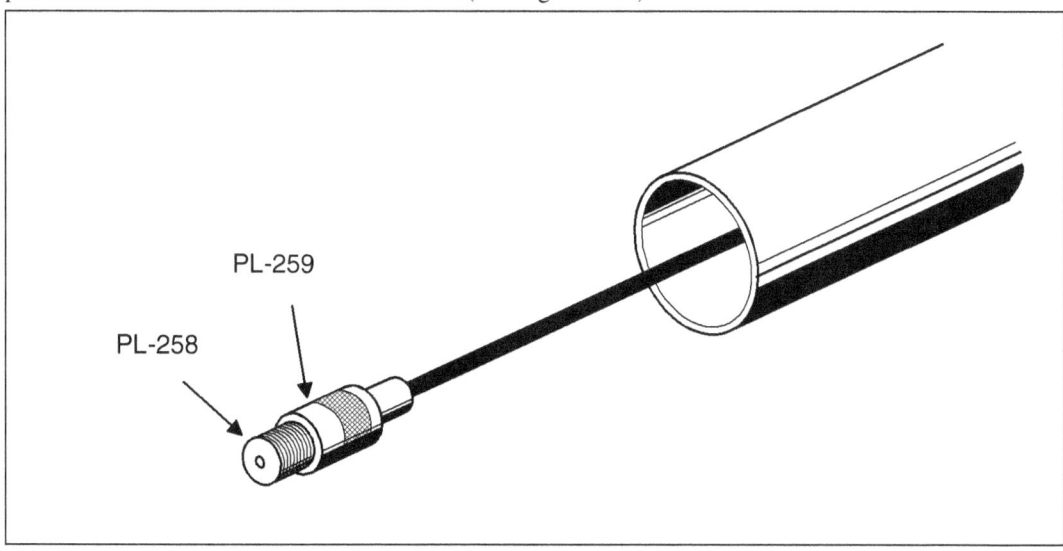

Figure 8.16   Connecting the PL-259 and PL-258 to the coaxial cable.

Finally, insert the PL-259, coaxial cable, and the mounting tube into the cone, starting from the top until the apex of the cone reaches just a tiny fraction of an inch below the plastic bushing holding the top disc element (see Figure 8.17).

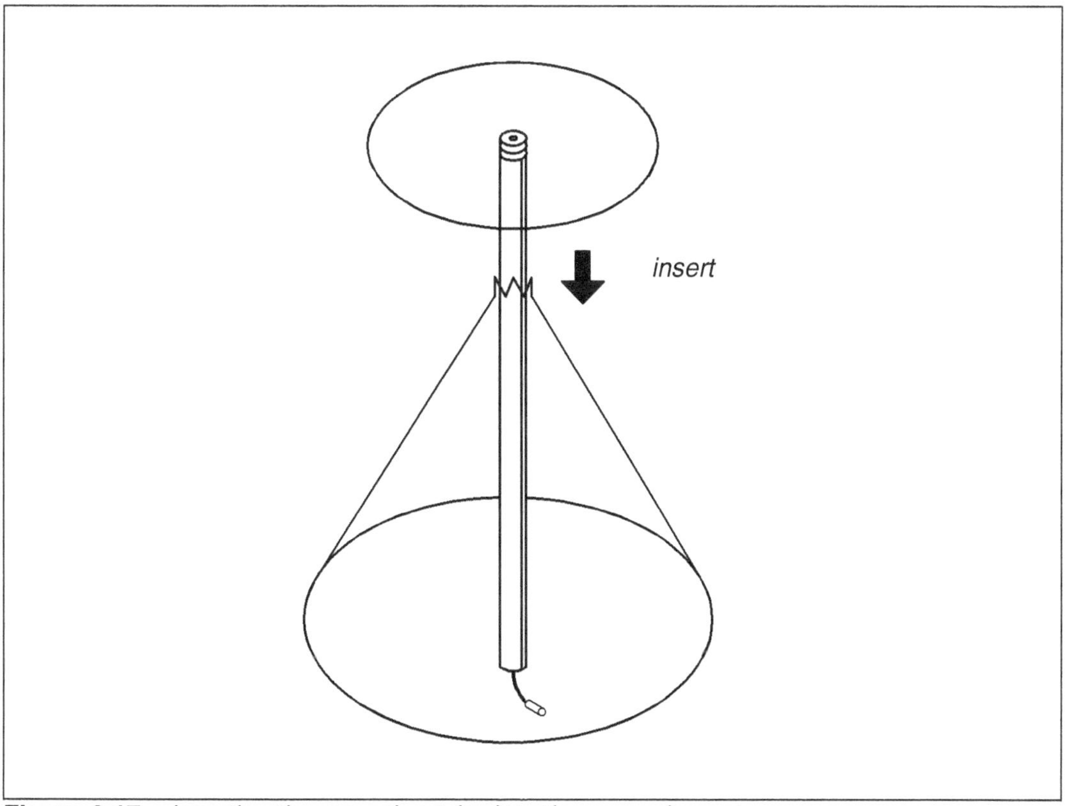

Figure 8.17   Inserting the mounting tube into the cone element.

Attach the stainless clamp around the upturned portion of the wire mesh just under the disc. Tighten the clamp to hold the cone in place. Trim the excess wire mesh protruding above the edge of the tube clamp (see Figure 8.18).

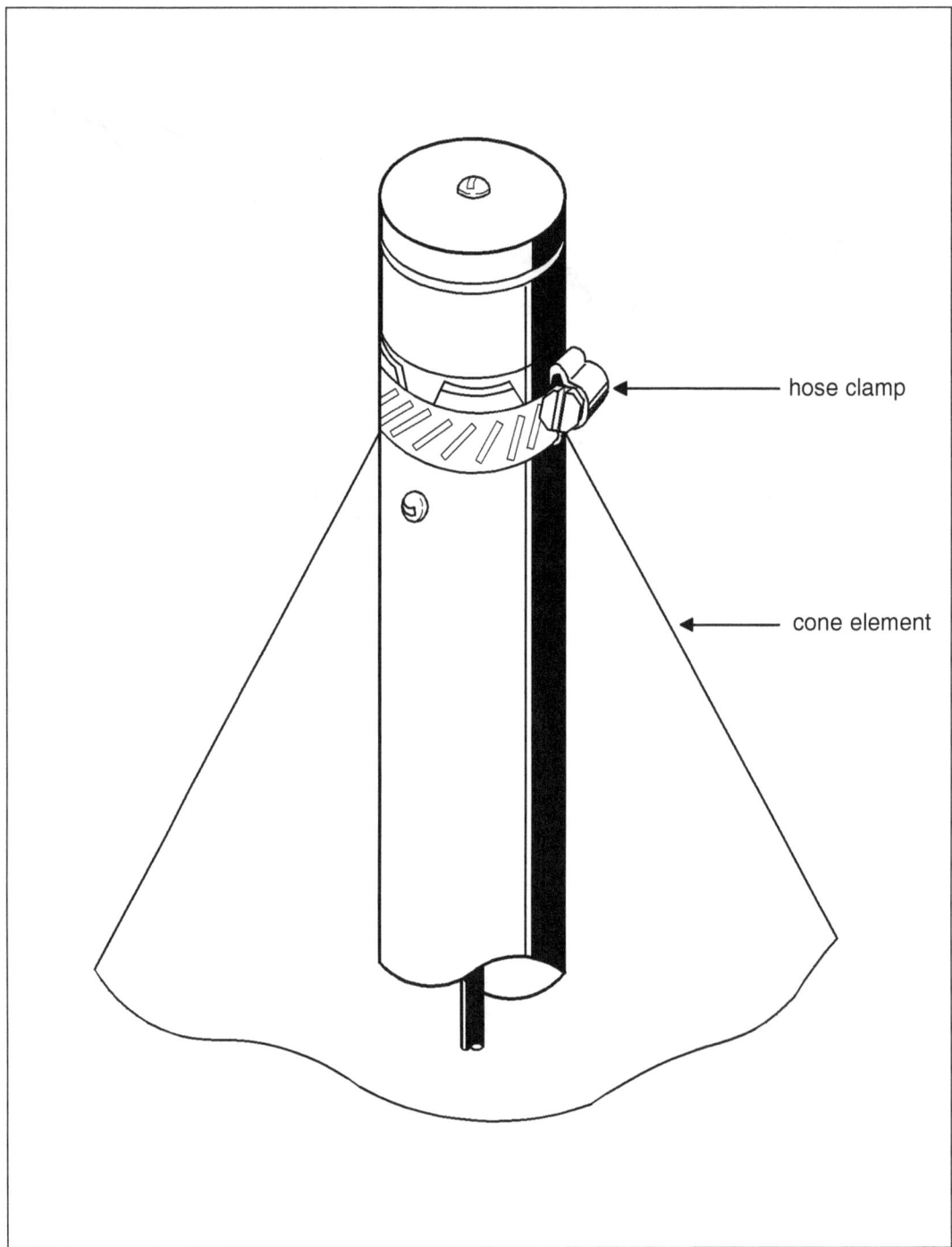

hose clamp

cone element

Figure 8.18   Clamping the apex of the cone element to the mounting tube.

Installation of CD-2W

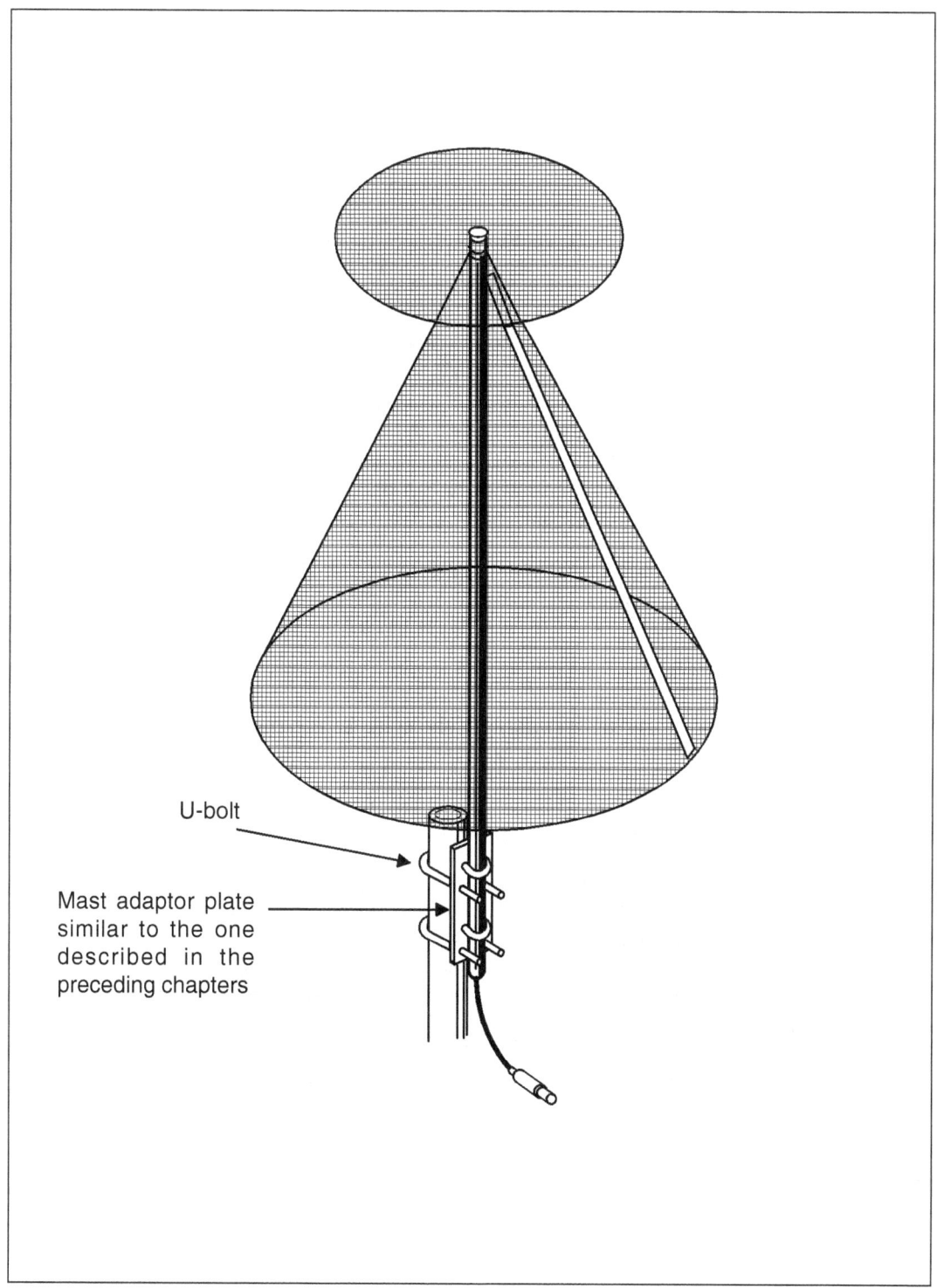

U-bolt

Mast adaptor plate similar to the one described in the preceding chapters

Figure 8.19   Mounting the CD-2W to the mast.

REVIEW QUESTIONS

1. How is the discone antenna able to operate over a very wide frequency band?

2. Why is it not good to use an antenna outside of its operational bandwidth?

3. What is the most important thing to consider in designing a discone antenna?

4. What is a cut-off frequency?

5. What is the most significant function of a discone antenna?

6. In this particular model, what is the reason for using a wire screen mesh?

7. Other than the use for wide band transceivers, what is the other popular use of the discone antenna?

8. In this particular model, what is the purpose of the aluminum tube under the cone element?

Table 8.1 Two-wire cable with plastic dielectric. Amphenol standard types.

	Impedance in Ω	Velocity factor	Wire diameter in mm	Attenuation in dB pro 1000 m long cable at:			
				7 MHz	30 MHz	150MHz	400MHz
14-080	75	0.68	7 x 0.32	68.62	137.24	311.82	467.30
14-023	75	0.71	7 x 0.7	16.50	49.50	160.70	
14-079	150	0.77	7 x 0.32	21.72	49.50	111.19	180.70
14-056	300	0.82	7 x 0.32	9.55	19.97	51.25	88.60
14-100	300	0.82	7 x 0.32	9.55	19.97	51.25	88.60
14-271	300	0.82	7 x 0.32	9.55	19.97	51.25	88.60
14-185	300	0.82	7 x 0.4	6.95	17.37	44.30	81.65
14-076	300	0.82	7 x 0.4	6.95	16.50	41.70	72.00
14-022	300	0.82	1.3	6.21	12.16	33.00	59.00

9 DISCONE ANTENNA

Model CD-2P

The discone antenna model CD-2P described in this chapter is functionally similar in most respects to the discone antenna in chapter 8. The only and obvious difference between the two models is the utilization of a metal plate for the disc and cone elements of CD-2P ('P' for plate).

The choice of using a metal plate becomes evident when the antenna is intended to be installed in areas with less than excellent weather conditions. For instance, if your region regularly experiences heavy rainfall or strong winds, then you should opt to construct and install this more robust model instead of the wire-screen version. Metal plate is more durable than wire-screen. The only trade-off is the total cost of the antenna, because metal plate is more expensive. In most occasions, a GI metal plate is satisfactory, but you can also use a more expensive aluminum plate if you desire. Aluminum is less susceptible to corrosion; so it is highly recommended if you plan to use the antenna near seashores or in places where there is a high level of salt present in the airborne moisture.

Paint has a negligible effect on the RF signal; so if you decide to paint the antenna to make it look attractive, just do it. But don't paint the aluminum mounting tube under the cone element, to ensure adequate grounding connection of the antenna to the mast or tower. This is a precautionary measure to avoid lightning striking your antenna and possibly causing damage to your transceiver or to you.

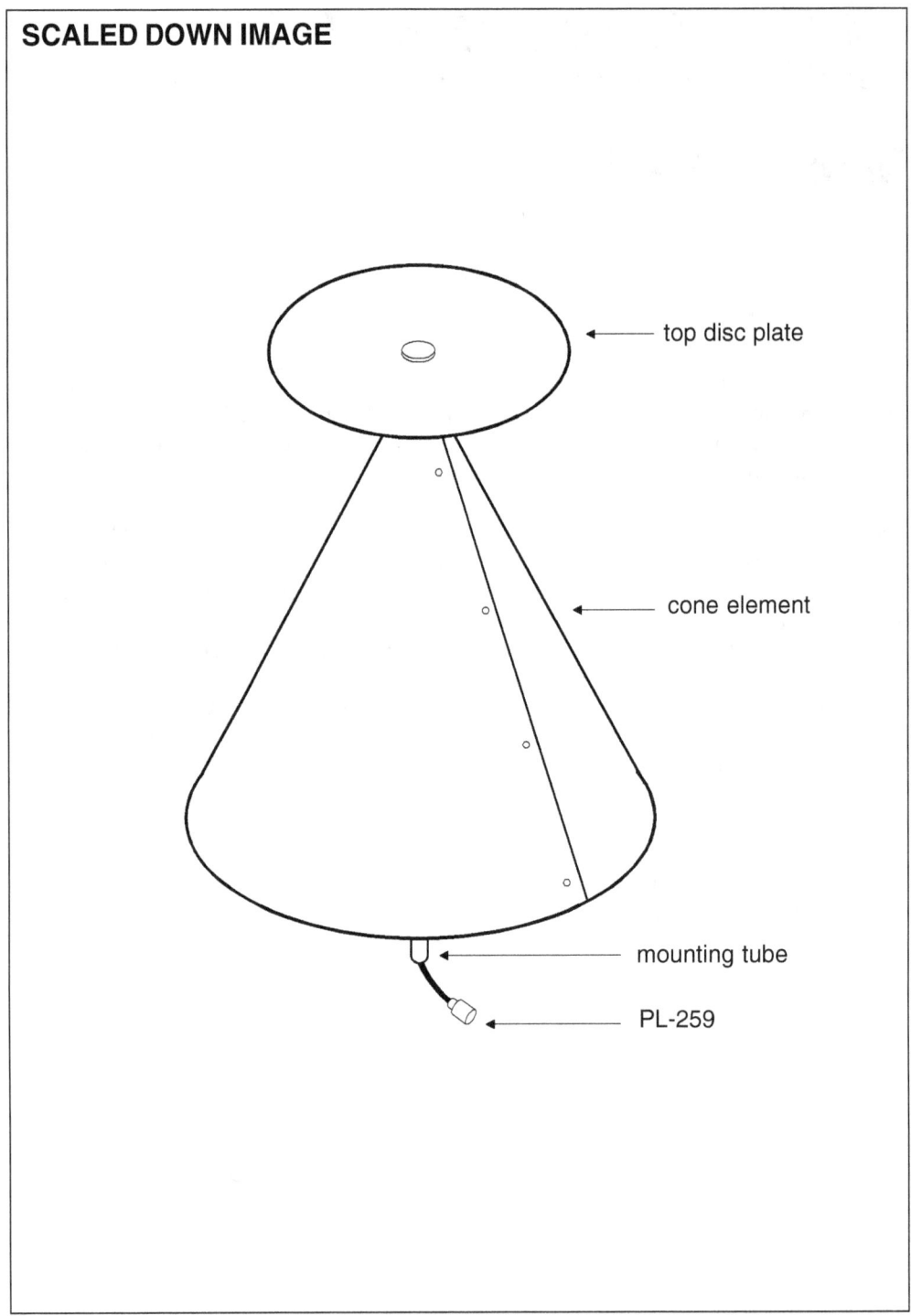

Figure 9.1 A discone antenna using metal plates.

Materials needed

The materials necessary for the antenna Model CD-2P are the same with those needed for the CD-2W, except for the wire mesh used for the cone and disc, which is replaced with a thin metal sheet in the CD-2P. Also, the two thin metal strips are not needed. However, you have to retain the blind rivets for the same purpose.

Construction

In constructing most of the parts of CD-2P, follow the instructions for the Model CD-2W, except those for the disc and cone elements. Cut the cone and disc elements out of the thin metal sheet following the dimensions shown in Figures 9.2 and 9.3.

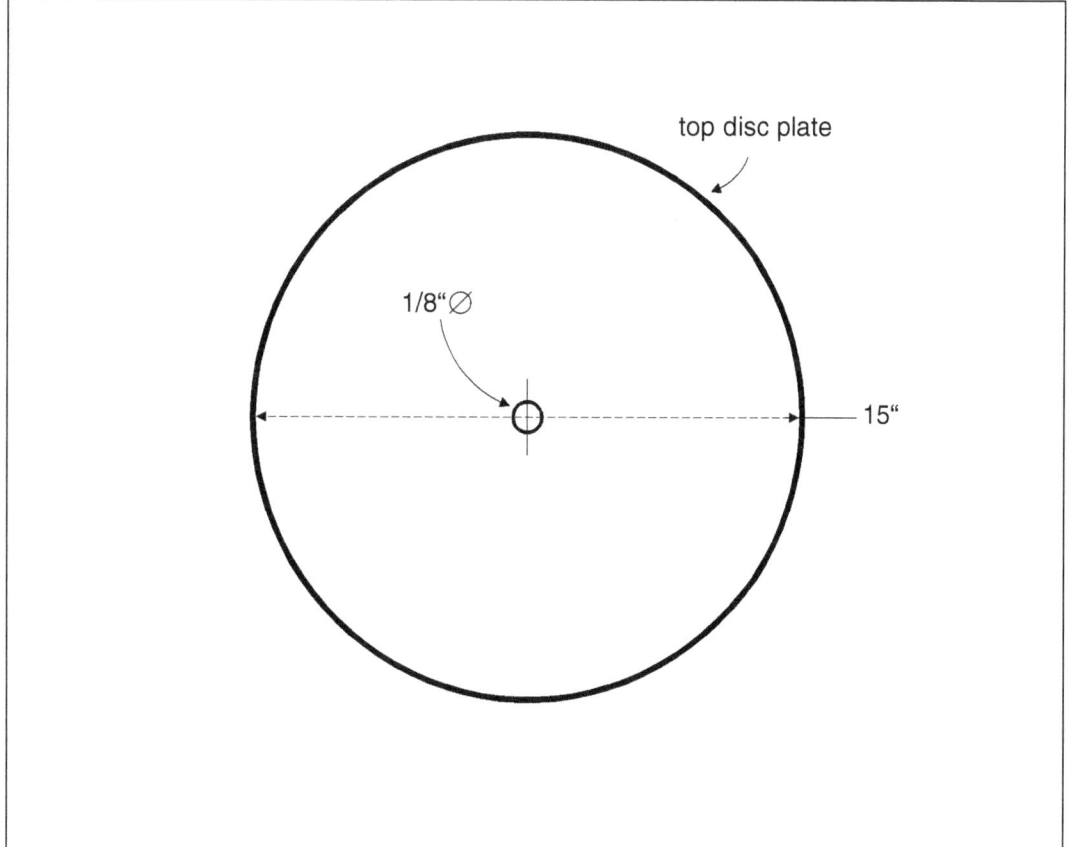

Figure 9.2   Disc element dimension.

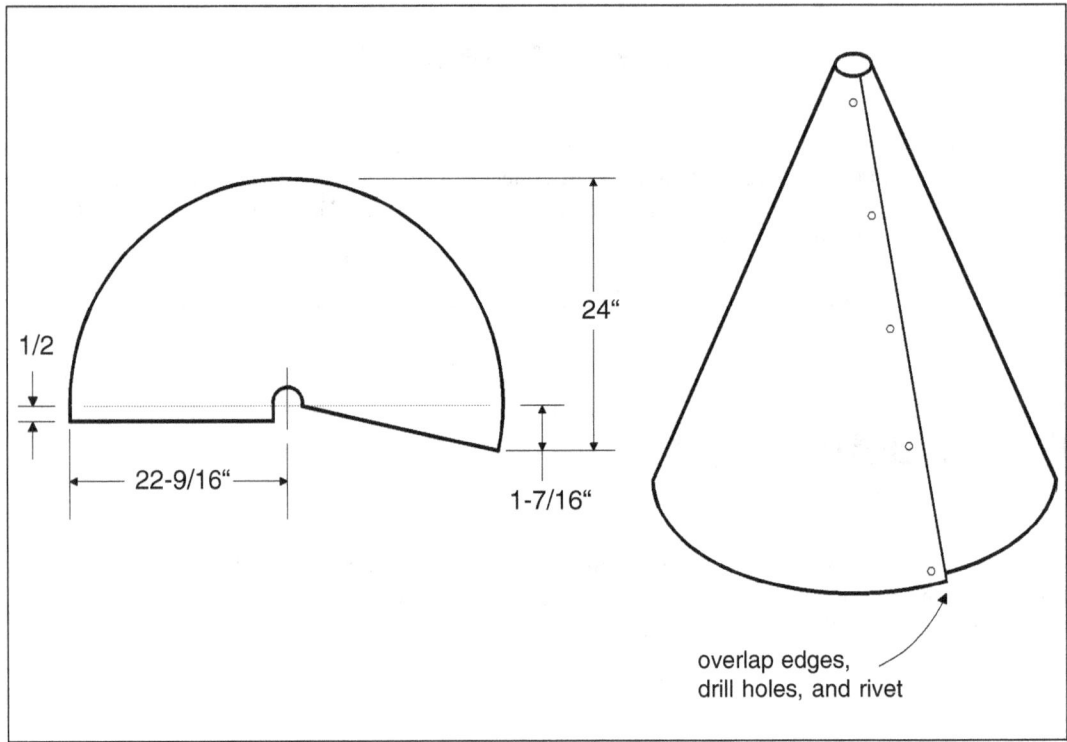

Figure 9.3　Cone plate dimensions.

Unlike the wire mesh cone of the CD-2W, the cone for the CD-2P must have a metal sleeve soldered to the opening at the apex. Cut a metal sheet and shape it to form a ring as shown in Figure 9.4.

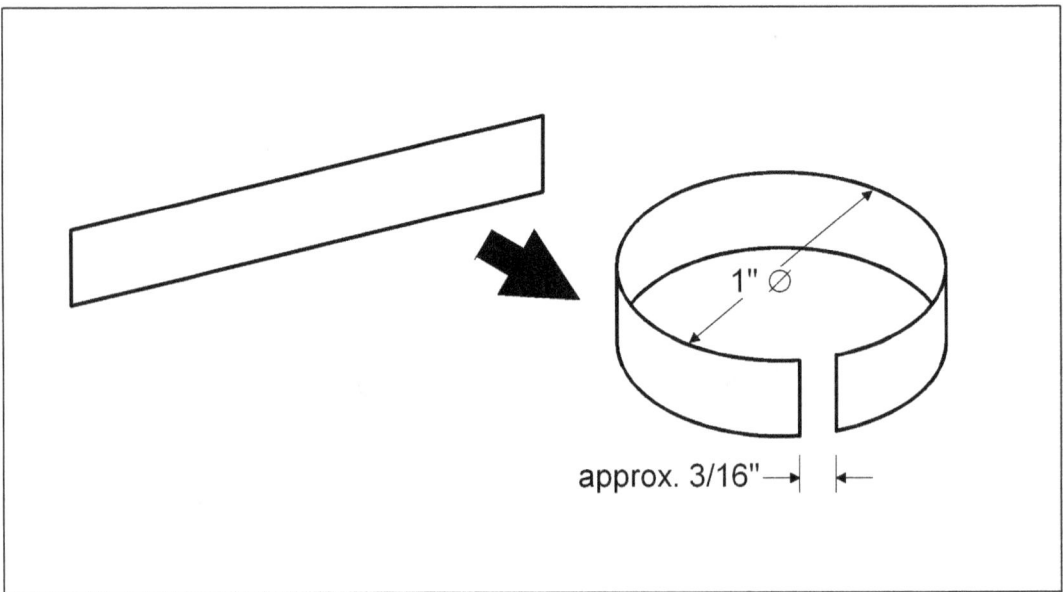

Figure 9.4　Preparing the sleeve.

Solder this sleeve to the rim of the apex opening, leaving a small gap between their ends (see Figure 9.5). If you are using an aluminum plate for the cone and sleeve, you need to electrically weld the two pieces together using a special technique for welding aluminum with protective gas.

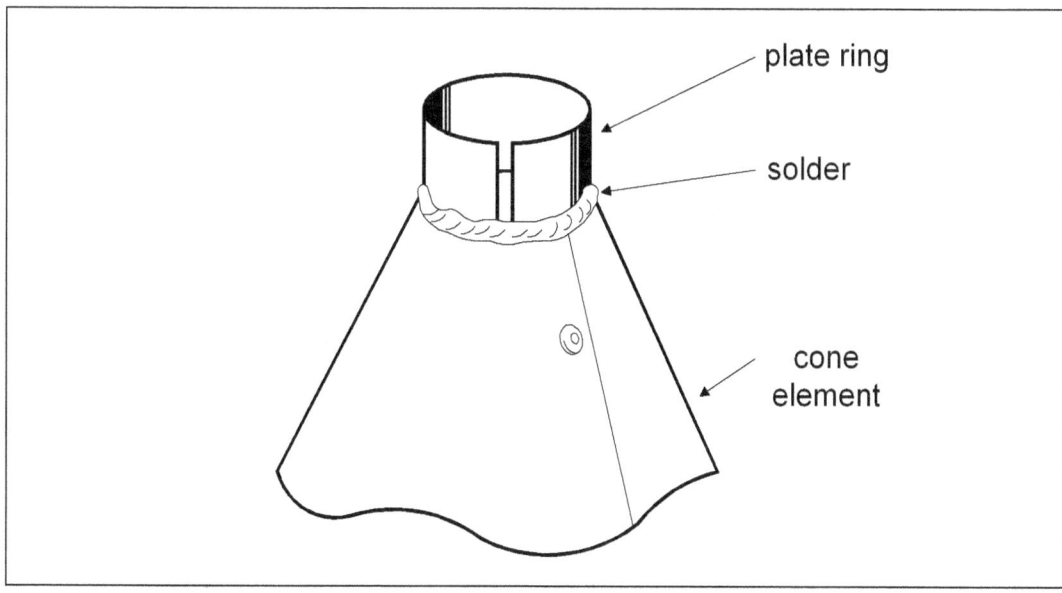

Figure 9.5   Soldering or welding the sleeve to the apex opening of the cone.

The assembly of CD-2P is similar to the steps for assembling the CD-2W (see Figure 9.6).

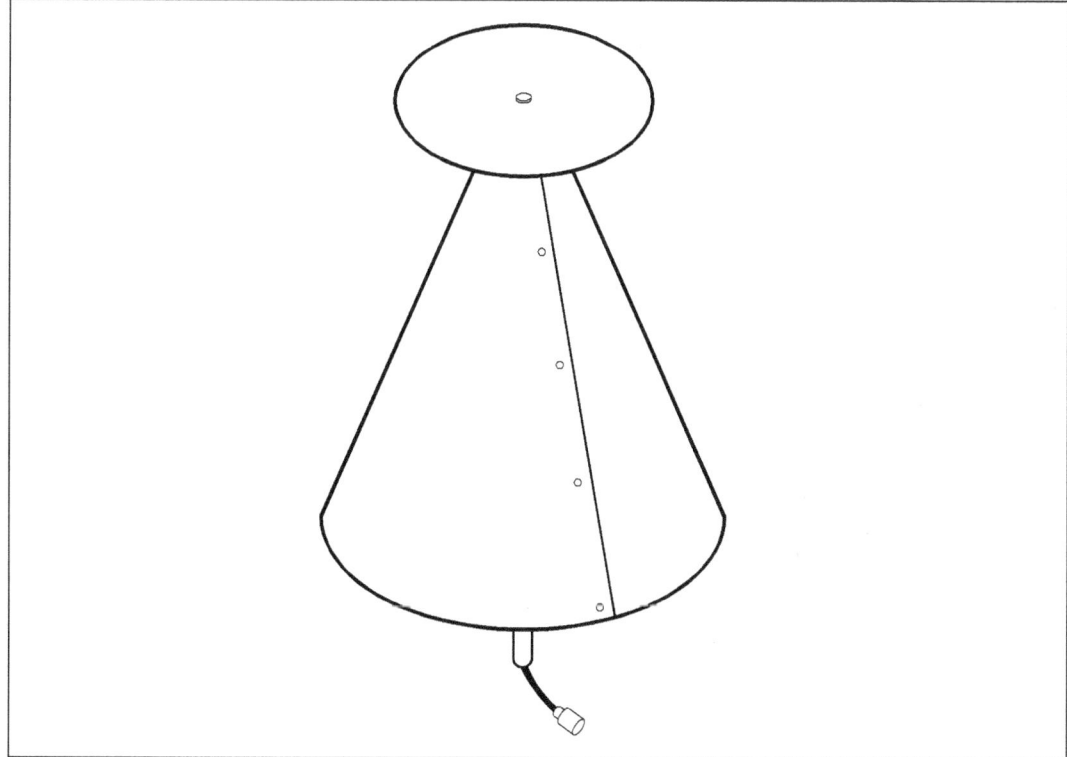

Figure 9.7   Assembled CD-2P.

REVIEW QUESTIONS

1. What is the advantage of using a metal plate for a discone antenna?

2. Why must the mounting tube be left unpainted?

3. What are the advantages of using an aluminum plate?

4. What is the reason for using an aluminum material for the bushing and washer?

Table 9.1

Coax cable MIL-C-17E equivalent types of the obsolete MIL-C-17 types

Obsolete Type	Equivalent	Obsolete Type	Equivalent	Obsolete Type	Equivalent
RG5/U... B/U	RG212/U	RG22/U ... A/U	RG22B/U	RG 10810	RGI08A/U
RG6/U	RG64/U	RG23/U	RG23A/U	RG111/U	RGl11A/U
RG8/U ... A/U	RG213U	RG24/U	RG25A/U	RG115/U	RG 115A/U
RG9/U ... A/U	RG214U	RG29/U	RG58C/U	RG116/U	RG227/U
RGI0/U ... A/U	RG215/U	RG34/U... A/U	RG34B/U	RG133/U	RG 133A/U
RG 1110	RG 11 NU	RG35/U... A/U	RG35B/U	RGl42/U...A/U	RG142B/U
RG12/U	RG12NU	RG58/U... B/U	RG58C/U	RG159/U	RG142B/U
RG 13/U ... A/U	RG216/U	RG59 ... A/U	RG59B/U	RG174/U	RG174A/U
RG14/U ... A/U	RG217/U	RG62/U... C/U	RG62A/U	RG178/U...A/U	RG178B/U
RG15/U	RG 11 NU	RG63/U ... A/U	RG63B/U	RG179/U...A/U	RG179B/U
RG17/U... B/U	RG218/U	RG65/U	RG65A/U	RG180/U...A/U	RG180B/U
RG18/U ... A/U	RG219/U	RG71/U ... A/U	RG71B/U	RG211/U	RG211A/U
RG191U ... A/U	RG220/U	RG74/U... A/U	RG224/U	RG228/U	RG228NU
RG20/U ... A/U	RG221/U	RG79/U ... A/U	RG79B/U	RG307/U	RG307A/U
RG21/U ... A/U	RG222/U	RG87/U ... A/U	RG225/U		

Obsolete MIL-Coaxial Cable types without equivalents:
RG16/U; RG36/U; RG54/U ... A/U; RG55/U ... B/U; RG57/U... A/U; RG72/U; RG-78/U; RG86/U;
RG94/U ... A/U: RG117/U... A/U; RG118/U... A/U; RG140/U; RG141/U ... A/U; RG143/U ... A/U;
RG147/U; RG148/U; RGl49/U; RGl50/U; RG181/U; RG187/U... A/U; RG188/U... A/U;
RG195/U...A/U: RG196/U... A/U: RG229/U; RG282/U; RG293/U... A/U; RG294/U ... A/U;
RG295/U; RG324/U; RG325/U; RG326/U: RG366/U; RG388/U; RG389/U.

10 DISCONE ANTENNA

Model CD-2T

In mobile operation, the problems related to antenna installation are much greater than those encountered in fixed stations. The problems are particularly worse for a one-man mobile-station that is transported on foot. Those seemingly small items like portable transceiver, spare batteries, coaxial cable, myriad of wires, solar panels, charging box, log books, scanning monitor, etc. could easily total up to more than 20 kilos of dead weight if crammed together inside a single backpack. Add to it the supply of food and few personal belongings, and it will surely feel like a nightmarish load when traveling across rugged terrain.

The over-all bulk of the load is another problem. Just imagine traveling while lugging a full-size metal plate discone at your back! Because of this, the tendency of mobile operators is to bring only the most important piece of equipment, which is usually a portable and lightweight version, to trim down the total weight and bulk of the load.

The antenna model described in this chapter is specially designed to satisfy the need for a lightweight and transportable discone antenna. The cone and disc elements are replaced with retractable telescopic rods, so that the antenna can be collapsed into a small unit and conveniently stored inside a backpack. The actual length of a packed discone is merely 8 inches! When the telescopic rods are extended to their maximum length and set in the proper angle, they approximate the function of a full disc and cone elements. Theoretically, the more elements used the better. Experience showed however that three elements for each disc and cone function are enough on most occasions.

This portable version of a discone antenna has the same electrical characteristics with the two full-sized models described in chapters 8 and 9. The only difference is in the mechanical construction. The chrome plated telescopic rods are quite expensive; so the total cost of this antenna is higher than the two preceding models.

If this antenna will be used solely for mobile operations, then the U-bolts intended for mounting may be discarded. Instead, the antenna can be tied with a thin nylon rope on its mounting tube near the feed point and hung under a post or branch of a tree. Never use a metallic wire to hang the antenna, because it will distort the radiation pattern of the signal or short out the disc and cone elements.

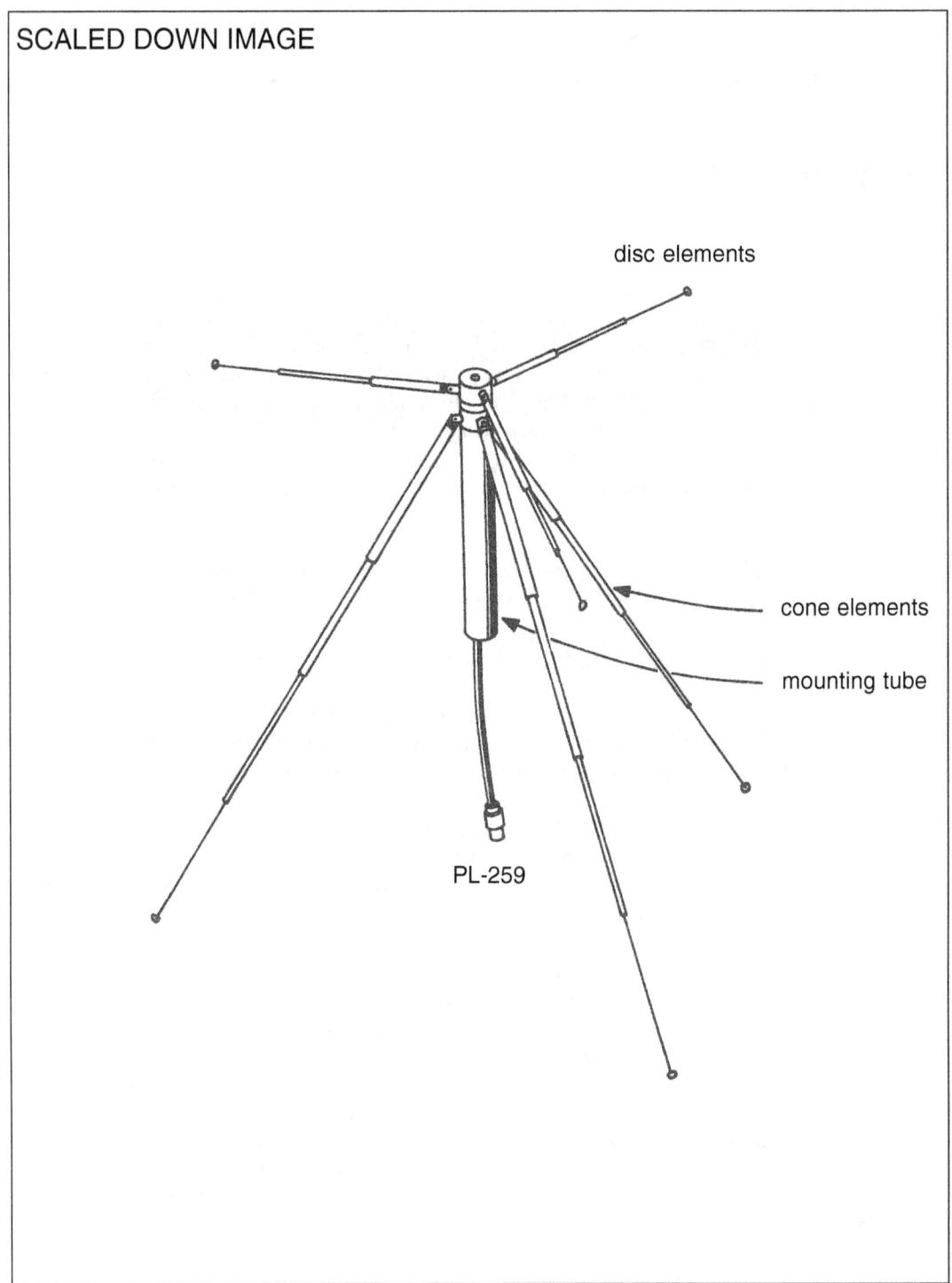

Figure 10.1   Discone antenna with telescopic elements.

Materials List

Quantity	Specification/Description	Dimensions
3	Telescopic antennas - with swiveling threaded base	7" fully extended 3-5" retracted
3	Telescopic antennas - with swiveling threaded base	22" fully extended 5-6" retracted
1	Aluminum disc base mount see text for exact dimensions	
1	Plastic spacer - see text for customized dimensions	
1	Aluminum cone base mount see text for exact dimensions	
1	PL-259 VHF male connector	
1	PL-258 VHF straight connector	
1	Aluminum plate 1/8" thick	3" x 6"
2	Eye terminals - no insulation	
4	U-bolts with accompanying hex nuts and lock washers	
1	Stove bolt - brass or GI	1/8" x 2"
6	Self-tapping metal screws	1/8" x 3/8"
1	Self-tapping metal screw	1/8" x 3/4"
1	Hex nut - brass or GI	1/8" id*

*id- inside diameter

Construction

First, prepare the top disc elements base mount by machining an aluminum rod to the necessary dimensions. Follow the dimensions shown in the illustration (Figure 10.2). The size of the holes (holes marked with an **a**) and their thread gauge must conform to the dimensions of the short telescopic antenna you intend to use.

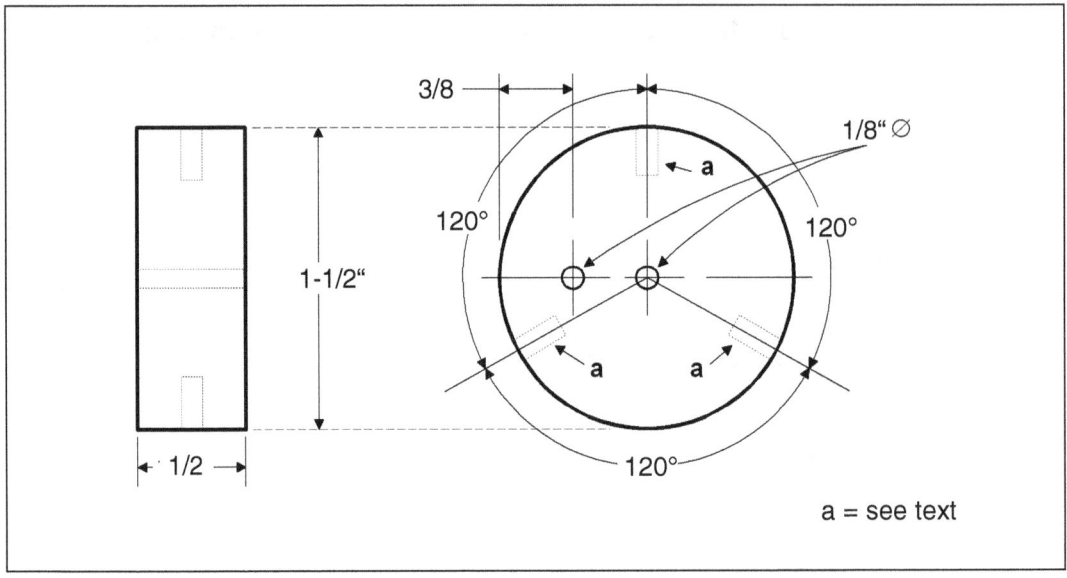

Figure 10.2   Disc base mount dimensions.

Second, prepare the plastic spacer/bushing from a piece of engineering plastic. Machine it to the form and dimensions shown in Figure 10.3. This plastic spacer insulates the disc base mount from the cone base mount.

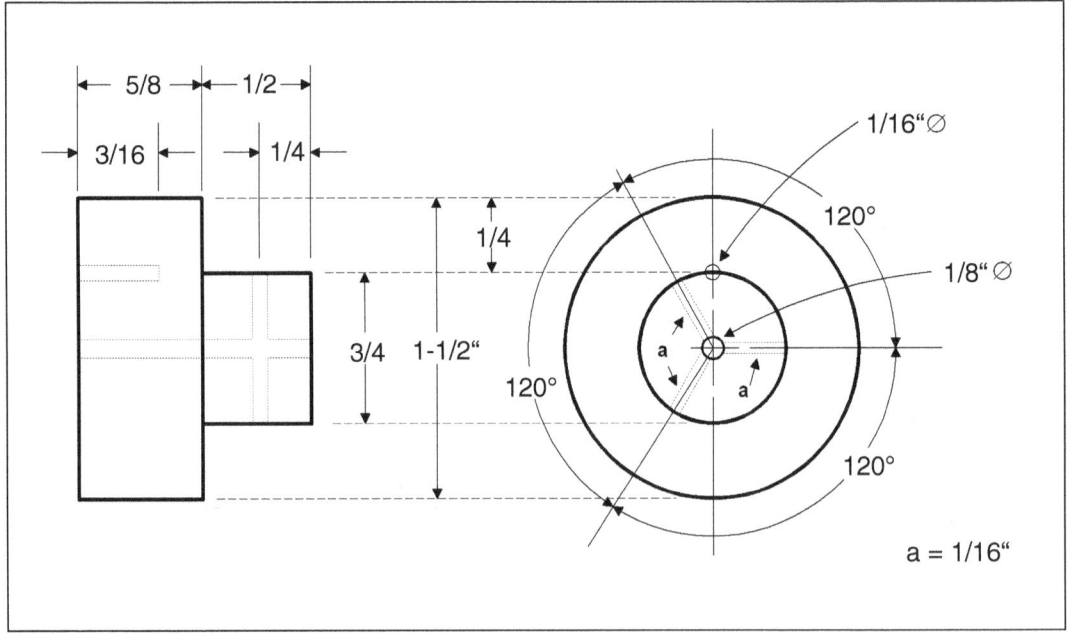

Figure 10.3   Plastic spacer dimensions.

The next step is to prepare the cone elements base mount from an aluminum rod with the necessary size. Machine it to the form and dimensions shown in the following illustration (Figure 10.4). The size, thread gauge, and deepness of the holes at the side (holes marked with an **a**) must conform to the base dimensions of the particular type of telescopic antenna intended for the cone elements.

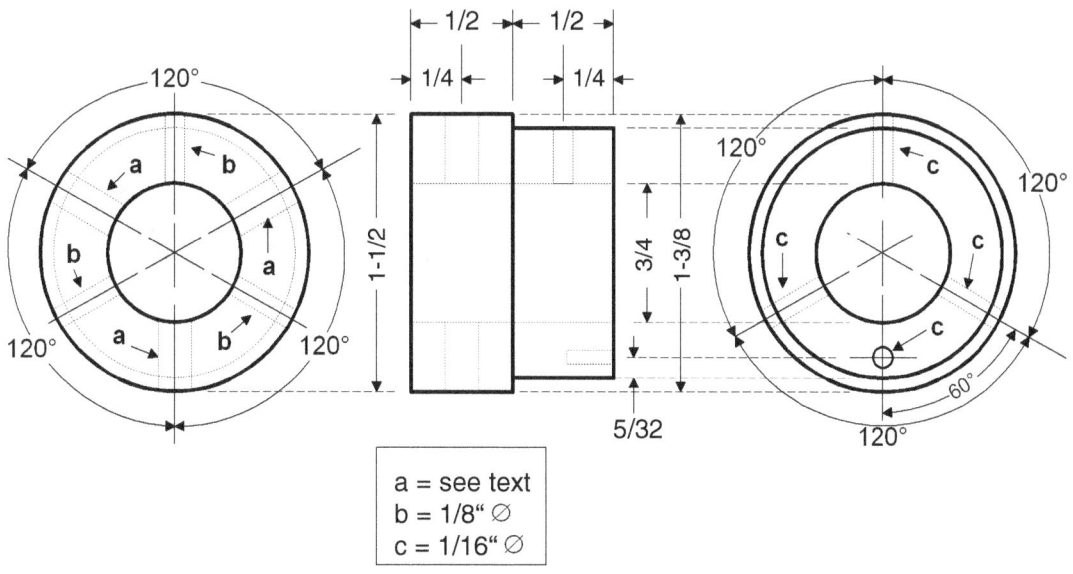

a = see text
b = 1/8" ∅
c = 1/16" ∅

Figure 10.4   Cone elements base mount dimensions.

Assemble the disc base mount, the plastic spacer, and the cone base mount together following the arrangement shown in Figure 10.5. Secure the assembly with self-tapping screws to the appropriate holes as illustrated.

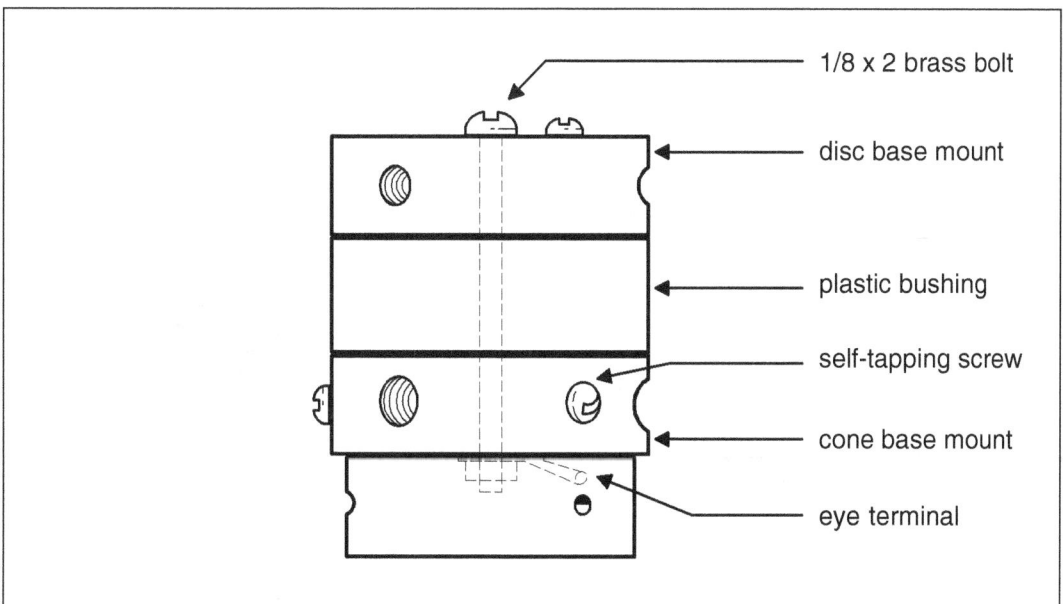

1/8 x 2 brass bolt

disc base mount

plastic bushing

self-tapping screw

cone base mount

eye terminal

Figure 10.5   Assembly of the elements' base mounts.

Attach the remaining eye terminal to the lone hole at the rim of the cone base mount (see Figure 10.6).

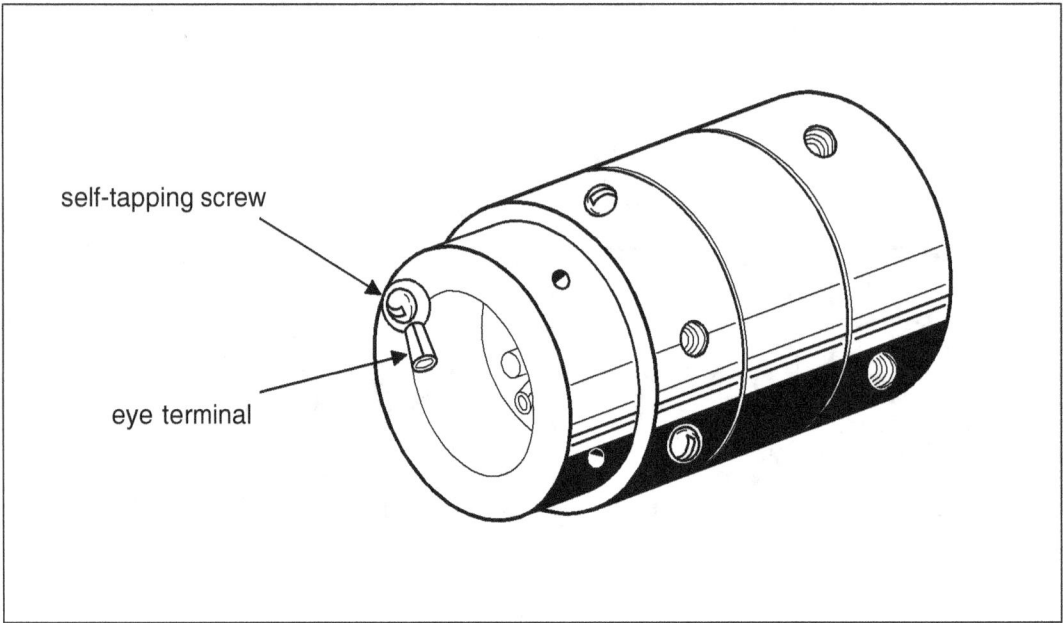

Figure 10.6   Eye terminal attached to the rim of the cone base mount.

Solder one end of the coax cable to the two terminals at the base mount assembly. The inner conductor must be soldered to the center terminal, and the braid must be soldered to the eye terminal at the rim (see Figure 10.7). *Never interchange the connection.*

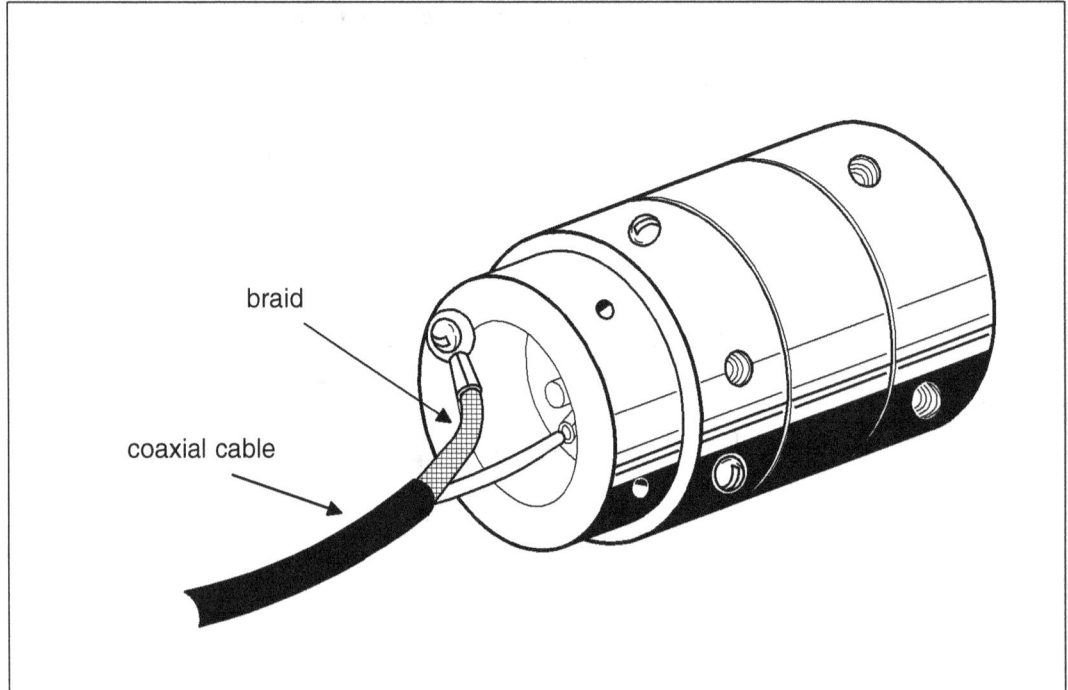

Figure 10.7   Connecting the coaxial cable to the terminals.

The next step is to prepare the aluminum mounting tube by cutting it to a length of 6 inches. Drill three holes (1/8" diameter) at one end (see Figure 10.8). The holes must be equally spaced from each other.

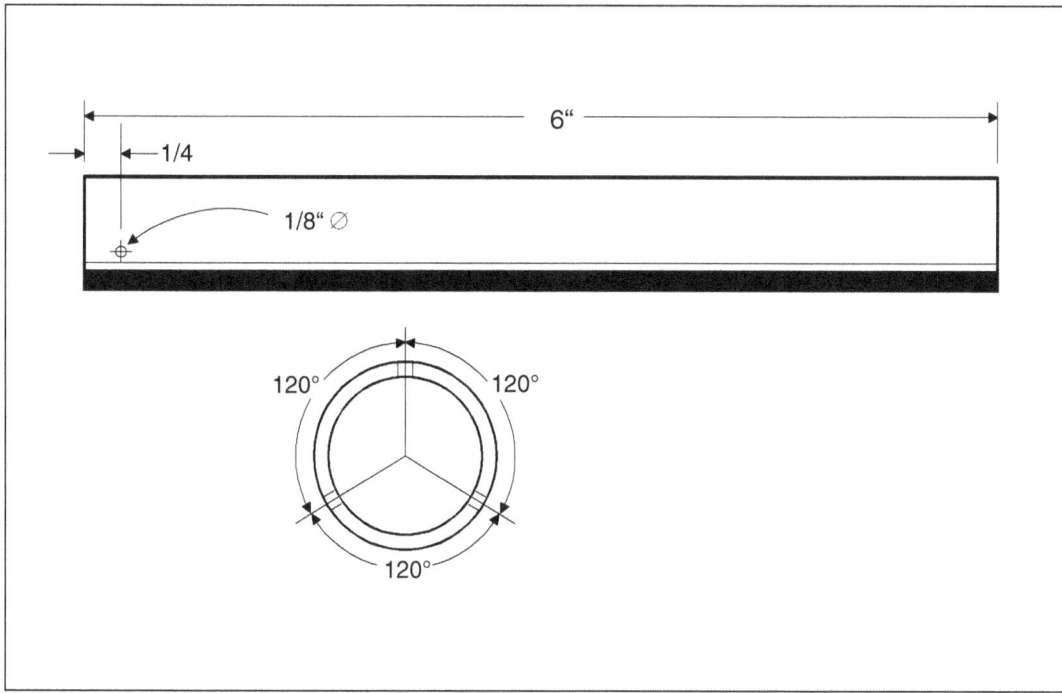

Figure 10.8   Preparing the mounting tube.

Insert the free end of the coax cable inside the aluminum tube starting at the end with side holes. Insert the aluminum base mount assembly into the tube and align the holes at the sides. Place screws through the holes to permanently attach the base mount assembly into the tube (see Figure 10.9).

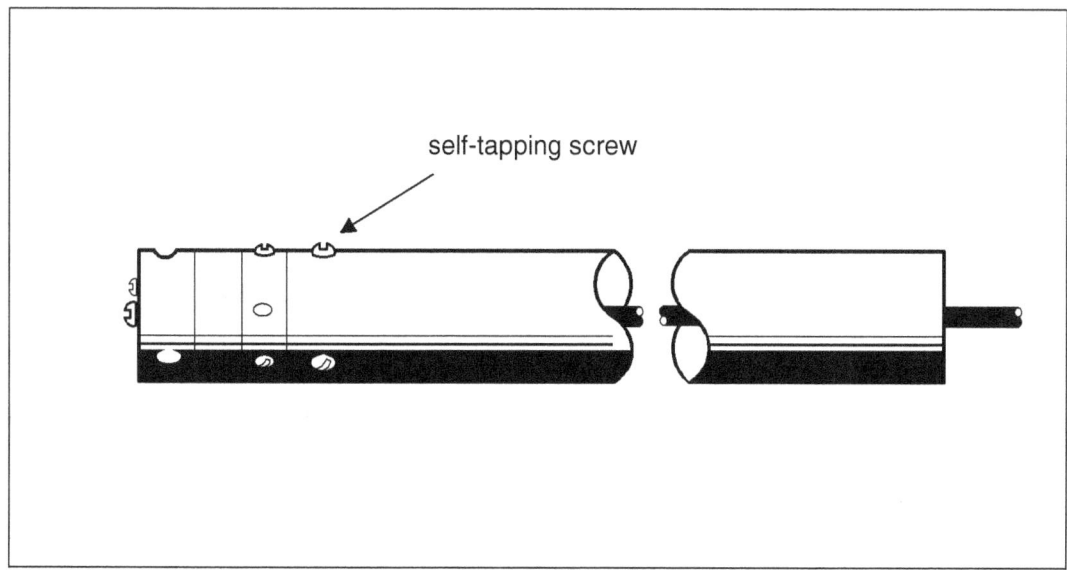

Figure 10.9   Fixing the base mount assembly into the tube.

Solder the PL-259 to the free end of the coaxial cable, and attach a straight connector (PL-258) into it (see Figure 10.10).

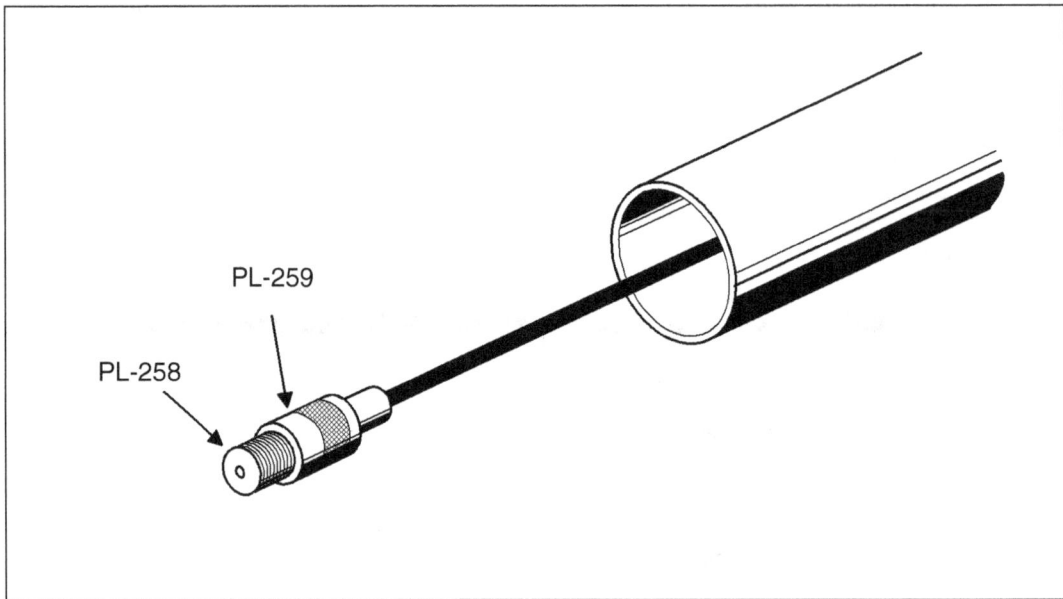

Figure 10.10 Connecting the PL-259 and PL-258.

Attach the three short telescopic antennas into the disc base mount (see Figure 10.11).

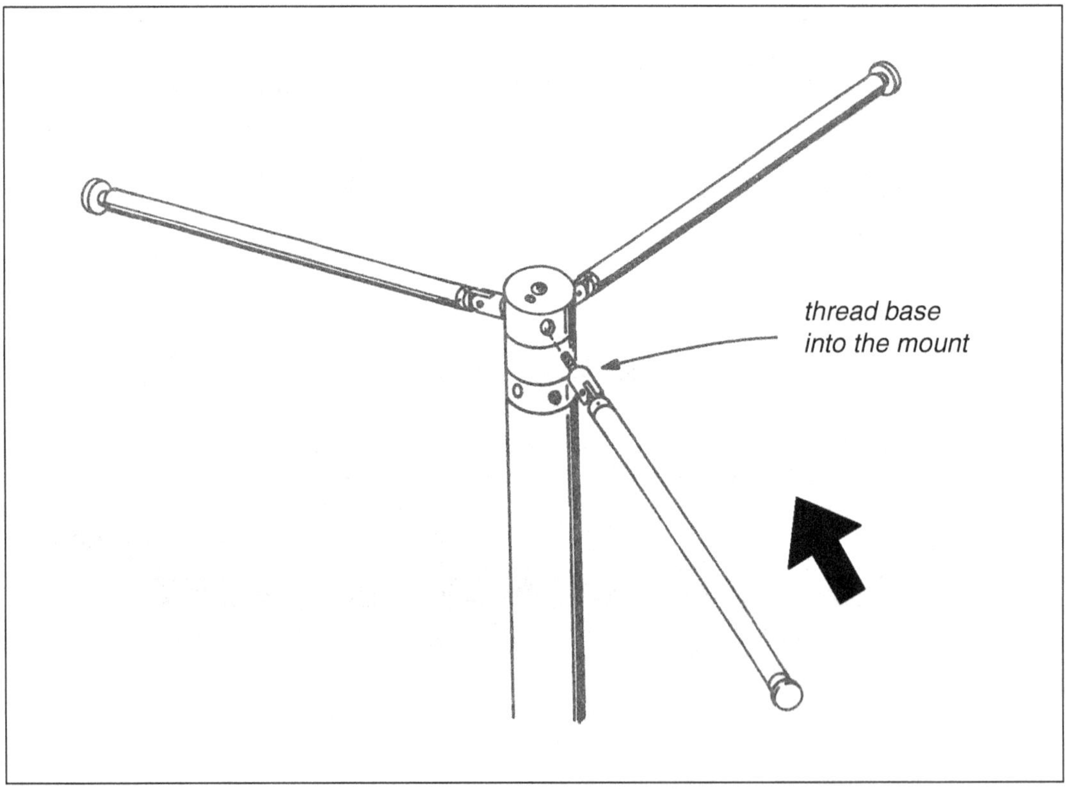

Figure 10.11 Telescopic antennas attached to the disc element mount.

Next, attach the three long telescopic antennas into the cone base mount under the first set of antennas (see Figure 10.12).

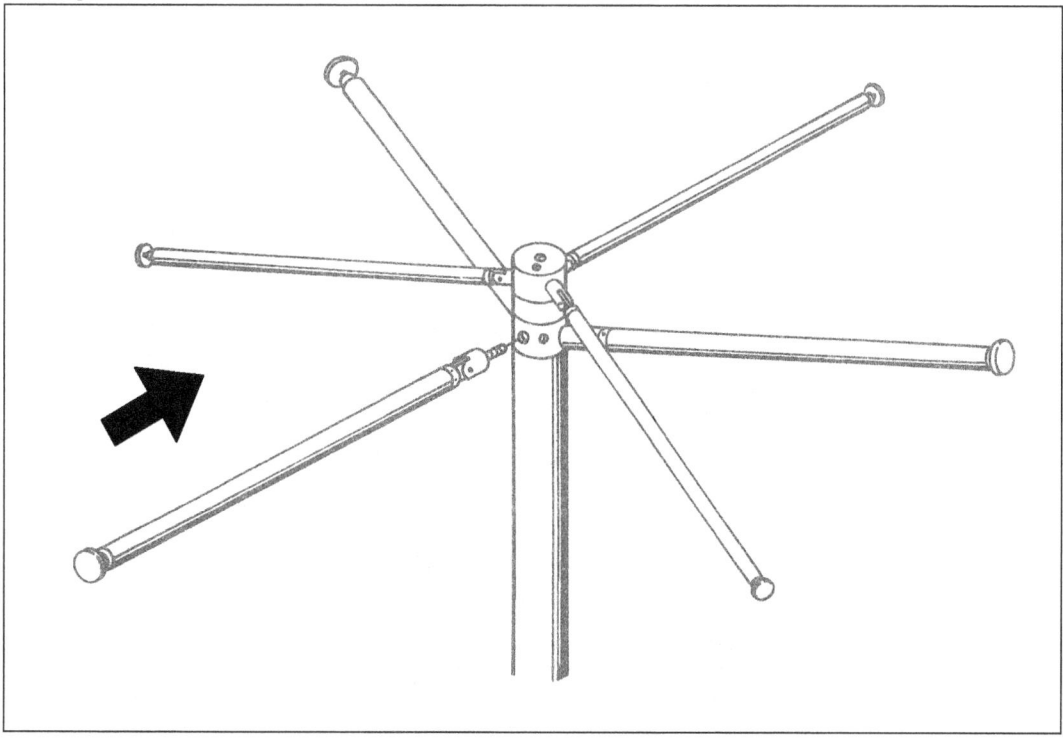

Figure 10.12   Attachment of long antennas to the cone element mount.

Attach the assembled antenna to the mast by using the aluminum mounting plate and one U-bolt. Extend the top telescopic antennas to their full lengths, maintaining them in a horizontal position. Similarly, extend the three long telescopic antennas to their full lengths; but they must be bent to about 60 degrees angle drooping downwards to the ground. See Figure 10.13.

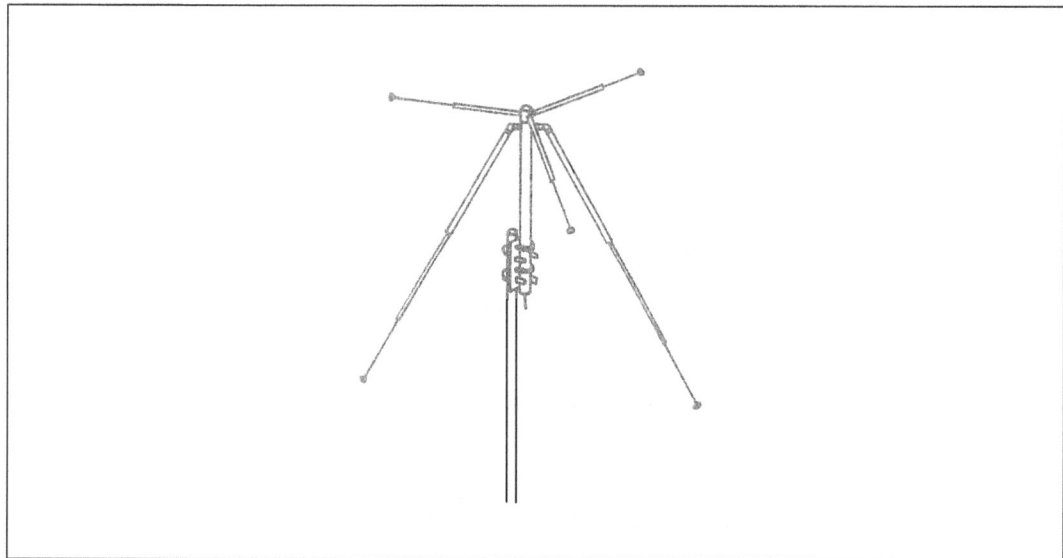

Figure 10.13   Mounting the completed antenna to the mast.

To carry the CD-2T in collapsed form for transportation, retract all the telescopic elements, and bend them towards the mounting tube (see Figure 10.14).

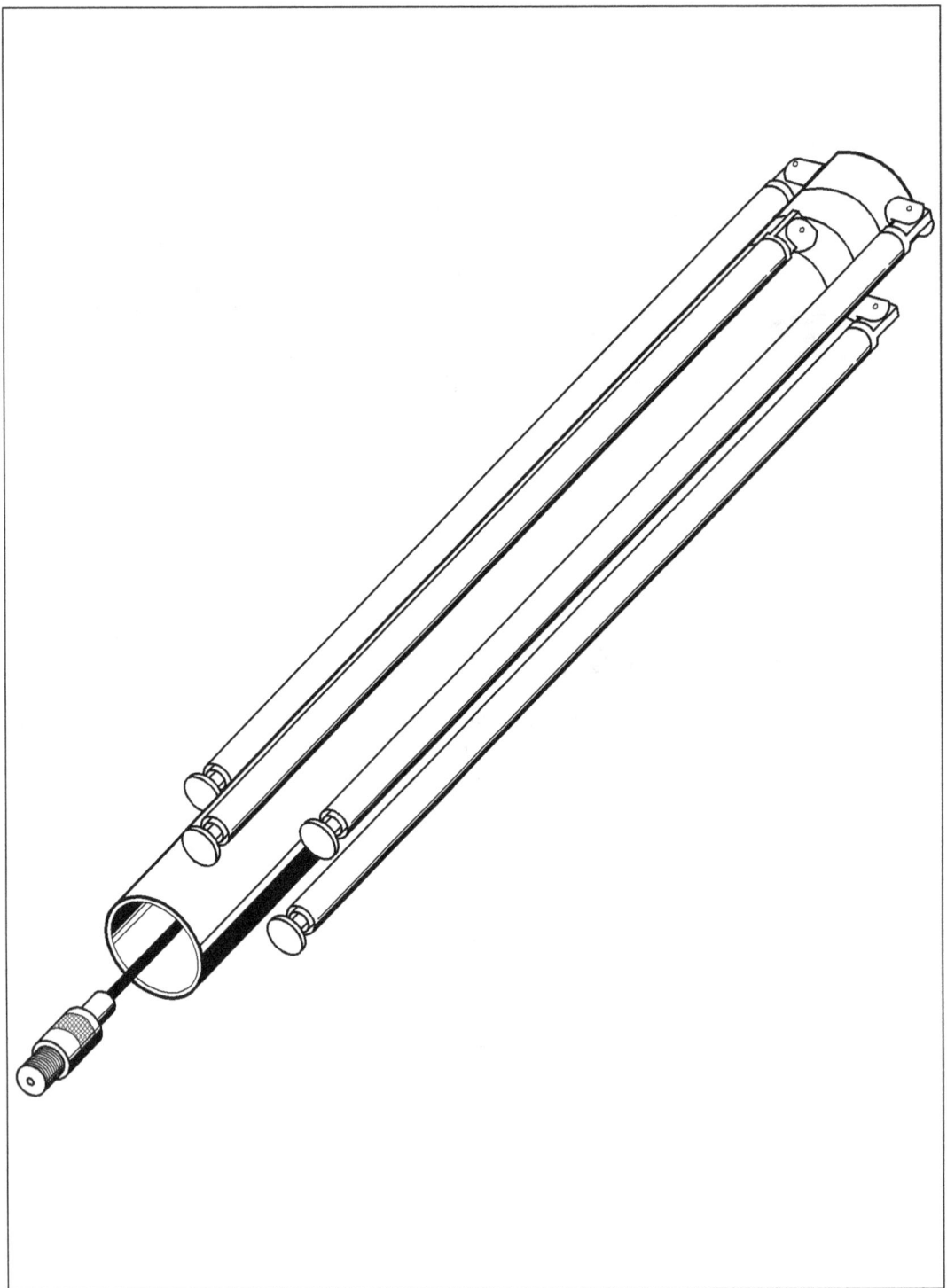

Figure 10.14 Antenna CD-2T in collapsed form.

REVIEW QUESTIONS

1. What is the main reason for using telescopic rods for this particular discone model?

2. How can the telescopic rods approximate the functions of disc and cone?

3. Would it be good to use four telescopic rods for each disc and cone function instead of three?

4. What are the distinct advantages of using this portable version of a discone antenna?

Table 10.1 American and English Wire gauges, diameter in inches and millimeter (Wire gauges 21 to 40)

The american standard wire gauge is based on the standards of the *Brown & Sharpe* company which uses numbers in identifying the wire size. In general, the abbreviation AWG (= *American Wire Gauge*) is used. In Great Britain, there are two standard wire gauges: *BWG* (= *Birmingham Wire Gauge*) and *ISWG* (= *Imperial Standard Wire Gauge*) or *SWG* (= *Standard Wire Gauge*). Both these standards also use numbers to identify the size of the wire.

Wire gauge Nr.	AWG diameter in inches	in mm	BWG diameter in inches	in mm	ISWG(SWG) diameter in inches	in mm
21	0.028	0.72	0.081	0.81	0.032	0.81
22	0.025	0.64	0.028	0.71	0.028	0.71
23	0.023	0.57	0.025	0.64	0.024	0.61
24	0.020	0.51	0.023	0.56	0.023	0.56
25	0.078	0.45	0.020	0.51	0.020	0.51
26	0.016	0.40	0.018	0.46	0.018	0.46
27	0.014	0.36	0.016	0.41	0.016	0.41
28	0.013	0.32	0.0135	0.356	0.014	0.36
29	0.011	0.29	0.013	0.33	0.013	0.33
30	9.010	0.25	0.012	0.305	0.012	0.305
31	0.09	0.23	0.010	0.254	0.011	0.29
32	0.008	0.20	0.009	0.229	0.0106	0.27
33	0.007	0.18	0.008	0.203	0.010	0.254
34	0.0063	0.16	0.007	0.178	0.009	0.229
35	0.0056	0.14	0.005	0.127	0.008	0.203
36	0.0050	0.13	0.004	0.102	0.007	0.178
37	0.0044	0.11	-	-	0.0067	0.17
38	0.0040	0.10	-	-	0.0060	0.15
39	0.0035	0.09	-	-	0.0050	0.127
40	0.0031	0.08	-	-	0.0047	0.12

NOTE: Values in millimeter were rounded off. AWG 0000 to 20 see Table 7.1 in page 88.

11   5/8 WAVE ANTENNA

Model WA-2

Probably one of the most popular vertical antennas for both mobile and fixed station installations is the 5/8 wavelength vertical, because it has some gain over a dipole. It is omnidirectional, and can be used either with radials or a solid-plane body (such as the one afforded by a car).

A version of a 5/8 vertical with radials is presented in this chapter. It is designed for fixed station installations. The common practice of radio operators is to install this antenna atop a tower with rotate-able Yagi arrays positioned a few feet below it. The two antennas are connected to a common transceiver via a switching box. Only one antenna is active at one moment. The 5/8 wave vertical is used as a monitoring antenna because of its omnidirectional characteristics. Once a contact has been established during operation, the operator quickly switches over to the Yagi antenna, and beams it towards the other station to optimize communications. When the contact is finished, the transceiver is again switched back to the 5/8 wave vertical antenna. This does not mean however that the average radio operator who cannot afford to erect a tower and a Yagi array should refrain from installing a 5/8 wave vertical. A properly constructed 5/8 wave vertical antenna if used singly works perfectly well!

Perhaps one advantage of constructing this antenna by the radio operator himself is the overall cost of the unit. All of the materials used in this model are readily available at hardware stores and can be bought cheap. In comparison, a commercial version of this antenna costs several hundred bucks!

This antenna model WA-2 is designed to operate in the 140-150 MHz VHF band. It exhibits an SWR of less that 1.5:1 over the entire band if properly tuned. It has a gain of 1.8 dB over a standard dipole reference.

SCALED DOWN IMAGE

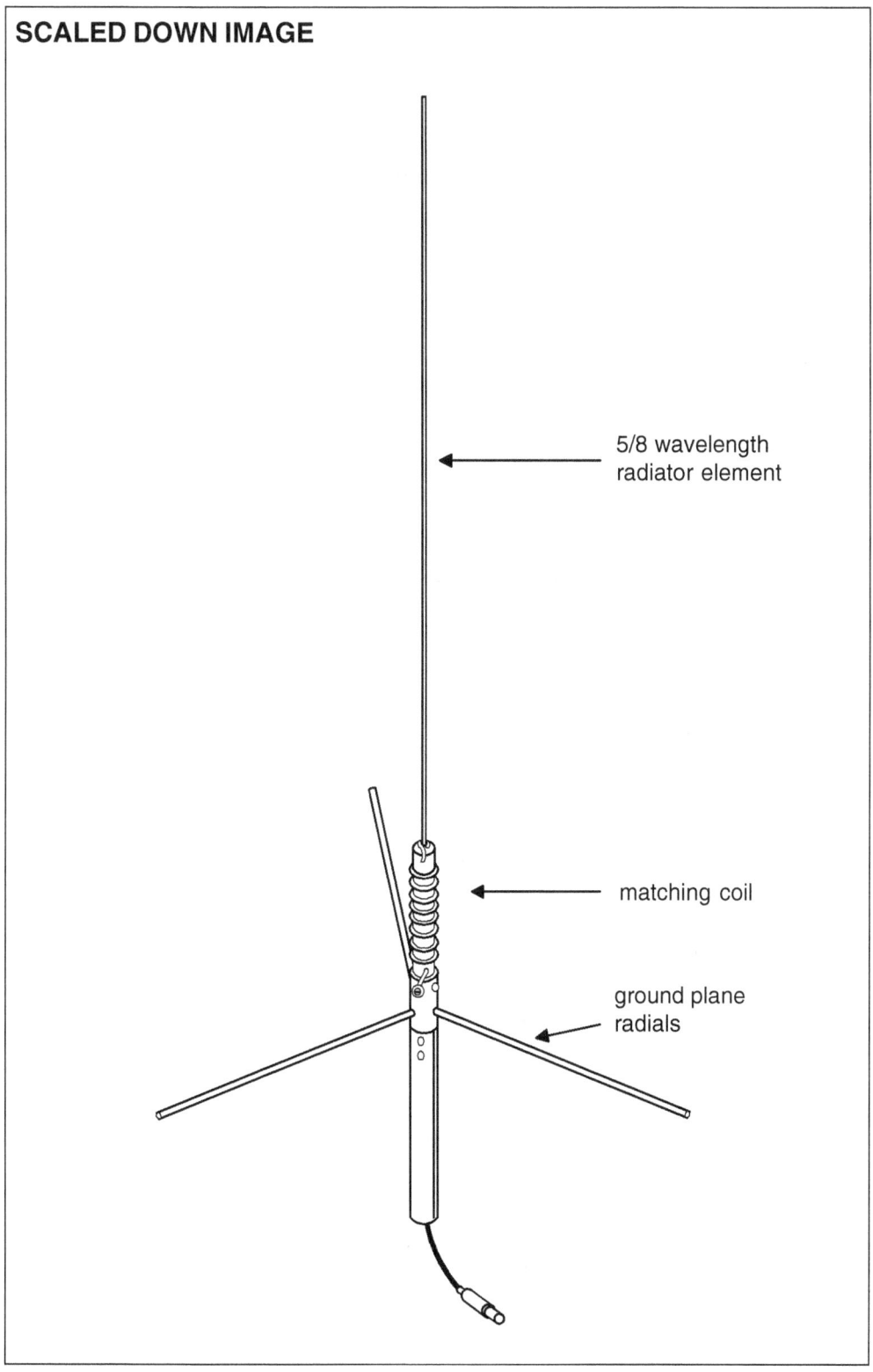

5/8 wavelength
radiator element

matching coil

ground plane
radials

Figure 11.1   5/8 Wave Antenna Model WA-2

Materials List

Quantity	Specification/Description	Dimensions
2	Brass rods the brass rod for acetylene welding is recommended	1/8" diameter
3	Brass rods 3/16" diameter	28" long
1	Engineering plastic rod see text for dimensions	
1	PL-259 VHF connector	
1	PL-258 VHF straight connector	
1	Aluminum bushing - see text for dimensions	
1	Aluminum tube	1" id* x 8"
1	Copper wire gauge no. 14	20" long
1	Aluminum plate 1/8" thick	3" x 6"
4	U-bolts with accompanying hex nuts and lock washers	
1	Coaxial cable RG-58/U	12" long
1	Stove bolt - brass or GI	1/8" x 3/8"
6	Self-tapping metal screws	1/8" x 3/8"
1	Eye terminal vinyl insulated	1/16" id*
1	Short hook-up wire	3"- 4" long

* id - inside diameter

Construction

First prepare the plastic coil form by machining the engineering plastic rod to the dimensions shown in Figure 11.2.

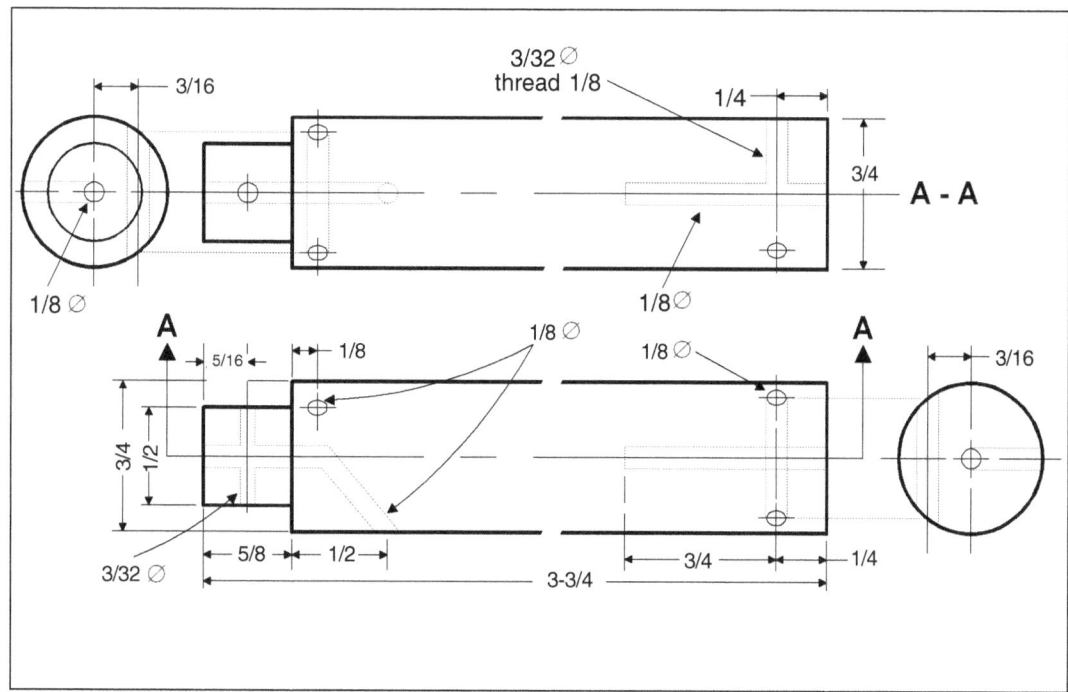

Figure 11.2   Coil form dimensions.

Next, prepare the bushing by machining the aluminum rod to the dimensions shown in Figure 11.3.

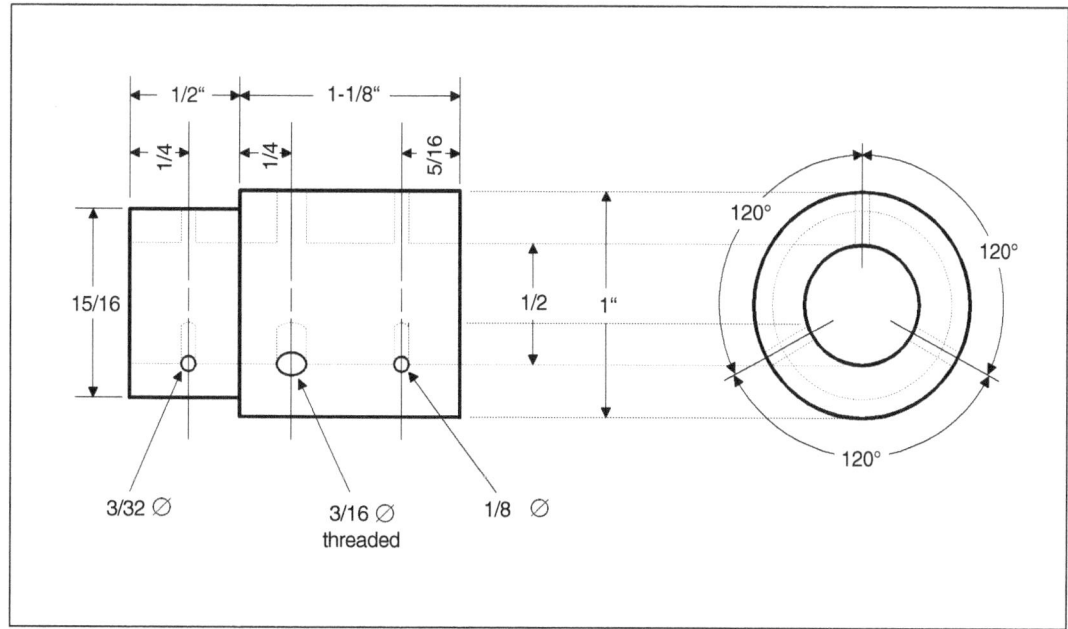

Figure 11.3   Aluminum bushing dimensions.

Assemble the plastic coil form and the aluminum bushing together. Secure the assembly with metal screws. Screw only through the two holes, and temporarily leave the third hole unscrewed (see Figure 11.4).

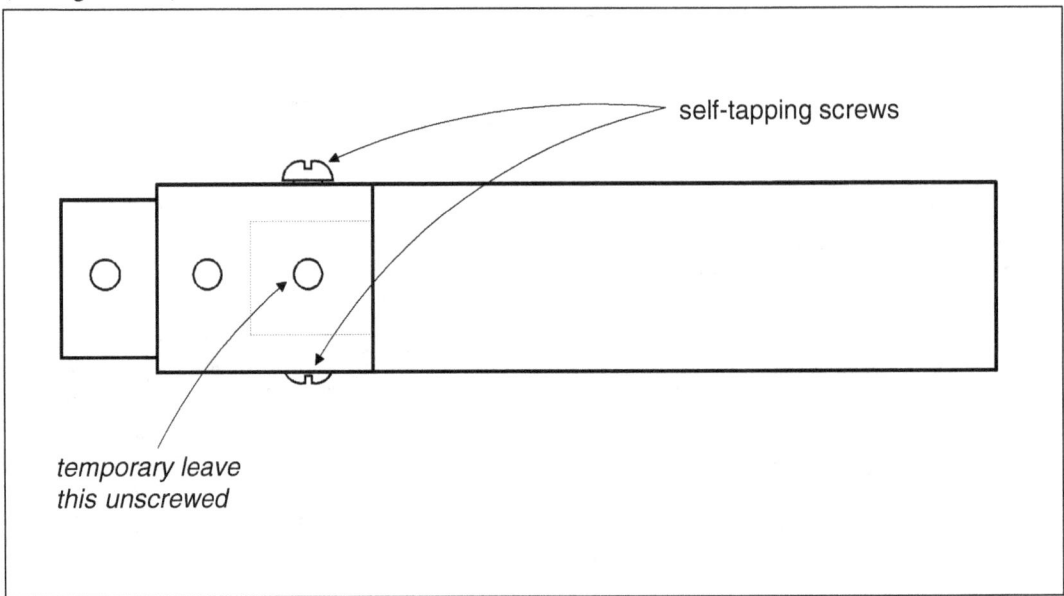

Figure 11.4   Assembling the coil form and the bushing together.

Prepare the radiator element. File a notch at one end of each 1/8" diameter brass rod. Join and solder the two notched ends together to make a long single rod (see Figure 11.5).

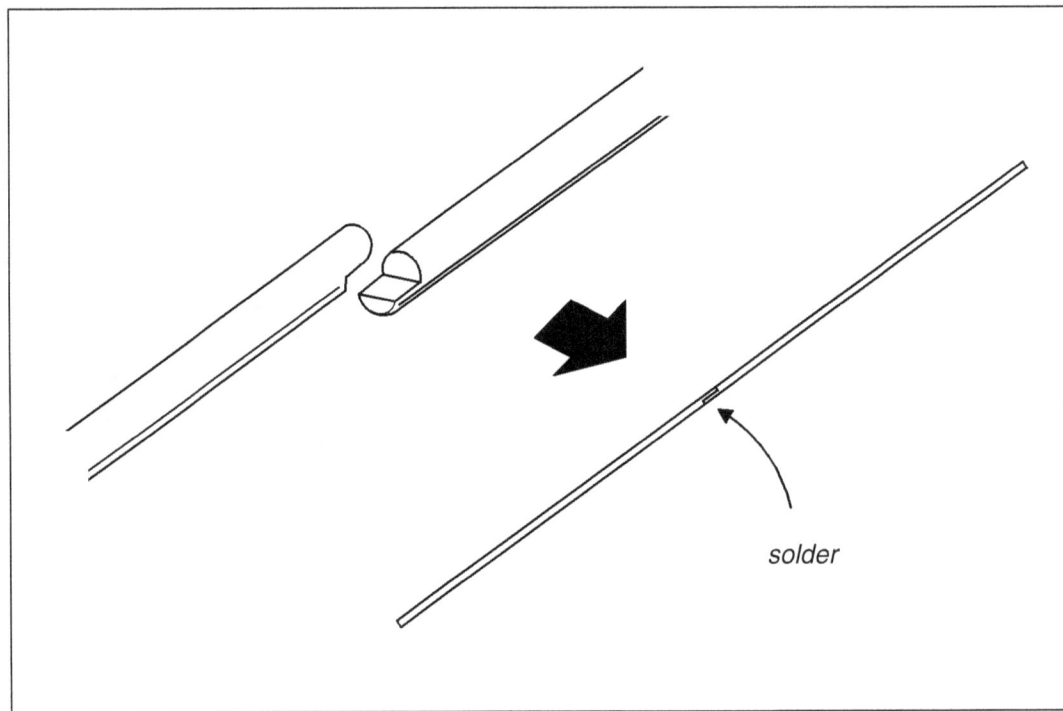

Figure 11.5   Joining the two brass together to make the radiator element.

Insert one end of the radiator rod into the top hole in the plastic coil form. Forcibly screw the 1/8" x 3/8" stove bolt through the hole at the side of the plastic coil form, pressing the brass rod inside to hold it firmly (see Figure 11.6).

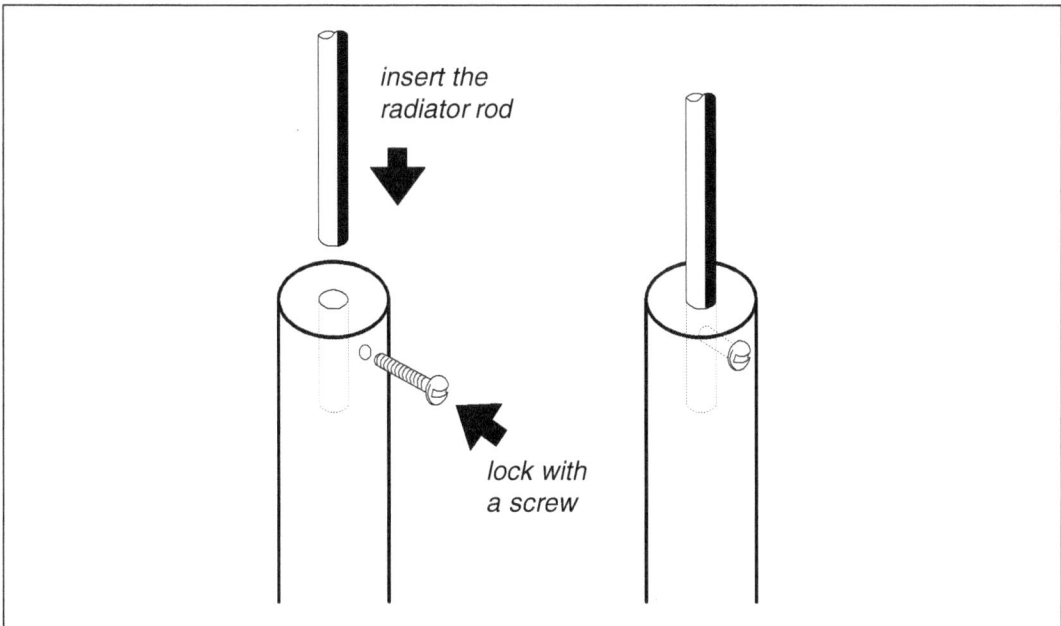

Figure 11.6 Securing the radiator element into the coil form.

Cut the brass rod to a length of 46 inches, measuring from the point where it emerges from the plastic form (see Figure 11.7).

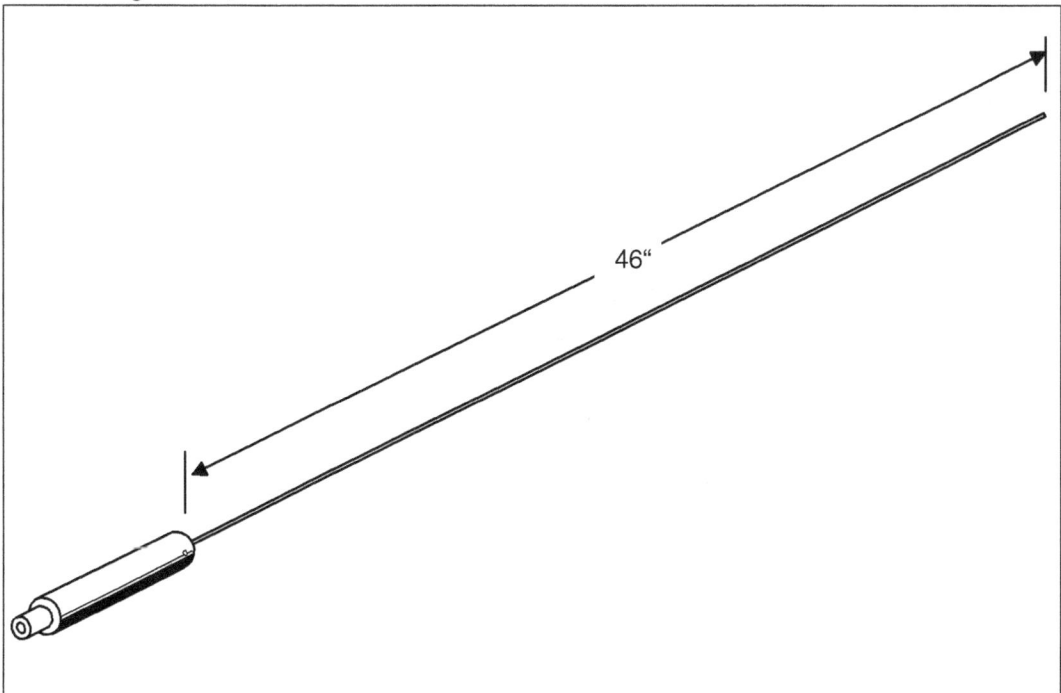

Figure 11.7 Cutting the radiator element to its proper length.

Wind the No. 14 copper wire around the coil form. Wind 10 and 1/2 turns evenly spaced and distributed to cover most of the length of the plastic form. Solder the top end of the copper wire to the base part of the brass rod (see Figure 11.8).

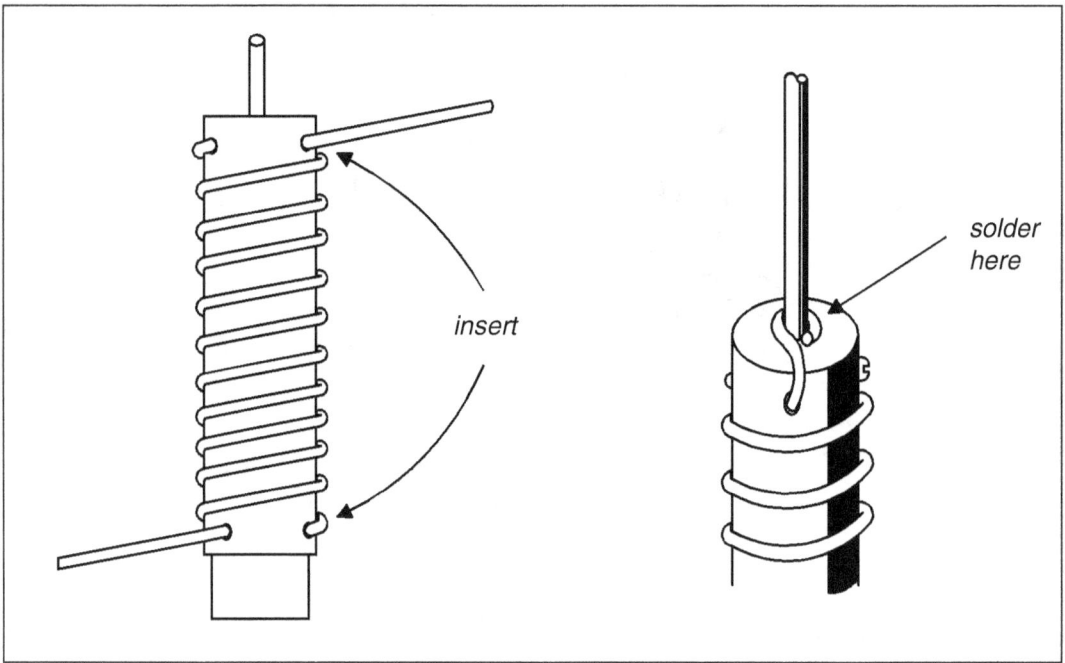

Figure 11.8   Winding the coil around the coil form.

Solder an eye terminal to the lower end of the copper coil. The eye terminal must be positioned in such a way that its eye is aligned with the unscrewed hole in the aluminum bushing. After you have soldered the eye terminal, attach it into the aluminum bushing with a metal screw (see Figure 11.9).

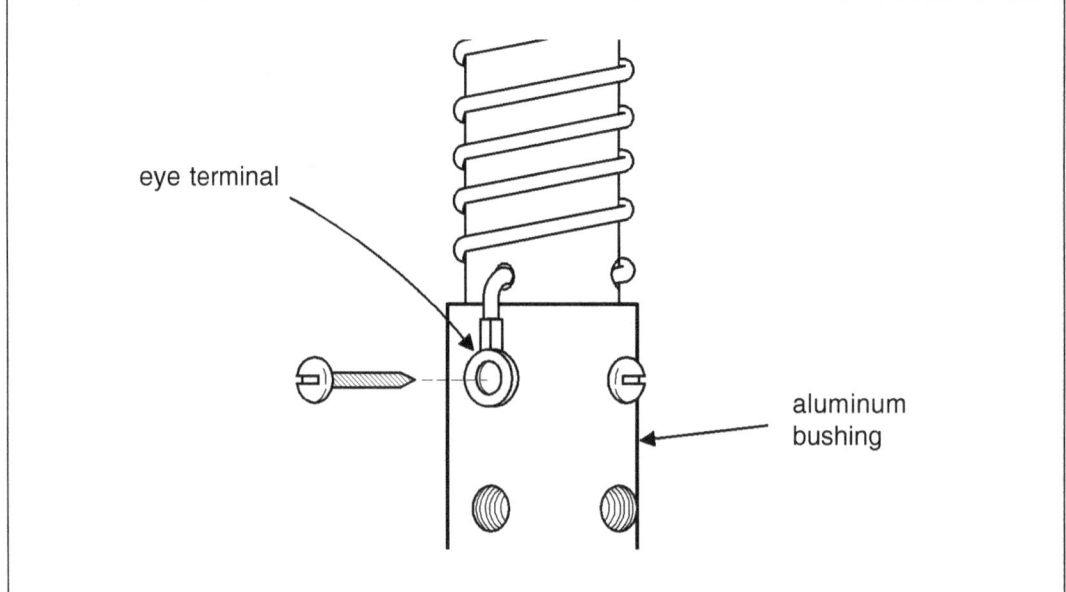

Figure 11.9   Securing the coil to the aluminum bushing.

Cut a short length of stranded hook-up wire (about 3 inches). Insert it into the hole in the plastic coil form until it protrudes from the center hole at the bottom. Solder the upper end of the hook-up wire to approximately 6 and 1/2 turns counting from the coil's lower end connected to the aluminum bushing (see Figure 11.10). This connection is temporary only, and it may be necessary to move the wire during the tune-up procedure.

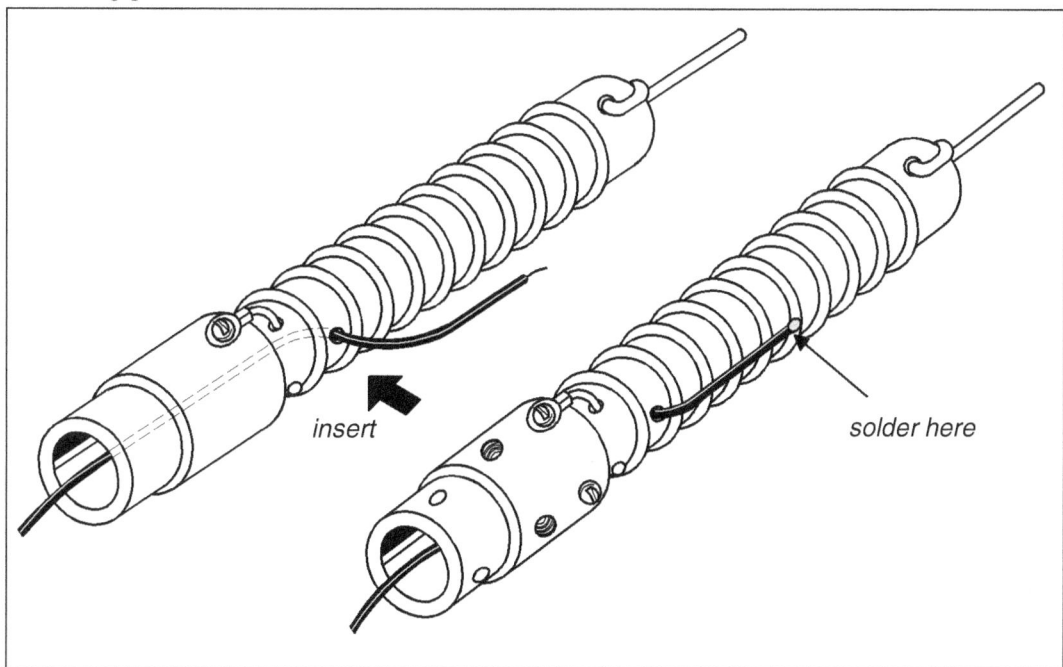

insert

solder here

Figure 11.10   Tapping the coil for feed point.

Cut 12" length of coaxial cable RG-58/U, and separate the braid from the inner conductor at one end (making a pig tail). Solder the inner conductor into the hanging end of the hook-up wire at the bottom of the plastic form. After joining the two wires, insulate the joint either with a shrinking tube or just plain vinyl tape. Solder an eye terminal into the braid of the coax cable (see Figure 11.11).

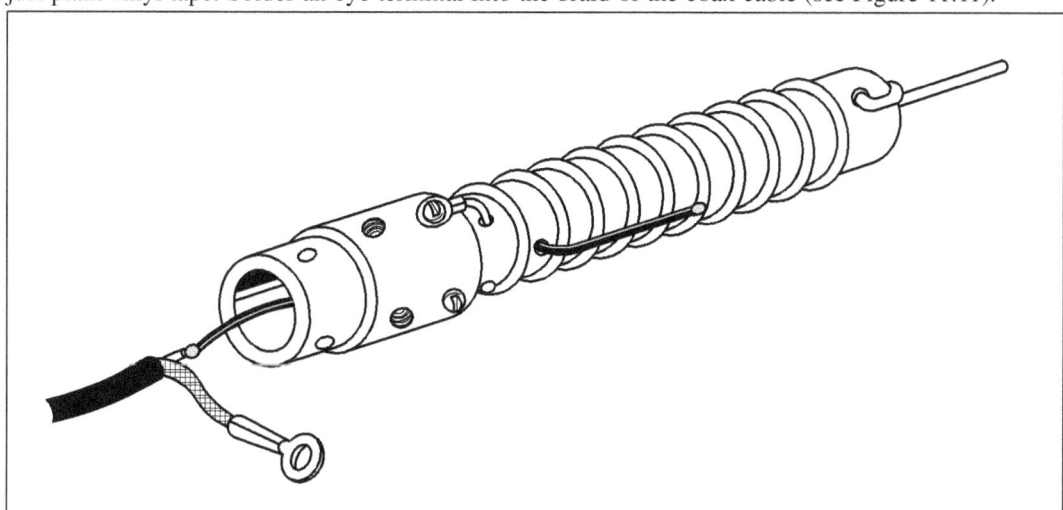

Figure 11.11   Connecting the coaxial cable to the hook-up wire.

The next step is to prepare the mounting tube. Cut 1" diameter of tube to a length of 12 inches, and drill three holes at one end. The holes must be 1/8" in diameter and equally spaced from each other. Drill a single hole at the same end, but slightly below one of the first 3 holes (see Figure 11.12).

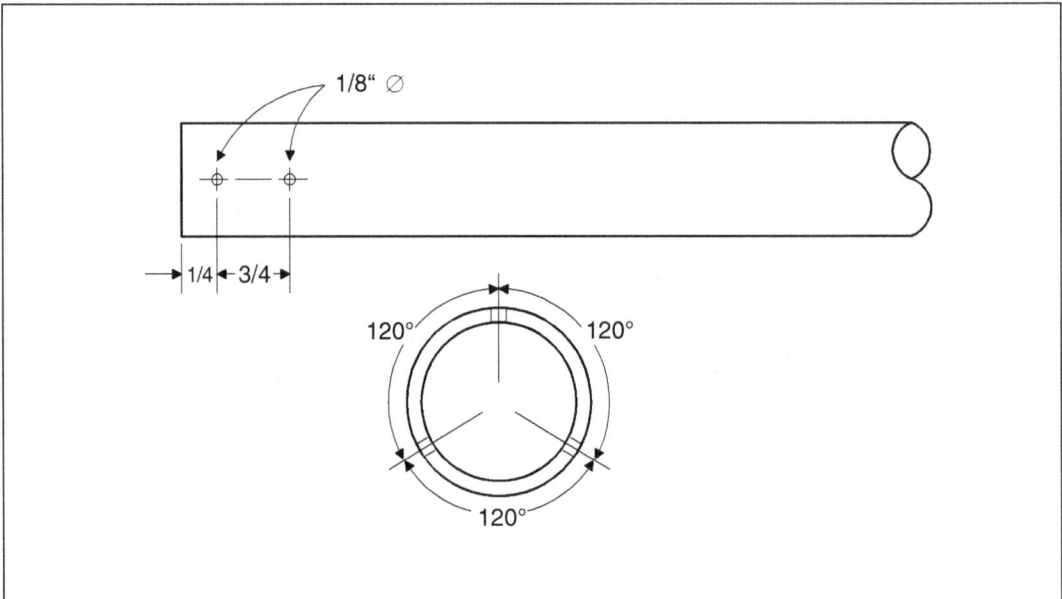

Figure 11.12   Preparing the mounting tube.

Next, insert the free end of the coaxial cable into the mounting tube, starting at the end with side holes. When the aluminum bushing and tube meet, insert the bushing inside the tube, and align the holes at their sides. Secure the bushing into the tube by screwing self-tapping screws into the holes (see Figure 11.13).

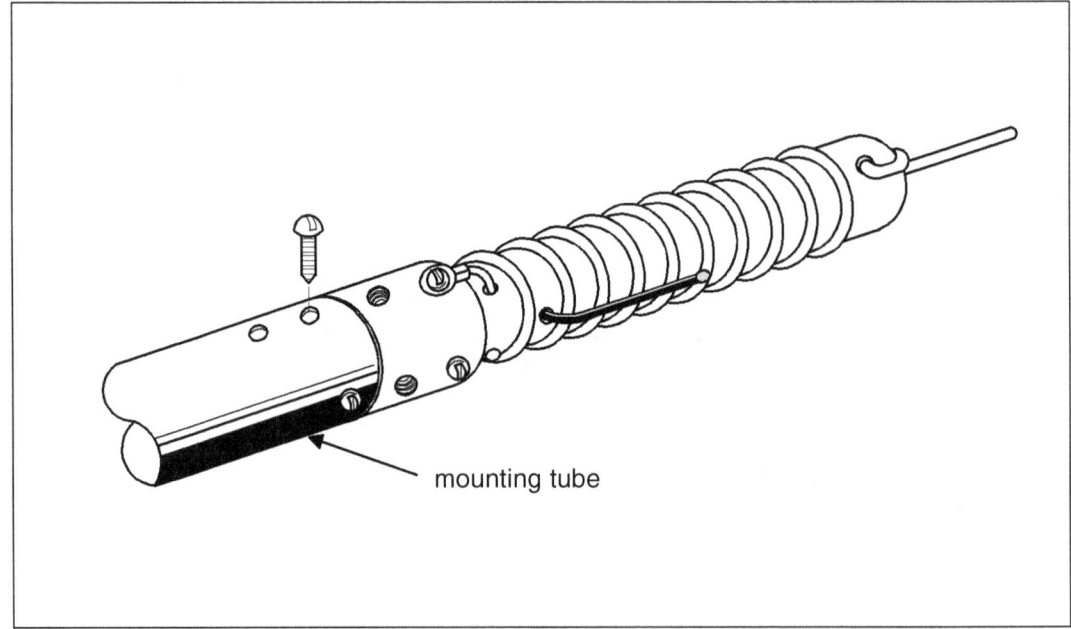

Figure 11.13   Securing the aluminum bushing into the mounting tube.

Secure the braid by inserting a self-tapping screw into the lone hole, and tapping into its eye terminal inside the tube. Tighten the screw to hold the eye terminal against the wall of the tube (see Figure 11.14). You may need to use a wooden stick inserted into the tube to position the eye terminal exactly under the hole.

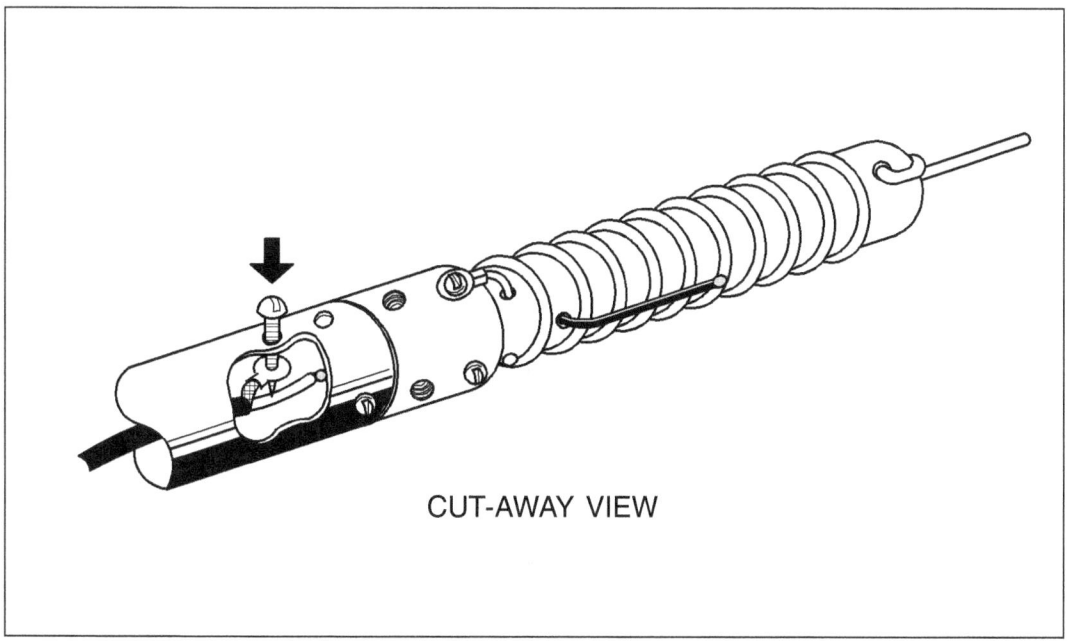

CUT-AWAY VIEW

Figure 11.14   Securing the braid inside the tube with a metal screw.

Next, prepare the ground plane radials. Cut three lengths of 3/16" diameter brass rods. Note that these rods are larger than the radiator rod. Each rod must be 28 inches long and threaded at one end (see Figure 11.15).

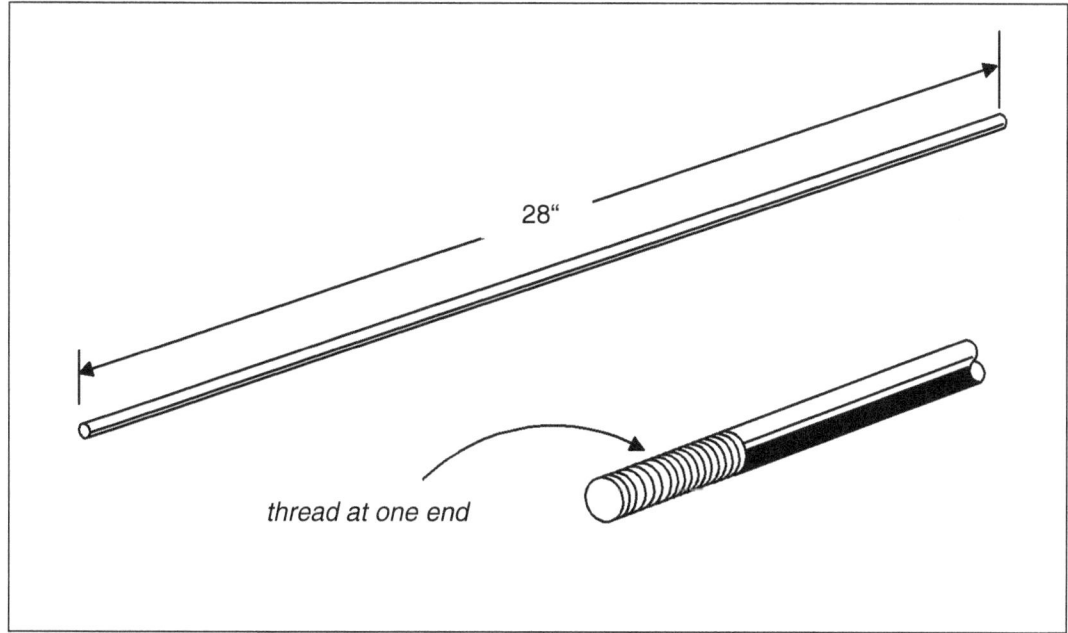

28"

thread at one end

Figure 11.15   Cutting the radials and threading one of their ends.

Final assembly and installation

Attach the three ground plane radials into their mounting holes at the aluminum bushing. Mount the antenna to the mast by using a metal plate adaptor similar to one described in the preceding chapters (see following illustration).

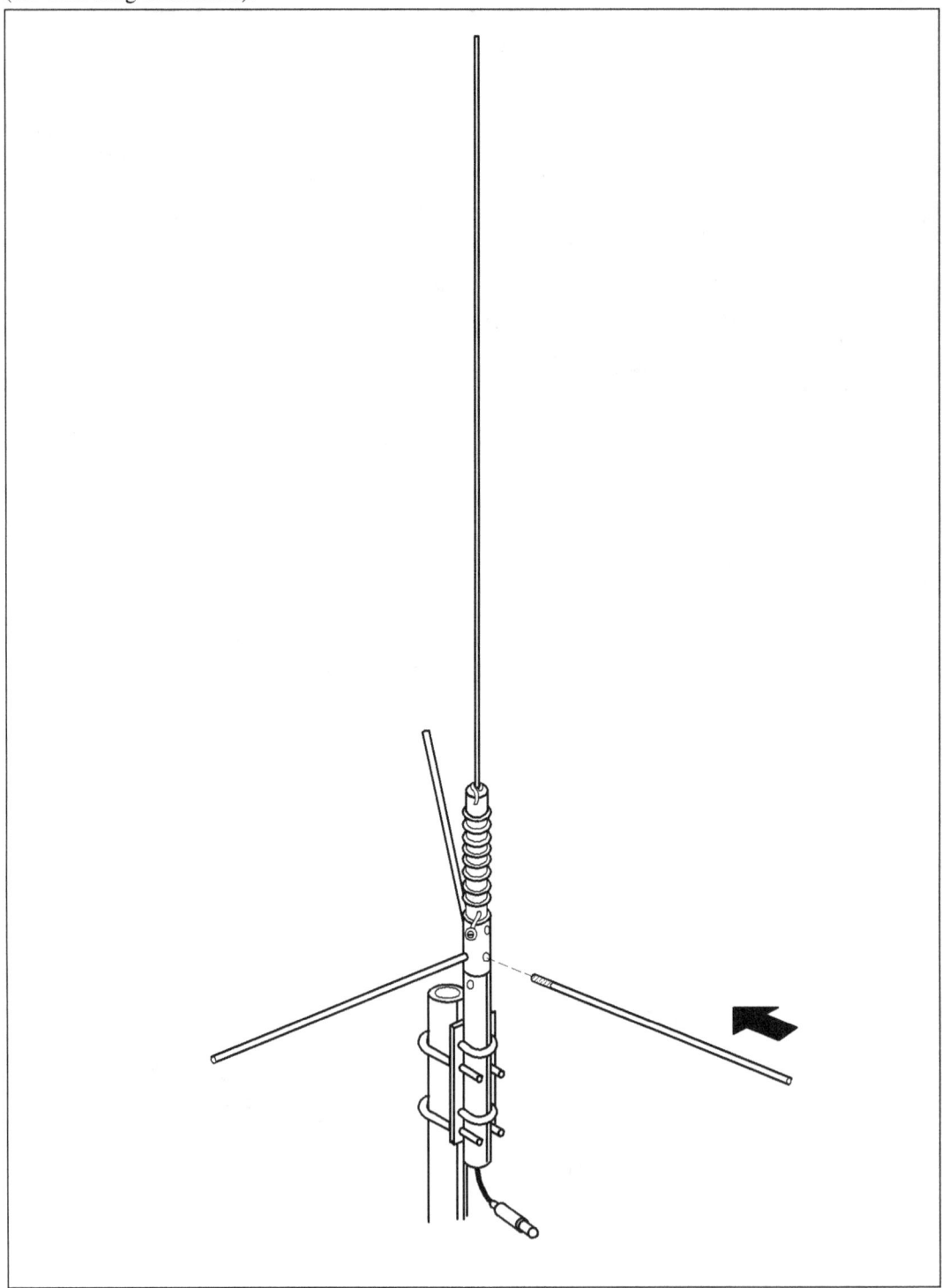

Figure 11.16   Mounting the WA-2 to the mast.

Tuning WA-2 to resonance

Mount the antenna to the mast as previously described. Connect a coaxial cable into the PL-258 connector of the antenna, and attach the other end into the output of an SWR meter (marked with 'antenna'). Attach also a short coaxial feeder into the input of the SWR meter (usually marked 'transmitter'); the other end of the feeder must be plugged into the output connector of your transceiver (see Figure 11.17). Set your transceiver's frequency to the center of the band, and key the PTT. Read the SWR response, and write it down in a chart similar to the one shown in Figure 11.18 on the next page.

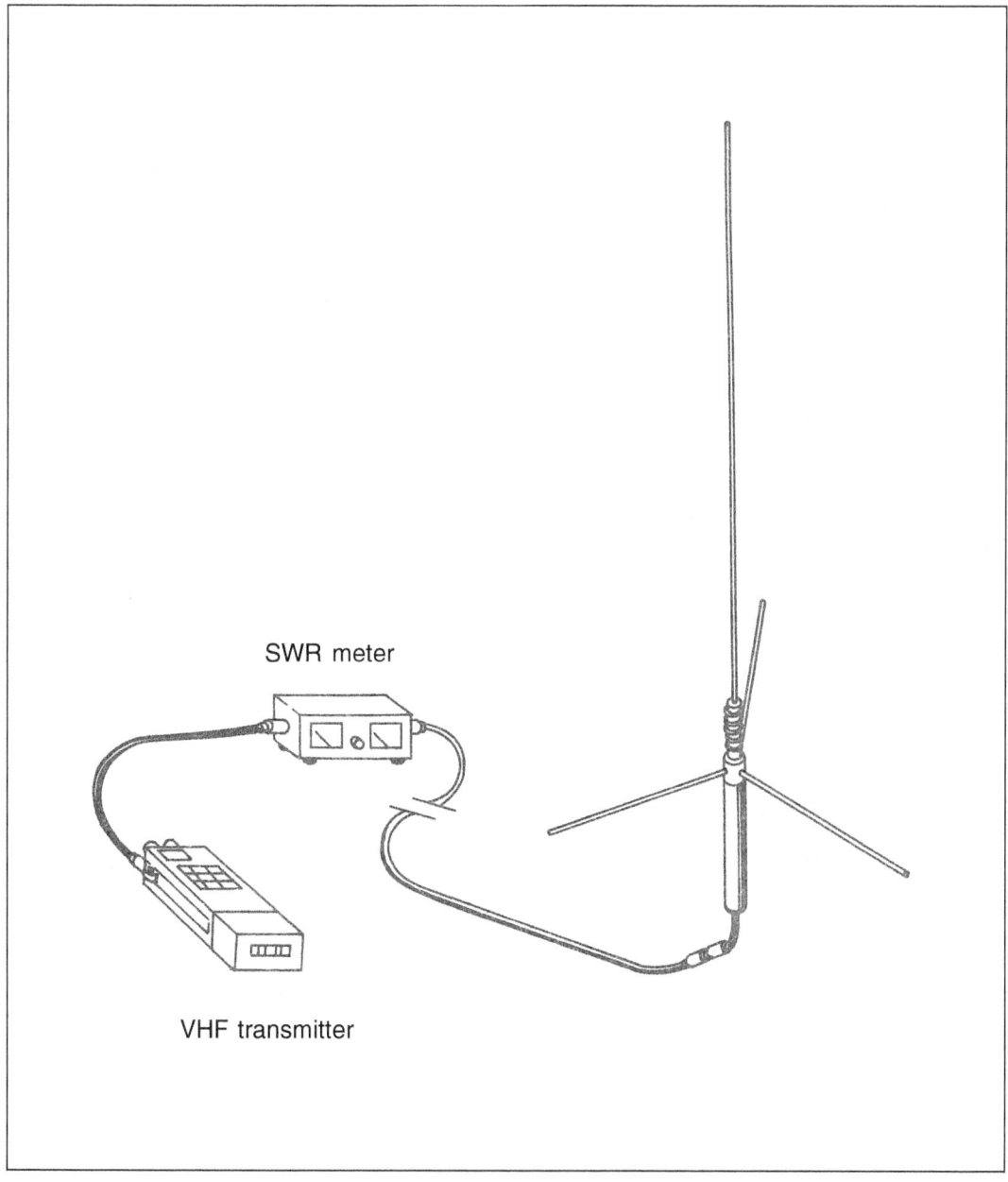

SWR meter

VHF transmitter

Figure 11.17   Preparing the antenna for tuning to resonance.

Set your transceiver's frequency to the center of the band, and key the PTT. Read the SWR response, and write it down in a chart similar to the one shown in Figure 11.18.

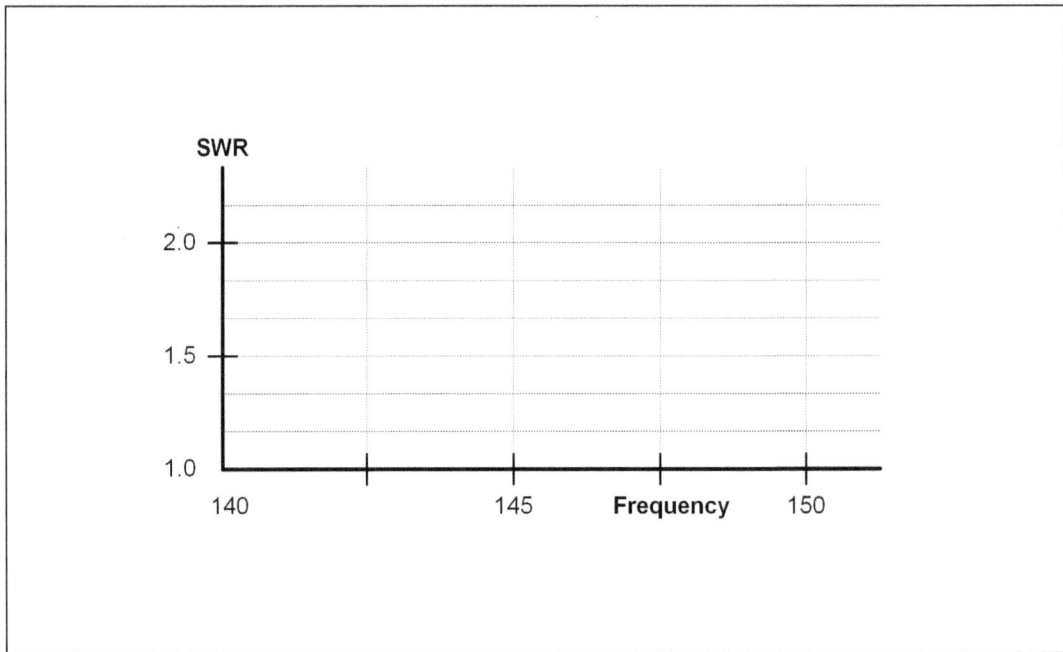

Figure 11.18   SWR chart.

Re-solder the hook-up wire to another point in the copper coil to get the lowest SWR response in the center frequency, and a relatively flat response over the entire band similar to the charted response shown below (see Figure 11.19).

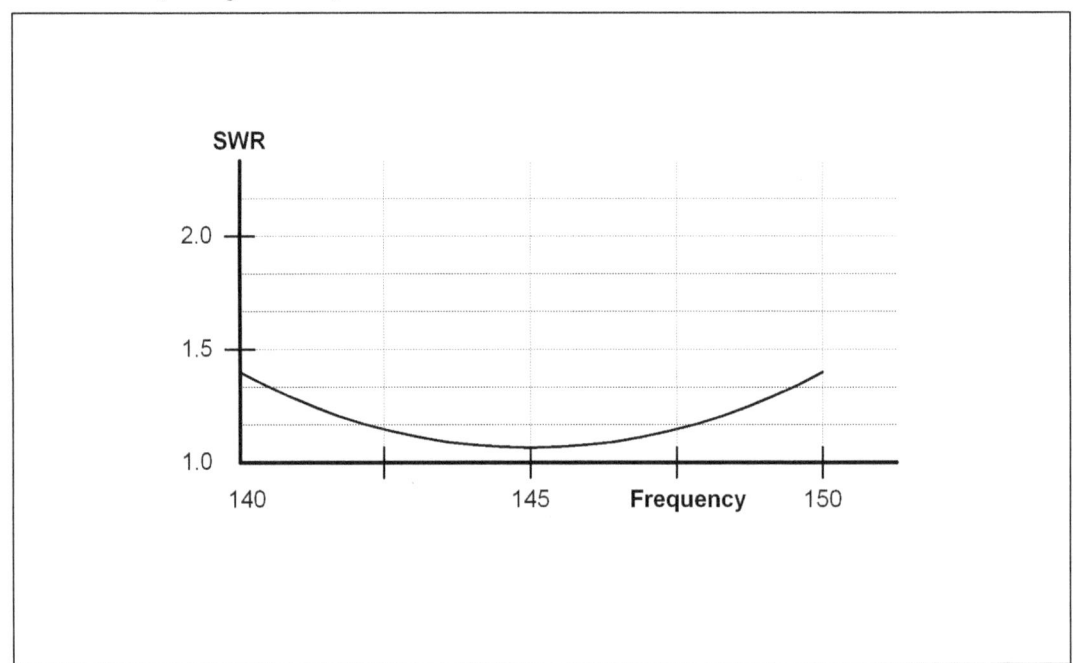

Figure 11.19   A sample of a charted SWR response.

You can move the soldered point or tap either way -- left or right -- depending on how the SWR responds. If you have moved the tap to the right and the SWR went higher, then obviously you must move the tap in the opposite direction -- to the left. You must check the SWR reading over the entire band every time you move the tap. Move the tap only about 1/4 inch farther each time. After you have found the best point in the copper coil, solder the hook-up wire permanently (see Figure 11.20).

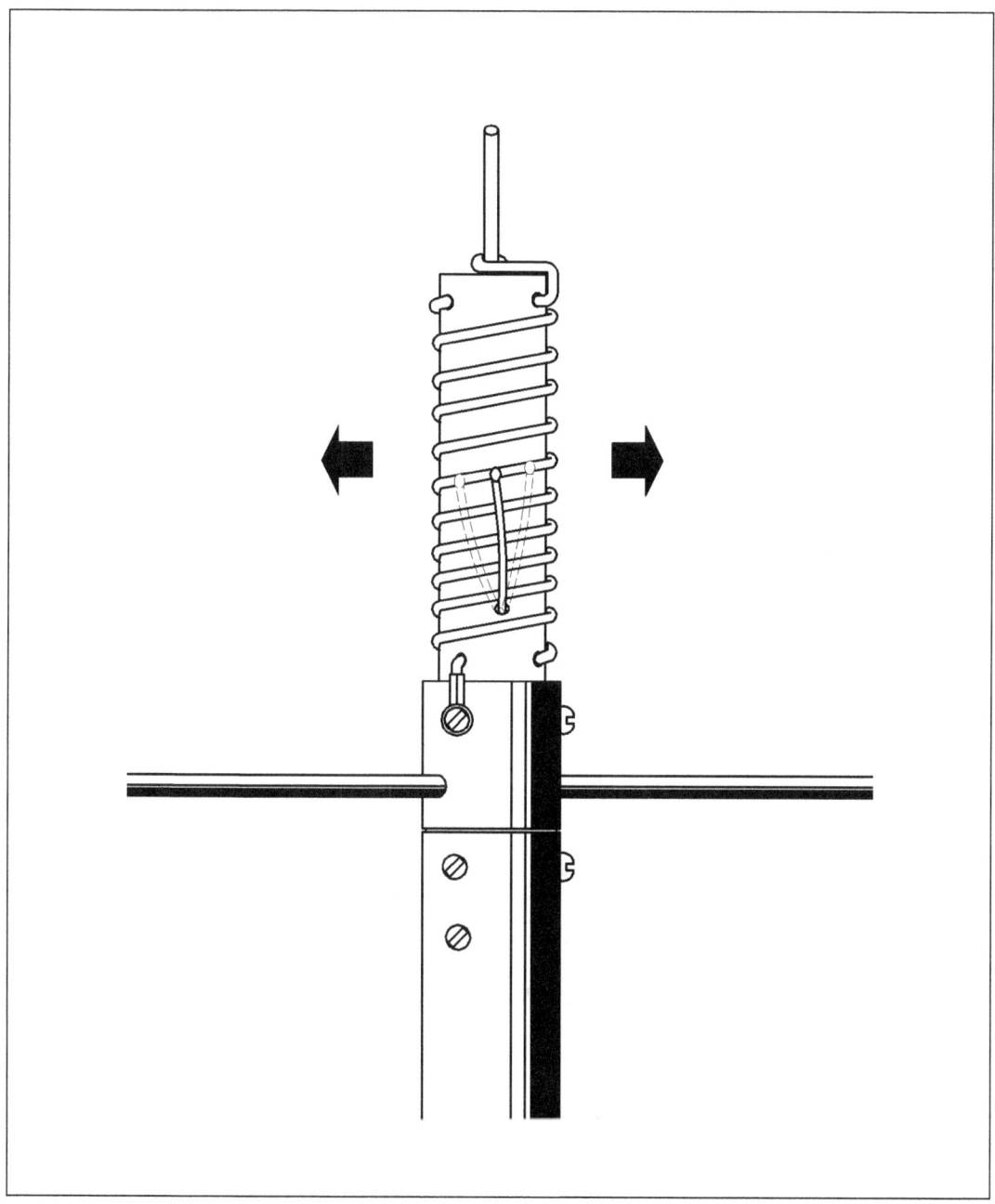

Figure 11.20   Resoldering the tap to a different point to find the best SWR response.

Dismount the antenna from the mast, and remove its three ground plane radials. Place the heat shrinking tube into the antenna, wrapping the entire coil form, and heat it over a flame or with a blow dryer. The coil form and shrinking tube must be rotated continuously over the heat to result in an even shrinking of the tube (see Figure 11.21). If you are heating the tube over the flame, don't let the flame touch the tube directly.

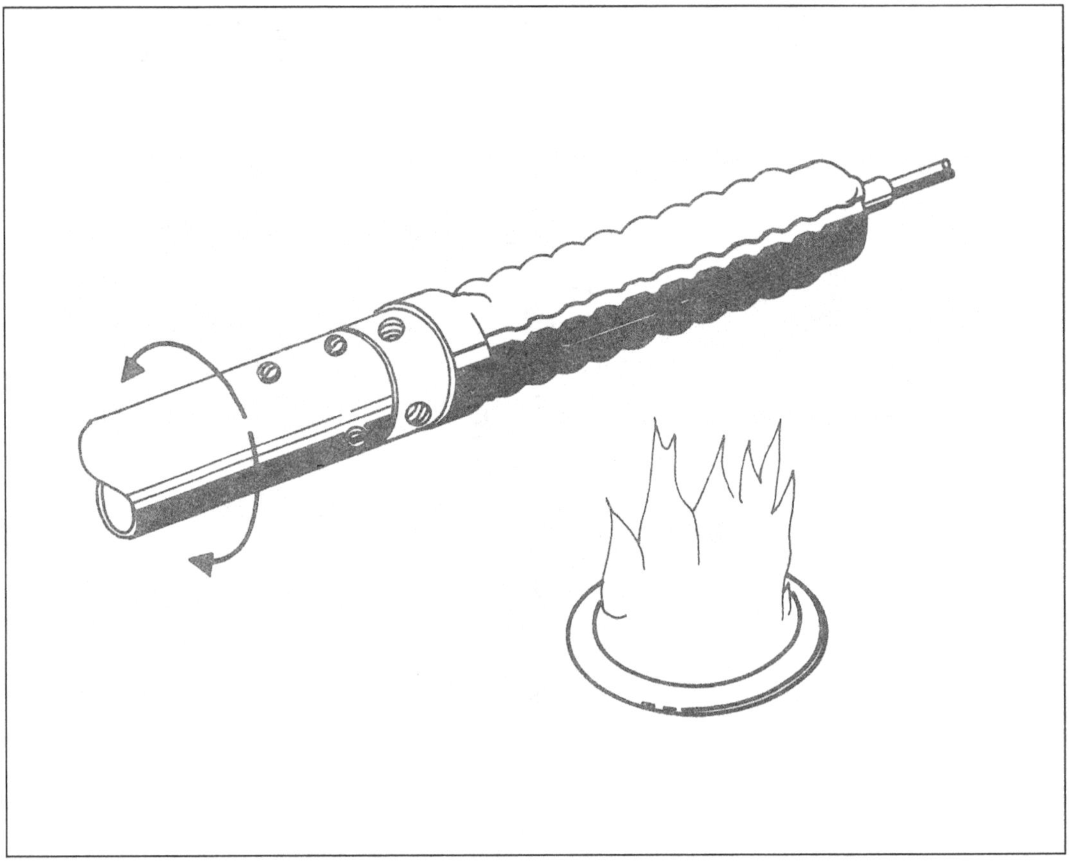

Figure 11.21   Heating the shrinkable tube.

REVIEW QUESTIONS

1. What is the advantage of using a 5/8 wave vertical antenna?

2. What is the function of the three rods connected to the base of the antenna?

3. What is the function of the coil at the base of the radiator rod?

4. What is the function of the shrinking tube?

5. How is the 5/8 wave vertical antenna tuned to resonance?

6. What is the advantage of using an SWR chart during antenna tuning?

12   5/8 WAVE ANTENNA

Model WD-2

This antenna is an improvement of the basic design of 5/8 wave vertical with radials. As can be clearly seen in the following illustration, it has two metallic cones attached to a long tube, which doubles as support for the radiator element. The cones are not intended for novelty, but serve a very important purpose: for a more efficient performance of the entire antenna system. Its function is to nullify the unbalanced coupling between the transmission and the antenna feed point, and prevent the unwanted current from flowing on the outside of the coaxial cable.

Why is this so? Consider some technical basics to understand this phenomenon. In a perfectly balanced antenna, the electrical current within each leg of the element is symmetrical. There will be no problem in coupling the RF signal to its feed point when a balanced feed line is used. However, if a coaxial cable is used to feed the antenna, the coupling action is inherently unbalanced because of the physical construction of the coaxial cable. Stated simply, the outside part of the outer conductor is not coupled to the antenna in the same way as its inner part is coupled to the inner conductor. The overall result is that current will flow on the outside of the outer conductor. This current is negligible in the HF frequencies, but must not be ignored in VHF or UHF frequencies. This problem is remedied by the metal cones described in this particular model -- it de-tunes the system for stray currents present on the outside of the line. The cones are also called "de-tuning sleeves" or "decoupling sleeves".

An antenna system with a properly decoupled line is commonly used in repeater systems, because by the very nature of its design, a repeater station is very sensitive to any kind of stray RF signal. A repeater station has both receiver and transmitter units simultaneously operating when used. Although the frequency of the transmitter unit is different from the frequency of the receiver unit, the very close proximity of the two units tends to blank out the generally weak signals from distant stations. This results in a phenomenon called "desensitization" or "desense" wherein the repeater cannot receive the signals from the user stations.

Feedback also results in loud squealing heard by the users. The entire system thus ceases to function as a repeater. Desense and feedback is avoided by using high-Q cavity filters inserted in the transmission line for the transmitter or receiver antenna, or both.

Additionally, the automatic switching electronics of the repeater is also protected against picking up unwanted RF by enclosing it in a metal box, and by extensive use of decoupling circuits in all the leads going in and out of the box. However all of these efforts could fail if the stray current that travels along the outside part of the transmission line is so strong that it penetrates all filters installed in the repeater system. Using a decoupled antenna system such as the one described in this chapter will save you from the trouble.

The model WD-2 is specifically dimensioned to operate in the 140-150 MHz band. It exhibits an SWR response of less than 1.5:1 over the entire band. The radiation pattern is omnidirectional. It has a gain of 1.8 dB compared to a standard dipole reference.

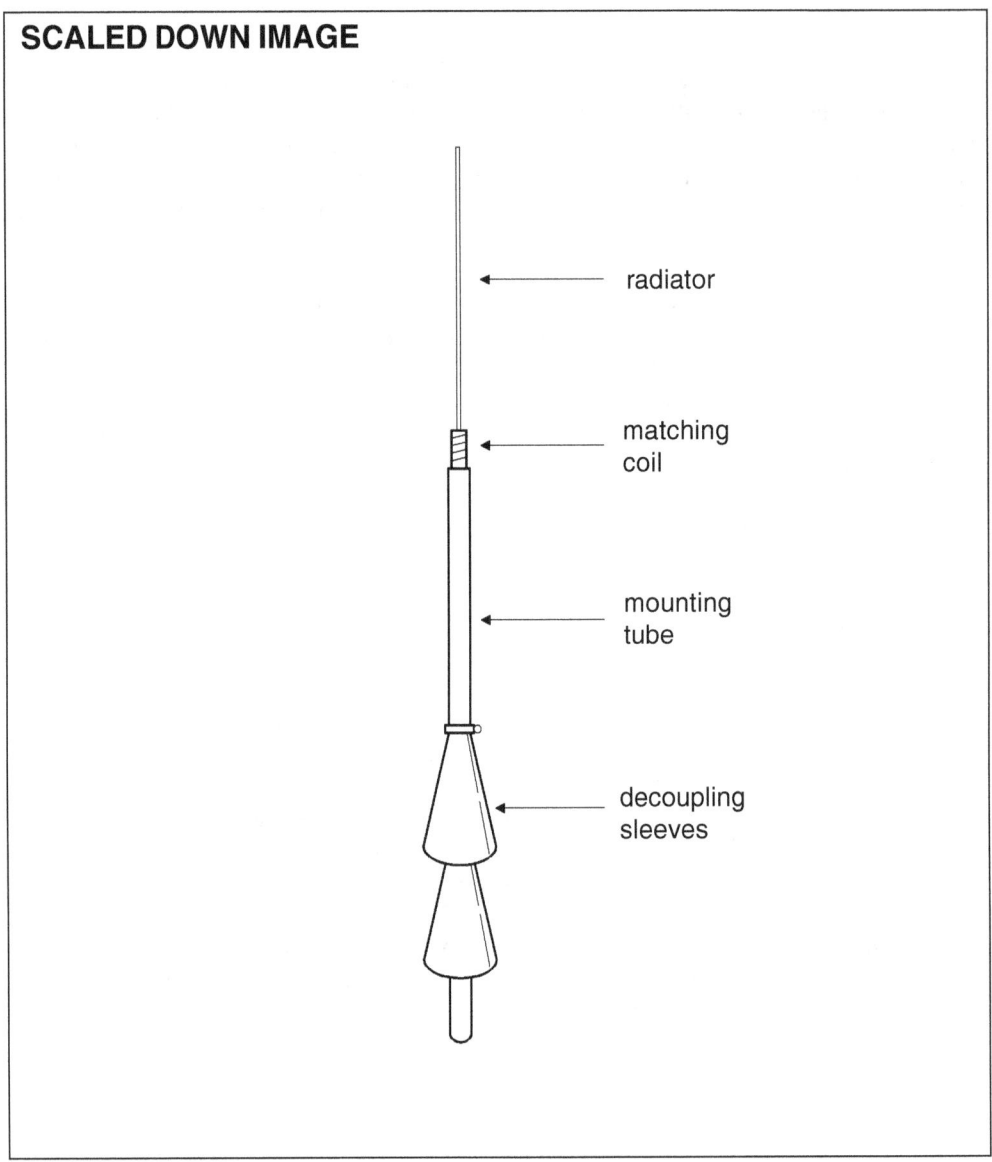

Figure 12.1   Decoupled 5/8 wave antenna model WD-2.

Materials needed

This antenna is basically the same as the antenna model WA-2. The difference between the two models is that the mounting tube for the model WD-2 is 104 inches long and has no ground plane radials, but instead it has two decoupling sleeves made of metal cones attached to the lower portion of the mounting tube.

Construction

Construct the antenna following the procedures described for the model WA-2 in chapter 11, except for the length of the mounting tube which is 104 inches long for the model WD-2. Also skip the procedure for preparing the ground plane radials; you don't need them for this antenna anyway. Furthermore, before you tune the antenna to its resonance, construct the decoupling sleeves and attach them to the mounting tube following the procedures described here.

Cut the cone form from a metal plate (GI sheet or aluminum) following the dimensions shown in Figure 12.2. In forming the cone, overlap its edges and drill holes along the edge. Rivet the overlapping edges through these holes.

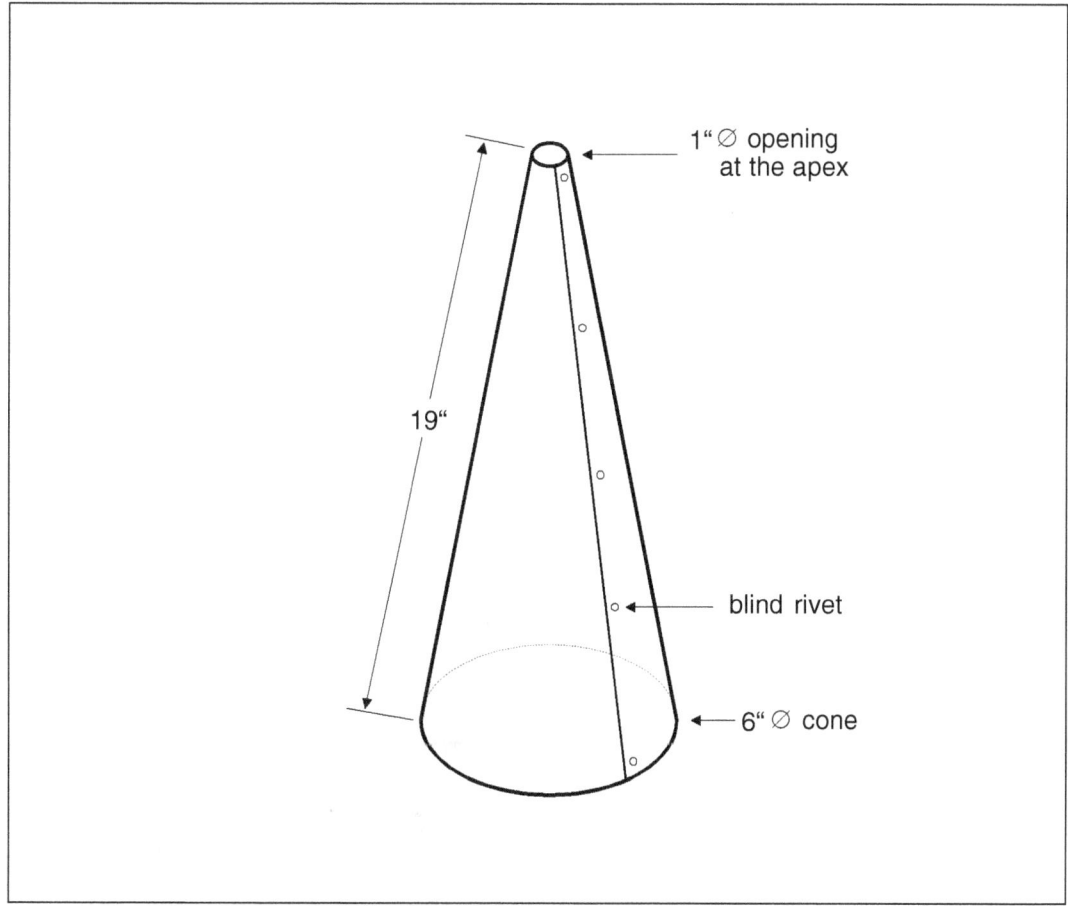

Figure 12.2   Fabricating the decoupling sleeve (or 'decoupling skirt').

Cut a narrow strip out of a similar material, and form it to a ring with a diameter of 1 inch as shown. Leave a small gap between the two ends. This ring will serve as a mounting sleeve so that the decoupling sleeve or skirt can be securely clamped to the mounting tube (see Figure 12.3).

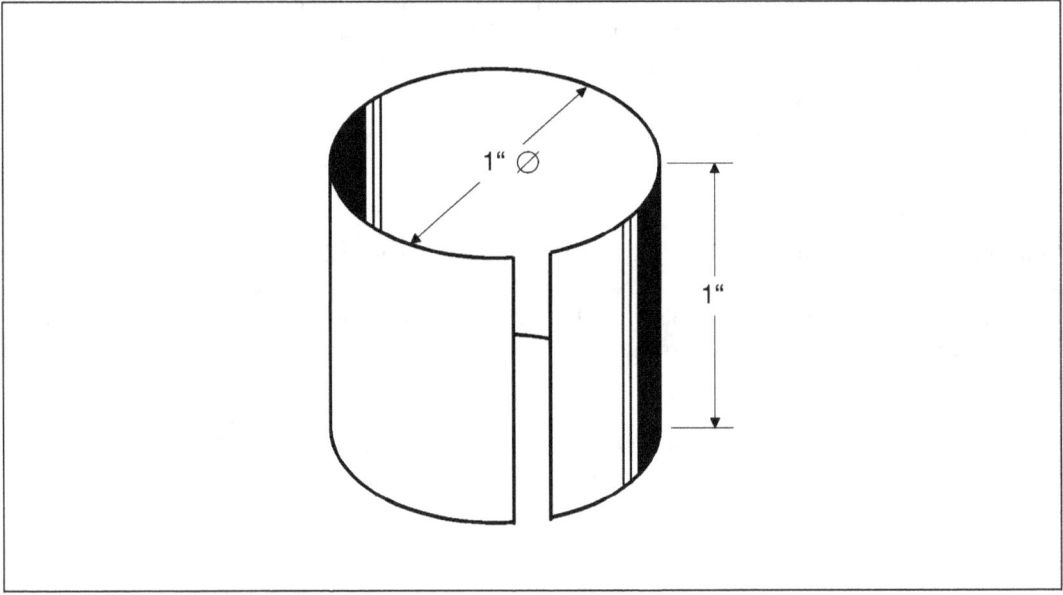

Figure 12.3   Preparing the metal ring.

Solder the ring to the apex of the cone (see Figure 12.4). If you use an aluminum plate you must electrically weld the two pieces together using a special welding technique with protective gas.

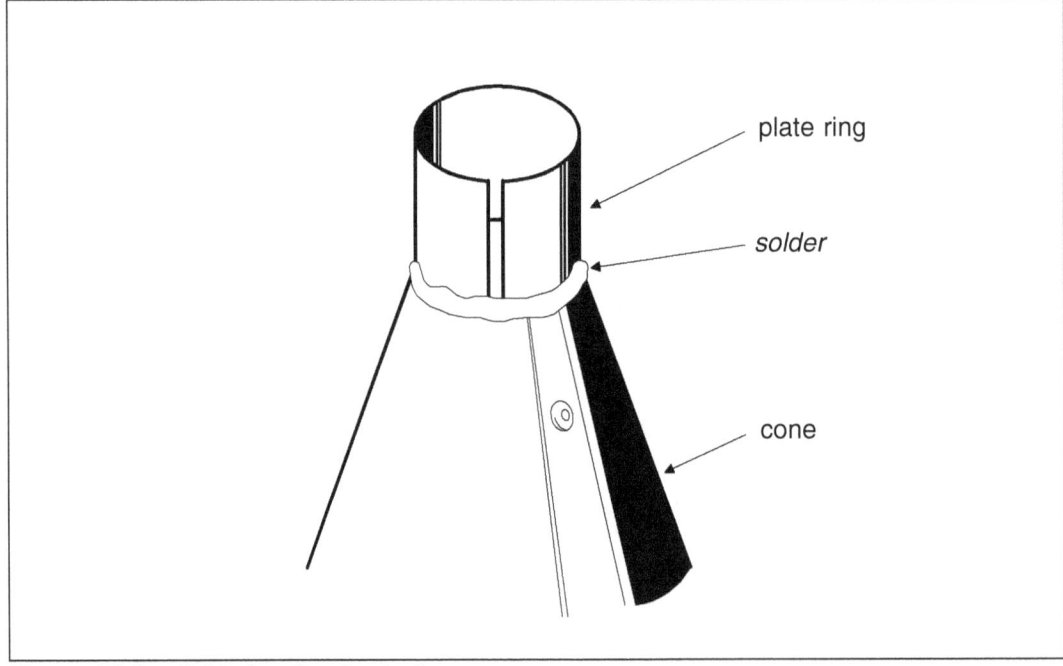

Figure 12.4   Soldering the ring to the cone.

Attach the two decoupling sleeves/skirts to the mounting tube, following the measurements shown in Figure 12.5. Place a tube clamp over each cone, and tighten it to secure the cones firmly to the mounting tube (see Figure 12.6).

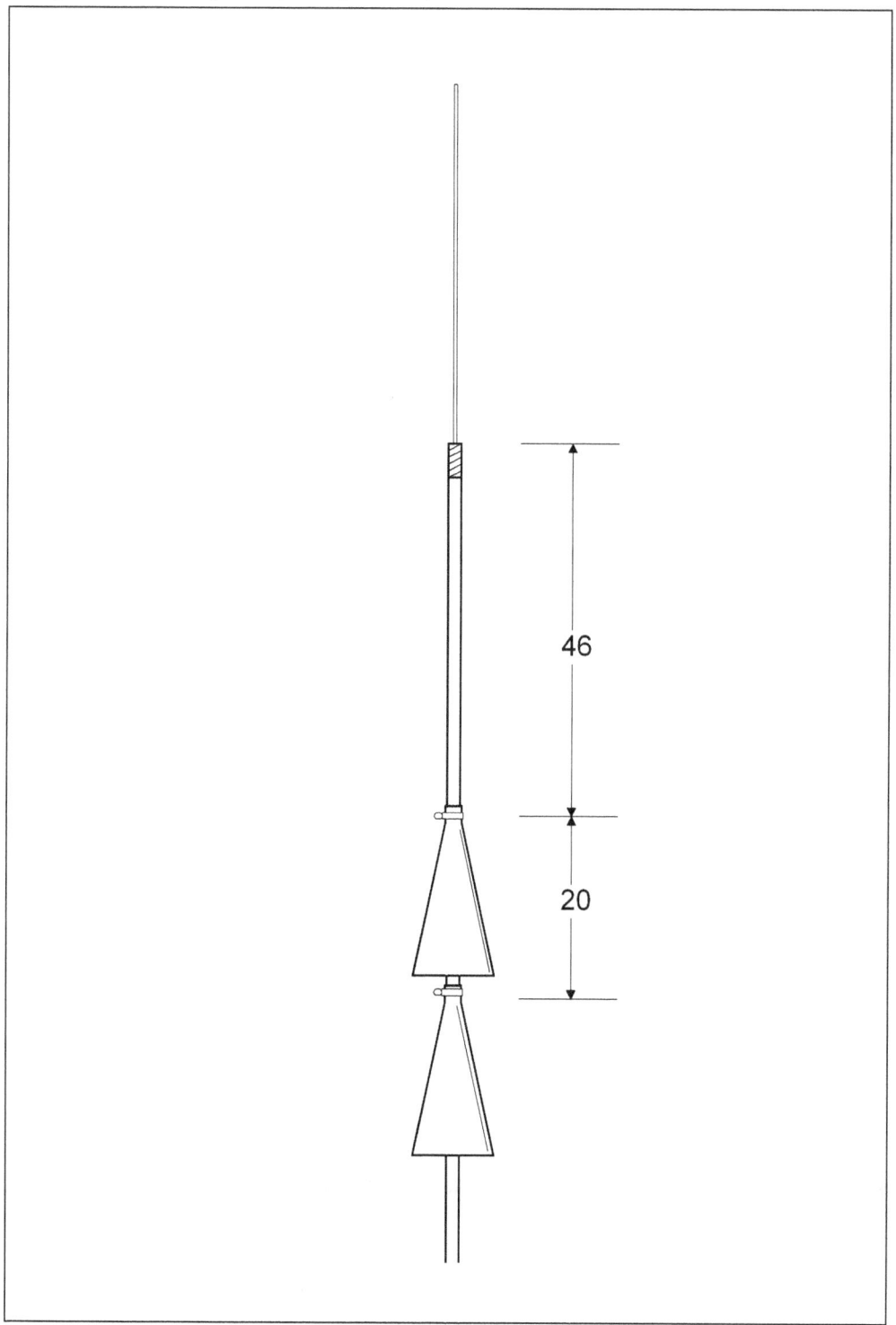

Figure 12.5   Mounting the decoupling sleeves to the antenna.

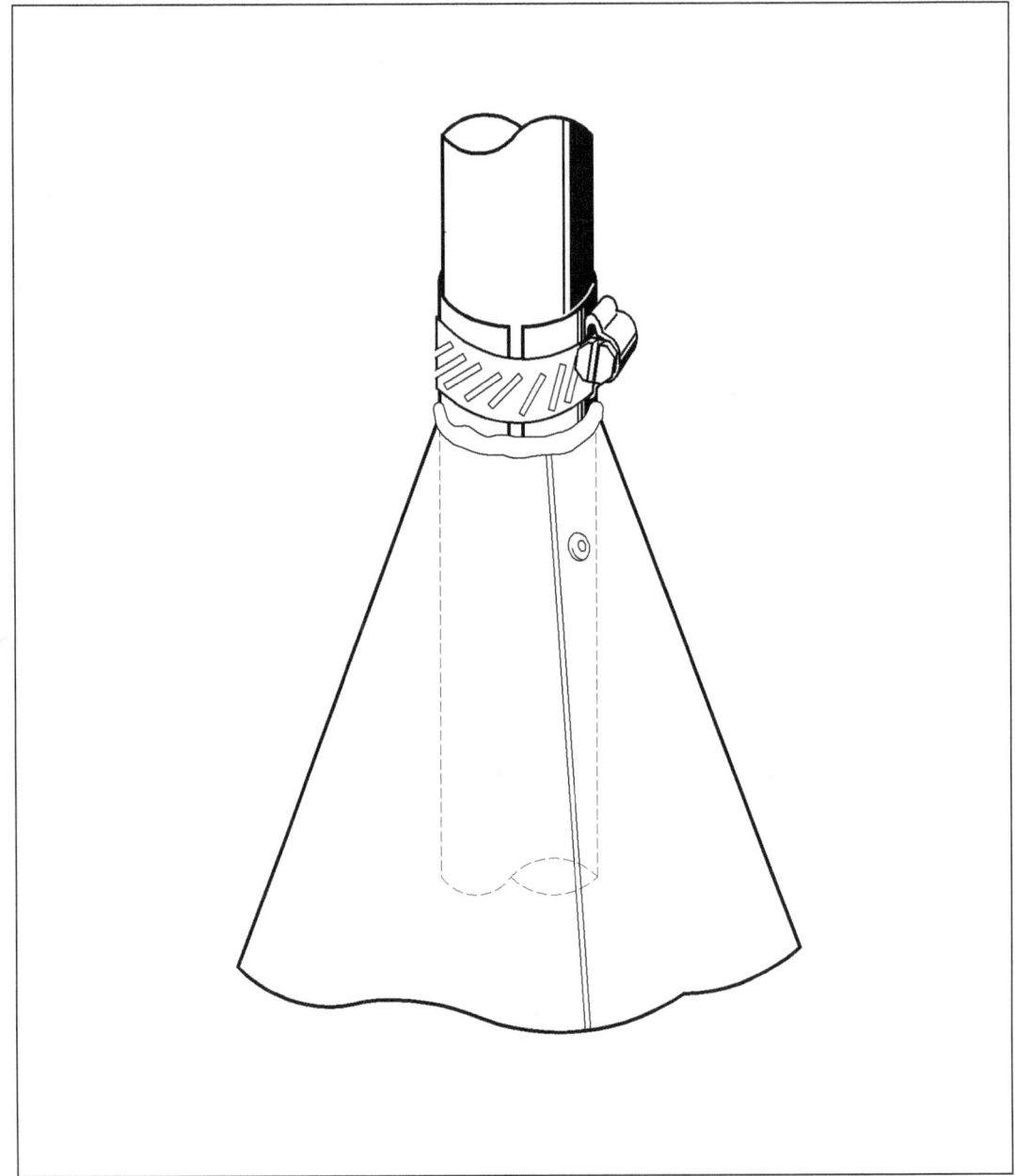

Figure 12.6   Securing the cone to the mounting tube with a hose clamp.

Tuning the antenna to resonance

The tuning procedure for this antenna is the same as the procedure for tuning the antenna model WA-2. Just follow the procedures described in 5/8 Wave antenna model WA-2.

INSTALLATION

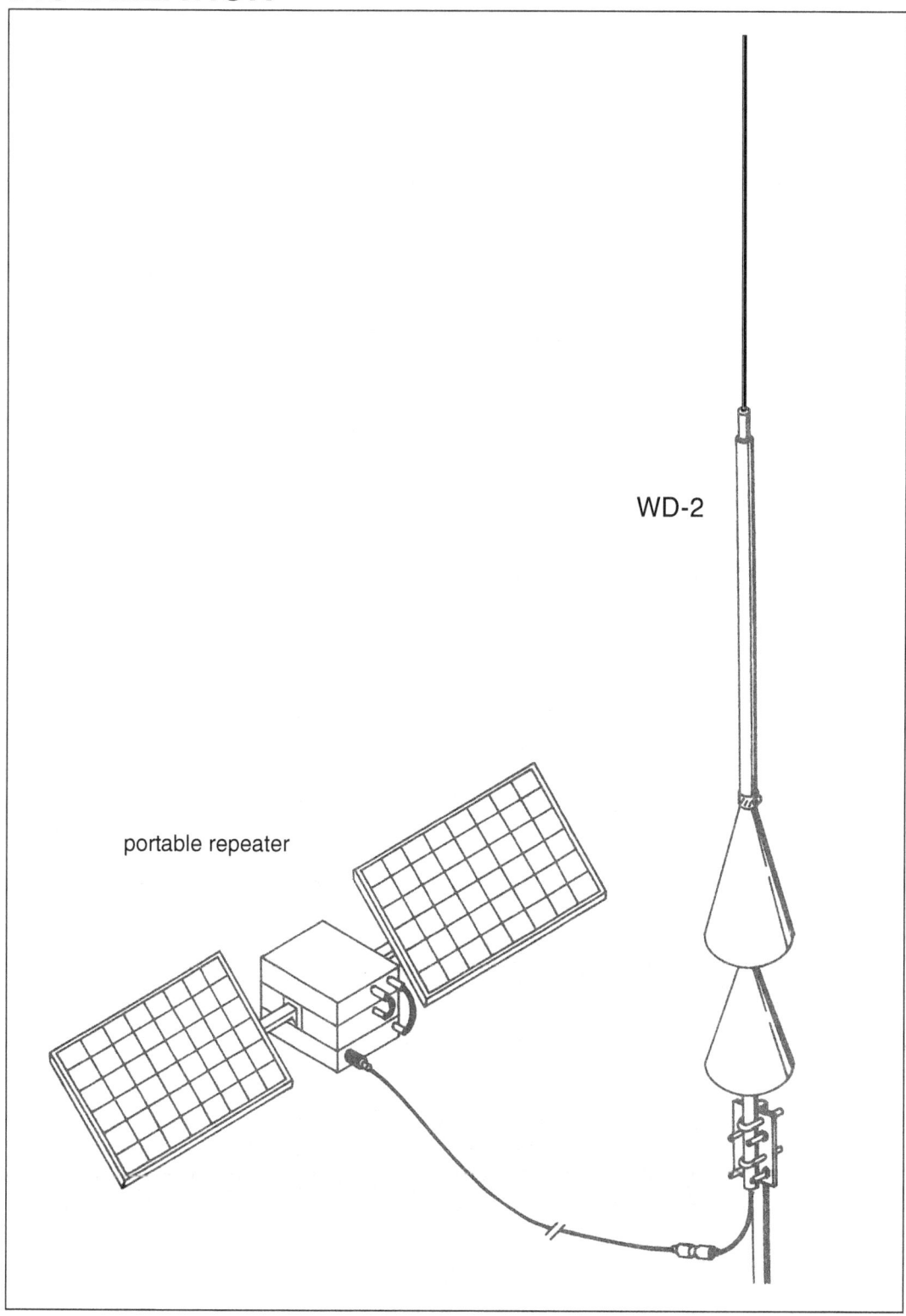

WD-2

portable repeater

Figure 12.7 Mounting the WD-2 to the mast.

REVIEW QUESTIONS

1. What causes an unbalanced coupling of antenna?

2. What results from an unbalanced coupling of antenna?

3. What is the function of the decoupling sleeves?

4. Why is a decoupling sleeve necessary in an antenna used for a repeater system?

5. What is desensitization?

6. By studying the design presented here, is it possible to replace the cones with metal rods cut to the same dimensions?

Table 12.1 American and english units in relation to metric units

USA and UK	Abbbreviation	Metric unit	conversion factor
1 inch = 10 lines = 1000 mils	('') in	2.54 cm	0.3937
1 foot = 12 inches	(') ft	30.48 cm	3.281×10^{-2}
1 yard = 3 feet = 36 inches	yd	91.44 cm	1.094×10^{-2}
1 fathom = 6 feet	fath	1.8288 m	0.547
1 into nautical mile = 6076 feet	naut. mile	1.852 km	0.54
1 statute mile = 1760 yards = 5280feet	stat.mile	1.6093km	0.6214
1 mile per hour	MPH	1.6093 km/h	0.6214
1 square foot	sqft	0.0929 m^2	10.7643
1 pound	lb	0.4569 kg	2.2046

To convert a metric unit into an english unit, use the conversion factor listed at the last column. For example : 40,000 km = 0.54 x 40,000 = 21,600 naut. miles.

13 | 5/8 WAVE ANTENNA

Model PF-2C

Most mobile operators use portable handheld transceivers, because these are lightweight and small. There are also available models today that equal the capabilities of their base station versions in terms of frequency coverage, sensitivity, computerized functions, PLL stability, and many other unique features. However, portable transceivers in general have low power transmitters because of obvious limitations in the type of batteries practical for mobile operations. The average transmitting power of handheld units ranges from 0.5 watts to 5 watts maximum. Because of this, most antennas used for portable sets are of gain type, to increase the effective radiated power.

The antenna described here is a portable version of a 5/8 wave vertical antenna. As stated earlier, an antenna of this length has a slight gain over a dipole. Approximately, a gain of 1.8 dB can be attained with this type of antenna. The radiator element of this model is made of telescopic rod, so that the overall length of the antenna can be reduced if desired. It is loaded at the base by a coil that doubles as a flexible spring supporting the telescopic rod. The telescopic element may be used while retracted or collapsed, and will function like an ordinary "rubber ducky" antenna that comes as a standard accessory for portable transceivers. Gain can only be realized if the antenna is used while the radiator is extended to full length.

Sometimes it is desirable to raise the height of the antenna to increase its effective range. Installing the antenna to a higher position clears it from most obstructions, such as houses or trees, and extends the horizon farther away, thereby increasing the area covered or "seen" by the antenna. This can be accomplished by using a length of coaxial cable to connect the antenna to the transceiver. The antenna is then mounted high up in a post or tower. It can also be hung under a tree by using a non-metallic material such as nylon or fish line.

This particular model is dimensioned to operate in the frequency band of 140-150 MHz. It exhibits an SWR of less than 1.5:1 over the entire band if properly tuned. Tuning is easy, as described in this chapter. The materials used for this model can be bought cheaply, and constructing it can save a lot of money. The total cost of the antenna is a mere fraction of the price of its commercial version. Furthermore, an invaluable knowledge can be gained during the actual construction of this antenna.

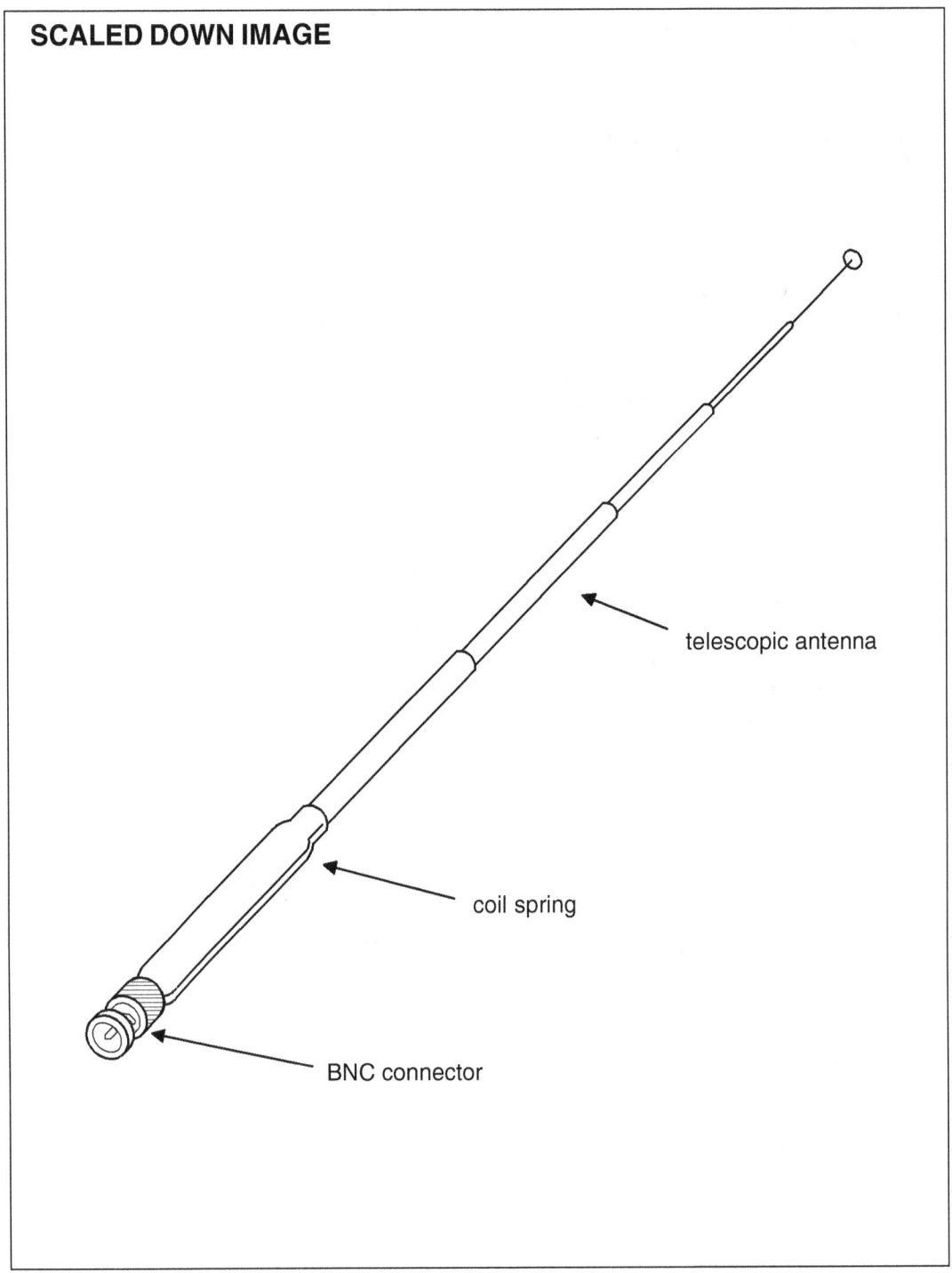

SCALED DOWN IMAGE

telescopic antenna

coil spring

BNC connector

Figure 13.1   5/8 wave antenna model PF-2C

Materials List

Quantity	Specification/Description	Dimensions
1	Telescopic antenna - approx. 7/16" od* of the base tube	46" fully extended 6" to 7" retracted
1	Brass wire no. 26 or approx. 3/32" diameter	24" or 60 cm.
1	BNC VHF male connector	
1	10 pF/150 Volts capacitor glass type	
1	Hook-up wire no. 22 stranded	4" long
1	Heat shrinkable tube	5/8" or 3/4" od* x 4" long

Miscellaneous: Epoxy glue

· od - outside diameter

Construction

Wind the brass wire into a spring-like coil form. Wind 13 turns of the wire with a pitch of approximately 4 turns per inch. The total length of the finished coil is approximately 3 inches (see Figure 13.2). The inside diameter of the coil spring must be force-fit to the outside diameter of the BNC connector, or approximately 3/8" id*.

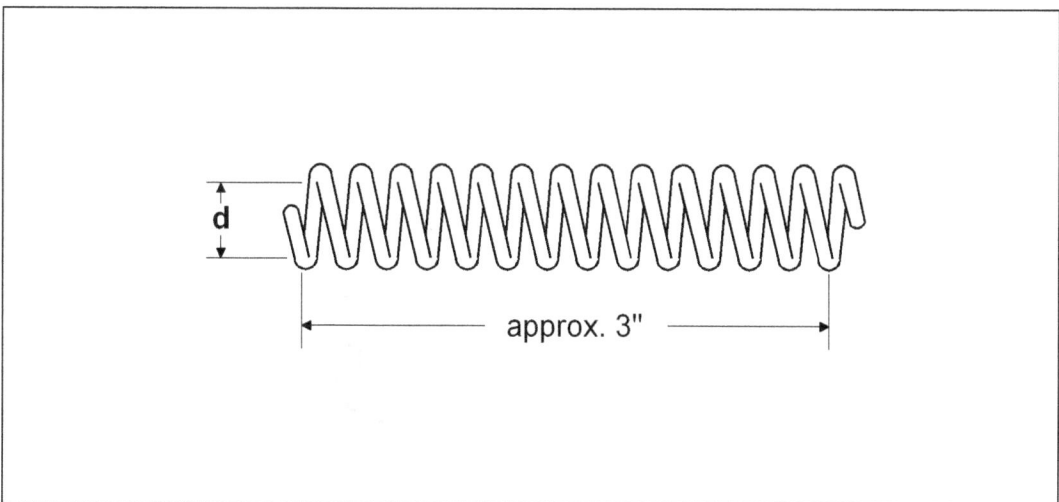

Figure 13.2 Constructing the spring coil.

*id-inside diameter

Solder about 2" long hook-up wire to the center pin of the BNC connector.

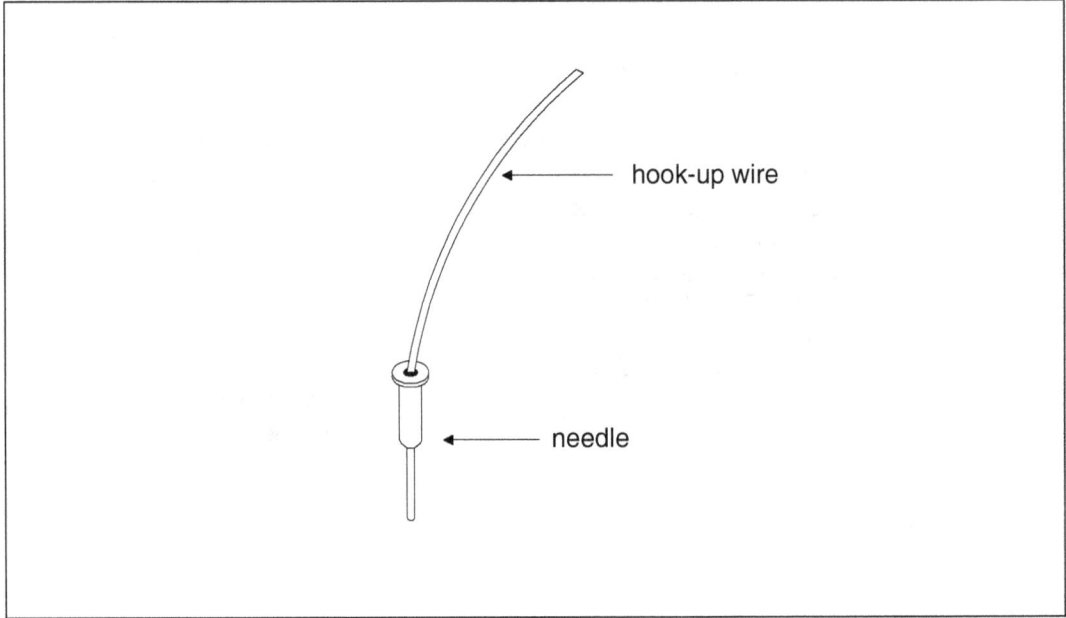

Figure 13.3   Soldering the hook-up wire to the center pin of BNC.

Place a moderate amount of epoxy glue around the soldered part of the needle. Avoid coating the epoxy around the body of the center pin. Insert the needle into the BNC connector, and cover the empty space inside with a liberal amount of epoxy (see Figure 13.4). Let the epoxy cure and harden.

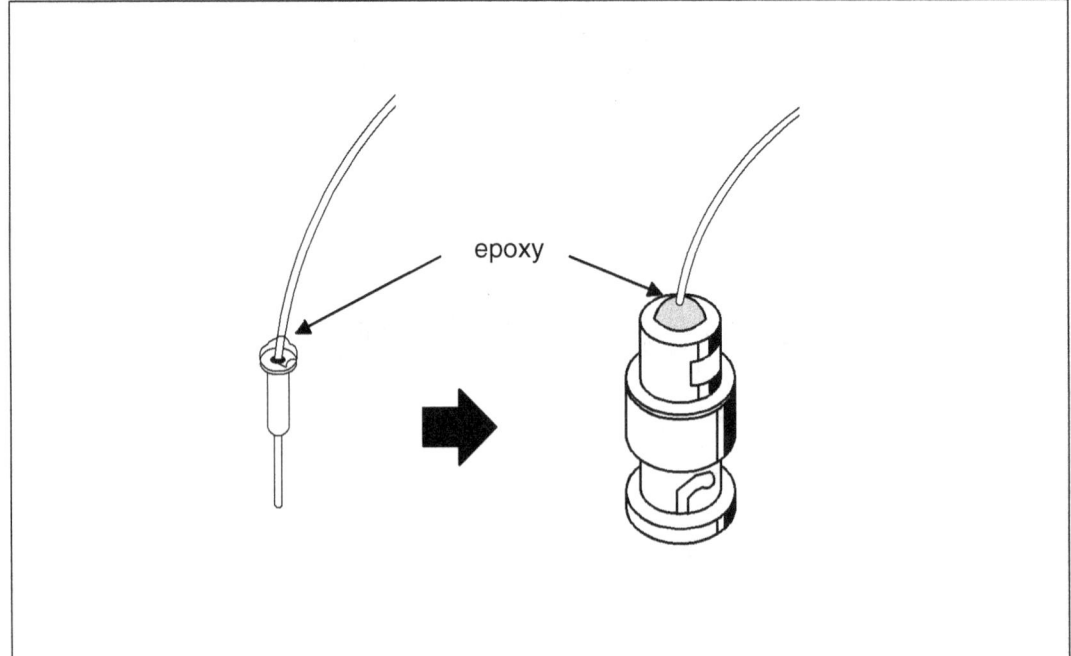

Figure 13.4   Fixing the center pin to the BNC connector with epoxy glue.

When the epoxy has hardened, insert the BNC connector into one end of the spring coil. Solder the part of the coil that wraps around the body of the BNC connector (see Figure 13.5).

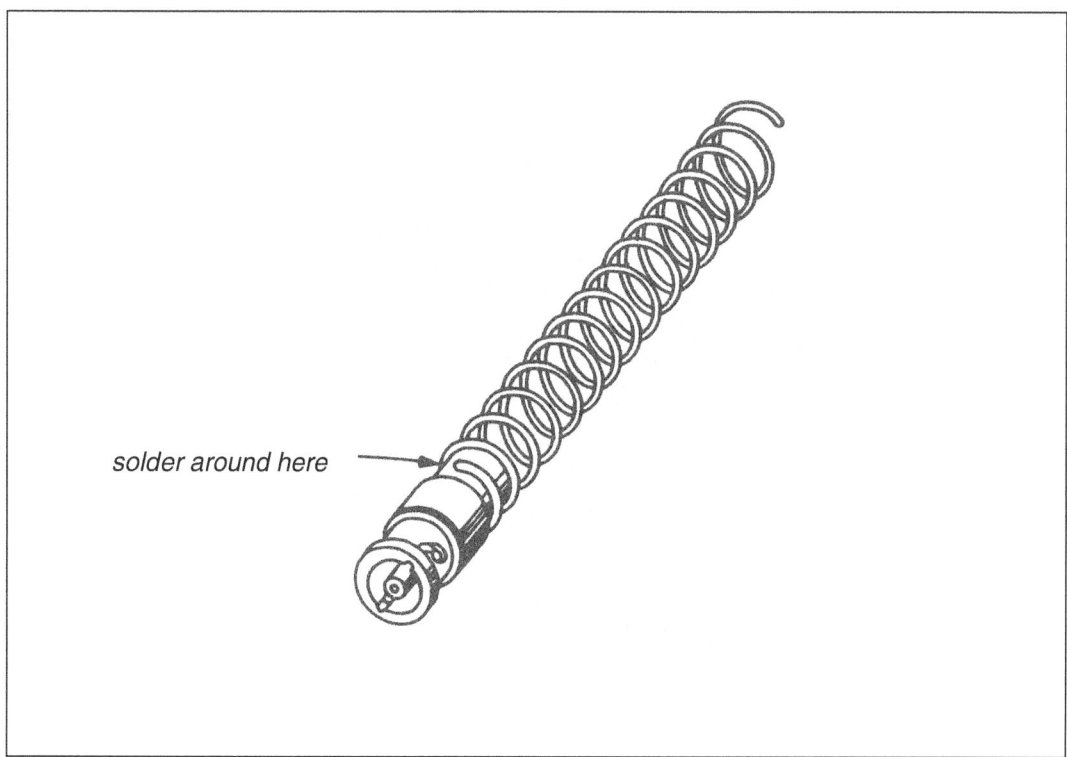

solder around here

Figure 13.5   Assembling the BNC connector and the spring coil together.

Pry out the free end of the hook-up wire inside the coil spring, and solder it to the point of the coil which is 1 and 1/2 turns counting from the ungrounded portion of the coil (see Figure 13.6).

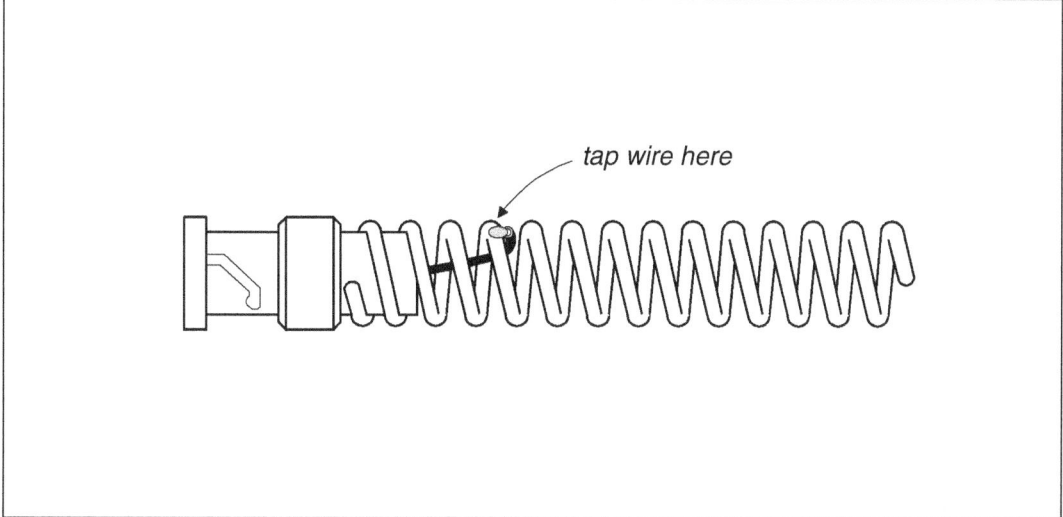

tap wire here

Figure 13.6   Soldering the hook-up wire to a temporary tap point.

Insert the 10 pF capacitor inside the spring coil and solder its lower lead to the grounded portion of the coil (see Figure 13.7).

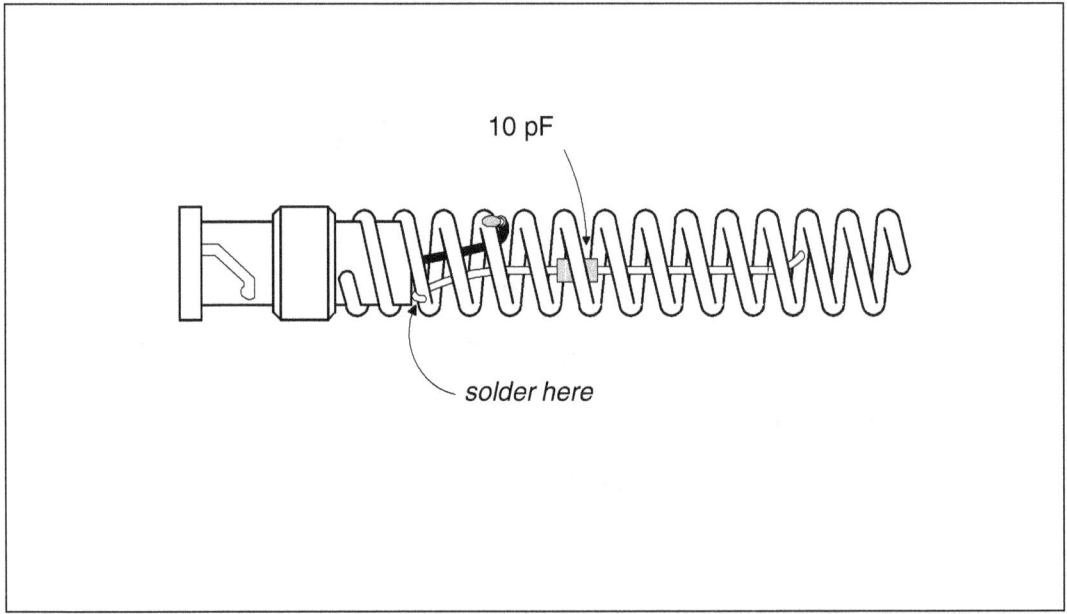

Figure 13.7   Soldering one lead of the capacitor to the grounded portion.

Next, solder the upper lead of the capacitor to the 6th turn of the coil spring, counting from the ungrounded portion. See Figure 13.8.

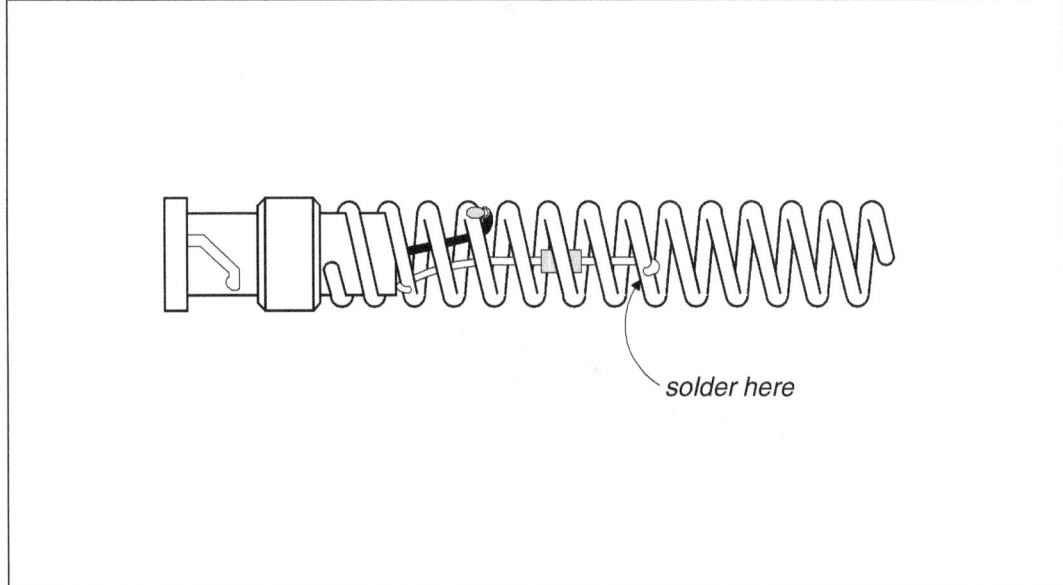

Figure 13.8   Soldering the upper lead of the capacitor.

The next step is to insert the base of the telescopic antenna into the open end of the coil spring. Let 2 turns of the coil hold the base of the antenna, and solder it to secure the two pieces together (see Figure 13.9). At this point the construction of the antenna is already finished; it only needs to be tuned to resonance for proper operation.

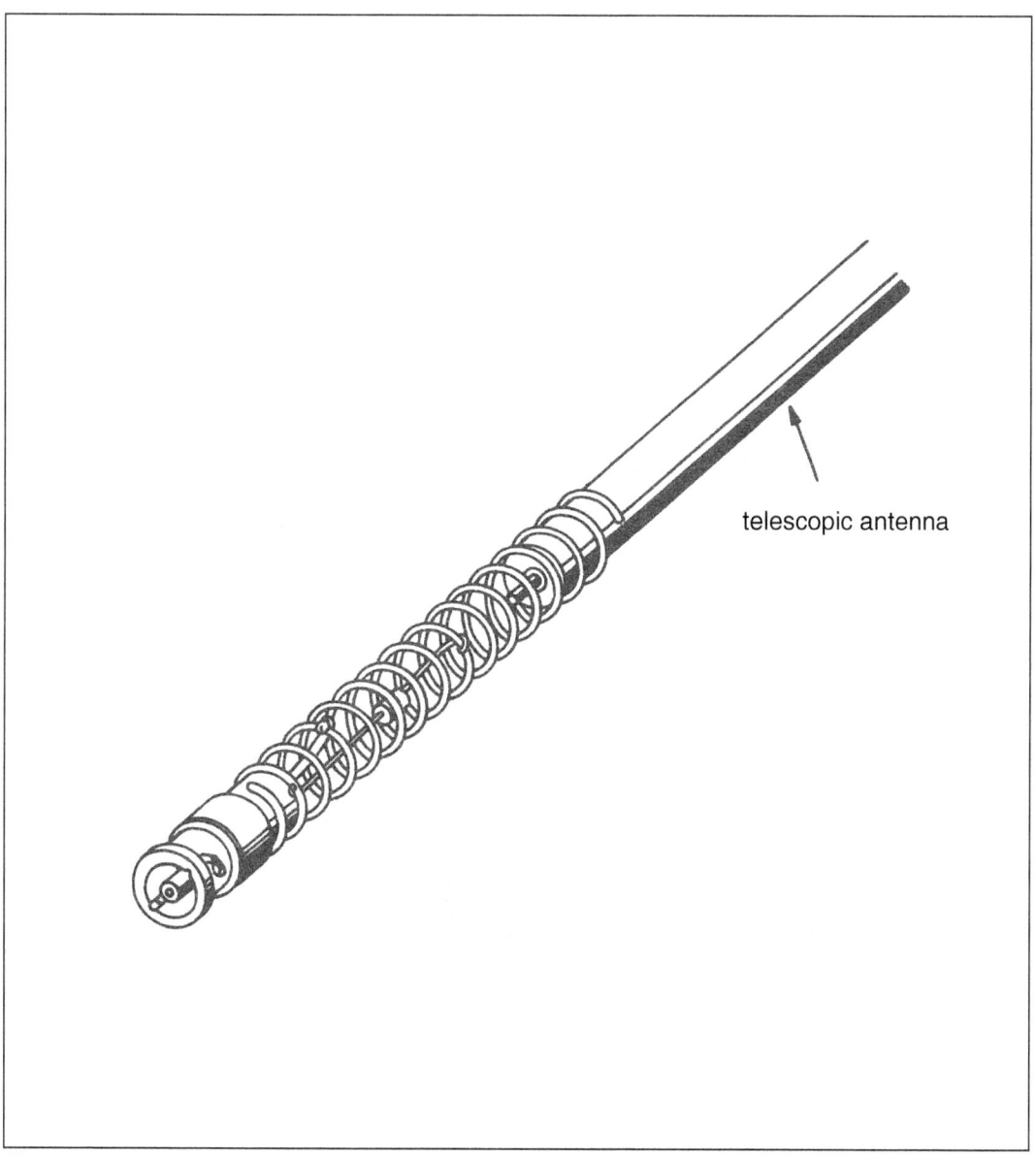

telescopic antenna

Figure 13.9 Final assembly of the antenna.

Tuning the antenna to resonance

Attach the antenna directly to the output connector of the SWR meter using the necessary adaptors. Similarly, connect the SWR meter to the transceiver using a short length of coaxial cable (see Figure 13.10).

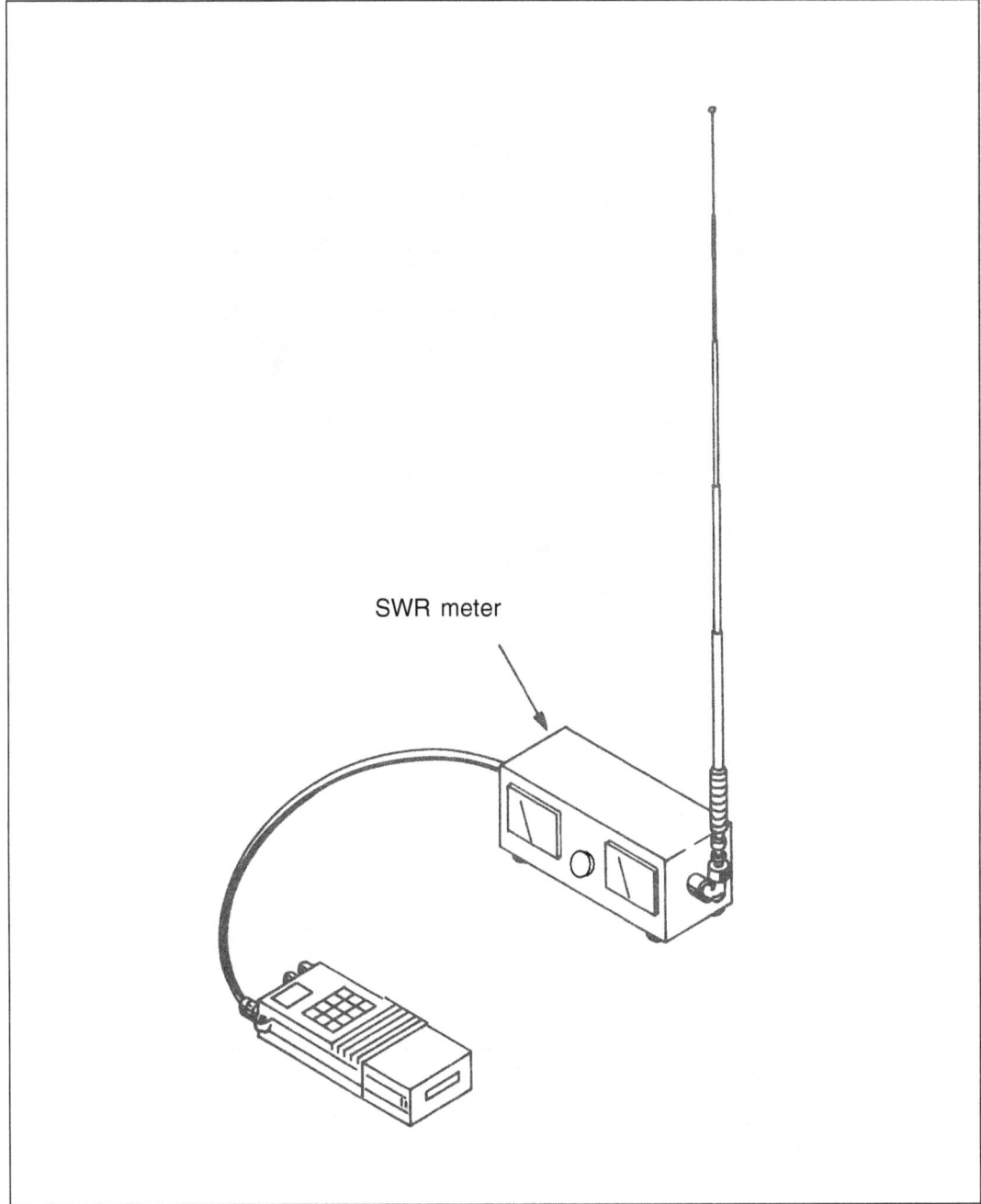

Figure 13.10   Preparing the PF-2C for resonance tuning.

Set the transceiver to the center frequency and key the PTT. Read the SWR response, and note it on a chart similar to the one shown in Figure 13.11.

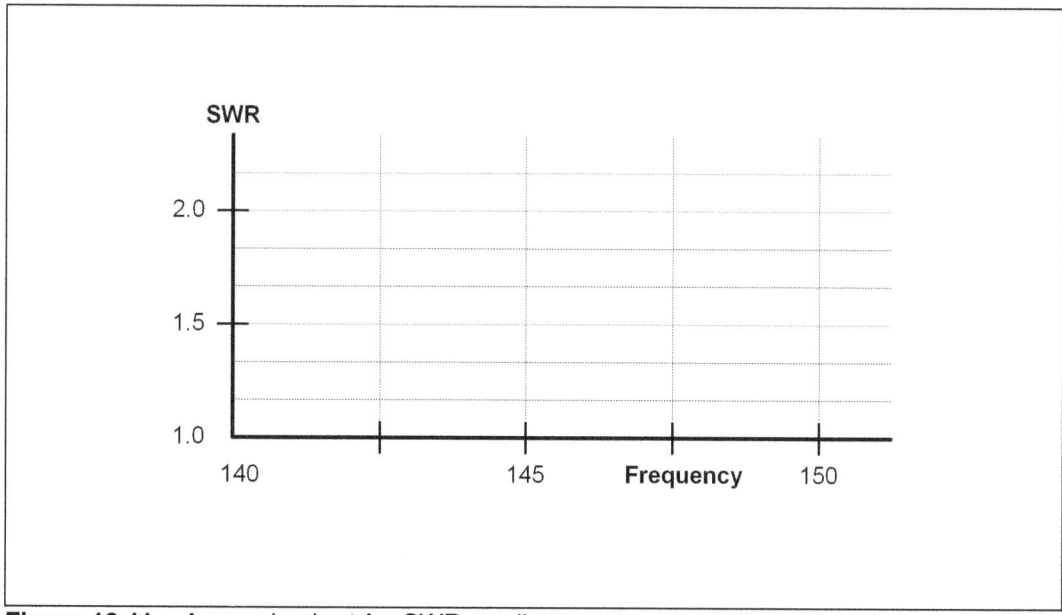

Figure 13.11   A sample chart for SWR readings.

Read all the SWR responses from the lowest frequency up to the highest frequency in the band, and mark all the results on the chart until you get a response curve similar to the one shown in Figure 13.12.

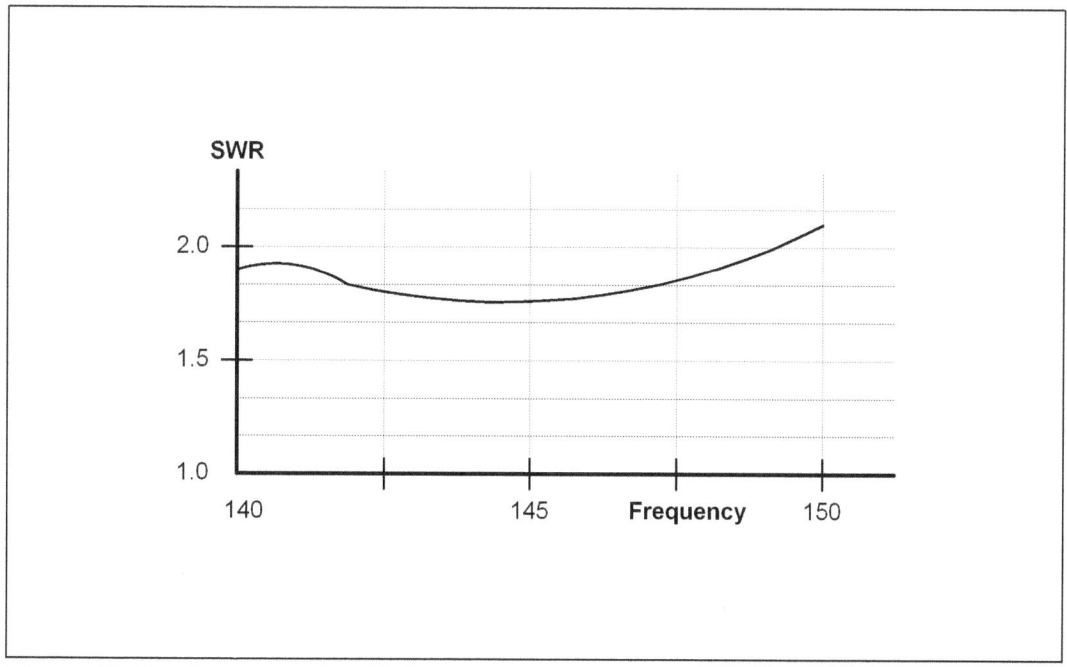

Figure 13.12   A sample of an SWR curve.

Re-solder the capacitor's lead to a different point, or tap either to the left or right of the original tap. If you have moved the tap to the right and the SWR went up, then obviously you must move the tap to the left. Key again the PTT, and mark the SWR responses once again on the chart. Move the tap about 1/8" farther at a time (see Figure 13.13).

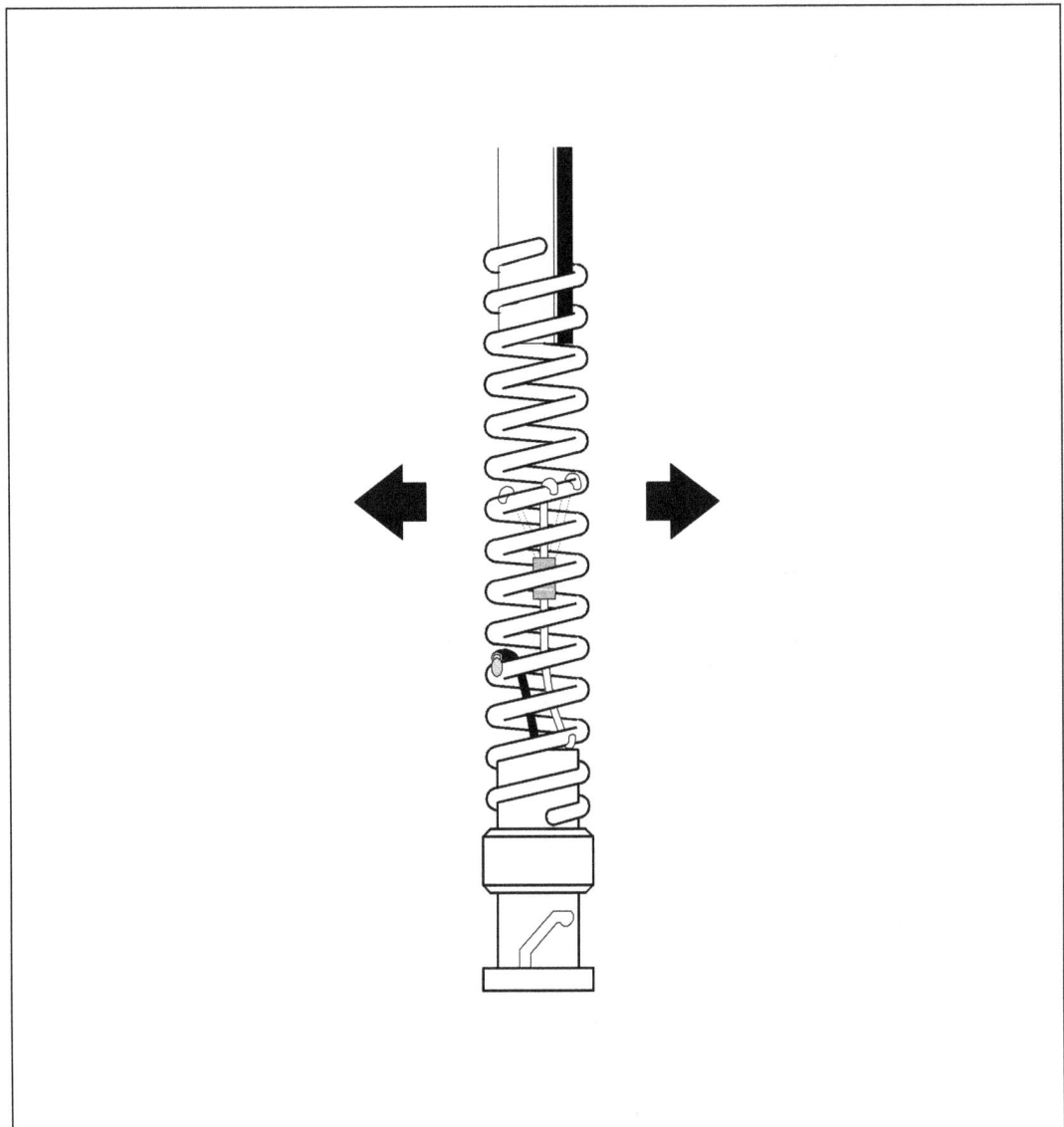

Figure 13.13   Re-soldering the capacitor's lead to find the right tap.

Repeat the whole process until you find the point in the coil that results in a very low SWR reading on the center frequency and relatively balanced responses on the extreme ends of the band. If you have followed the instructions in constructing this antenna carefully, it is possible to get an SWR response of 1.1 at the center frequency and 1.5 at extreme ends of the band, similar to the response curve shown in Figure 13.14 on the next page.

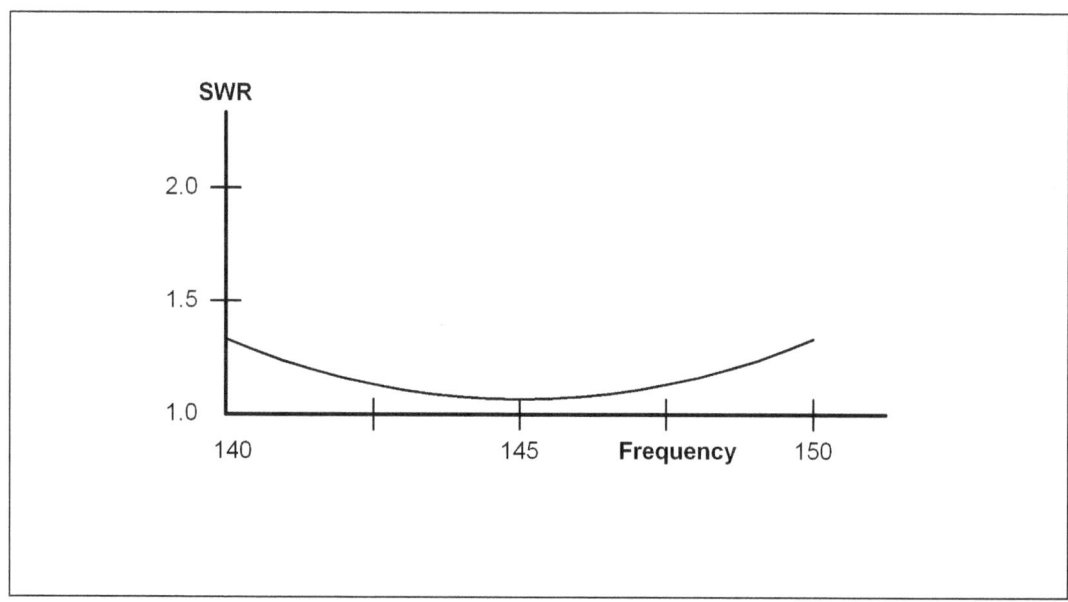

Figure 13.14 A sample of a good SWR response.

After you have found the right tap, solder it to the coil spring permanently. Insert the spring coil into the heat shrinkable tube, and heat the tube over a flame or with a blow dryer. Rotate the antenna and the tube continuously while being heated, to get an even shrinking of the tube (see Figure 13.15).

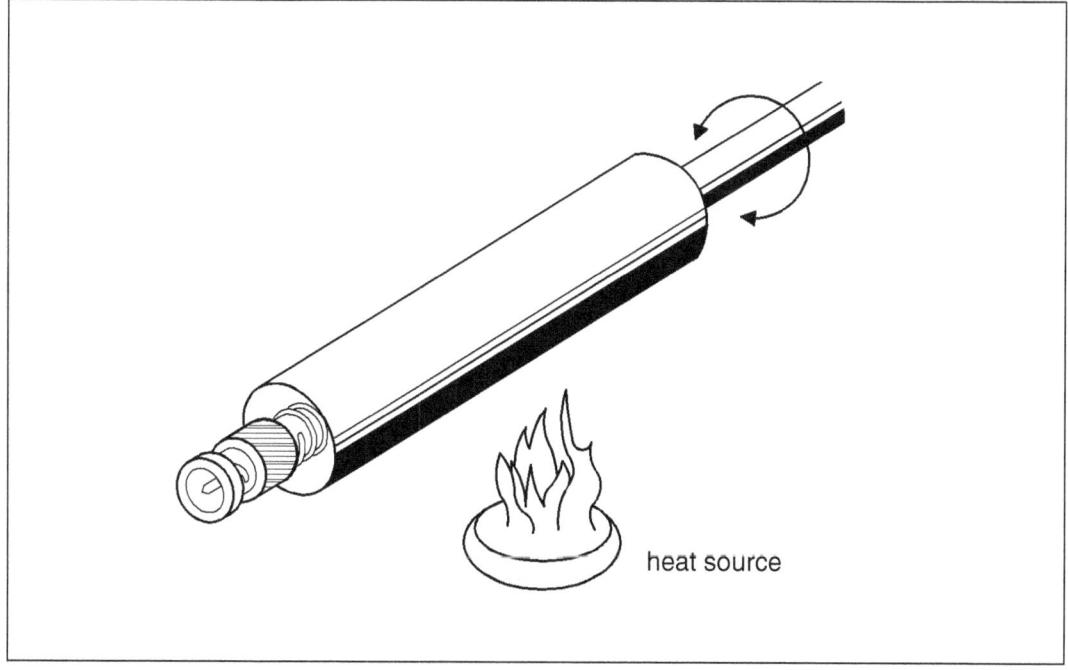

Figure 13.15 Heating the shrinkable tube.

REVIEW QUESTIONS

1. What are the advantages of using this antenna?

2. What is the function of the coil at the base of the telescopic element?

3. What is the function of the capacitor inside the coil?

4. How can the effective coverage of this antenna be increased?

5. How is the antenna tuned?

6. Would the antenna exhibit gain if its radiator element is retracted?

Power loss in relation to the SWR figure in the transmission line:

SWR	Power loss in %
1 :1	0 %
1,3 :1	2 %
1,5 :1	3 %
1,7 :1	6 %
2 :1	11 %
3 :1	25 %
4 :1	38 %
5 :1	48 %
6 :1	55 %
10 :1	70 %

COLLINEAR ANTENNA

Model SD-22 (2 stacked dipoles)

A collinear antenna is made up of a multiple number of dipoles mounted in a common structure with their axis arranged in one straight line. The dipole elements are always driven in phase; otherwise, the array simply becomes a harmonic type antenna. A collinear array is a broadside radiator, meaning the direction of maximum radiation is at right angles to the line of the antenna.

When mounted vertically, it radiates an omnidirectional pattern. One advantage of this design is its ability to attain high gain. When dipole elements are stacked collinearly, the power gain increases in direct proportion to the number of dipoles used. Obviously, this type of antenna is limited to fixed installation only because of its mechanical construction.

An actual working design of a collinear array is presented here. It has two identical dipoles fed with a coaxial phasing line or 'harness'. Each dipole element is tuned by a gamma matching system, similar to that described in chapter 6. In fact, it is the same design of dipole except 'doubled' and fed simultaneously. This configuration gives a gain of 3 dB compared to a single dipole.

This model is dimensioned to operate in the frequencies of the 140-150 MHz band. It has an SWR response of less than 1.5:1 over the entire band. The tuning procedure is similar to that described for dipole model DP-2.

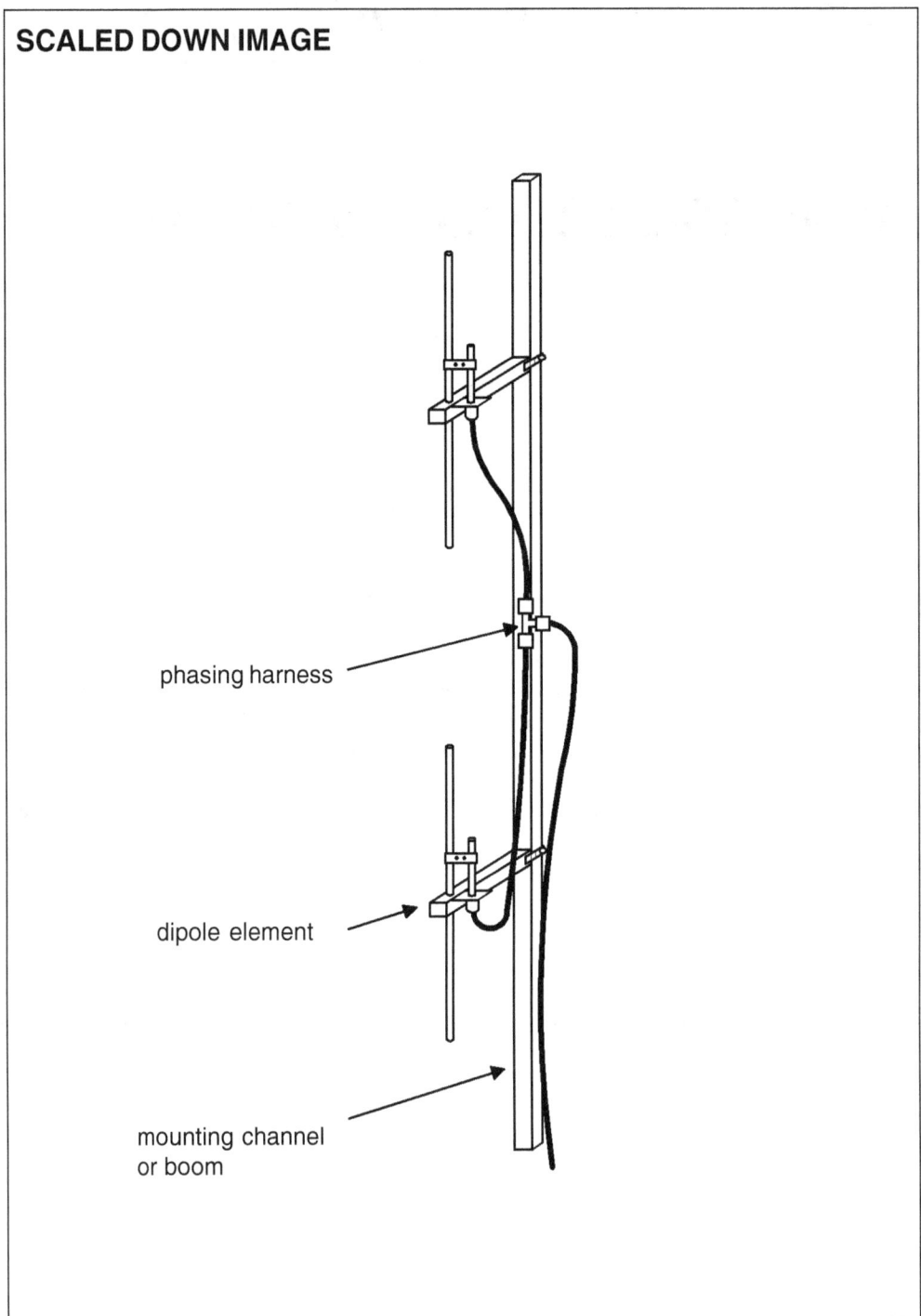

SCALED DOWN IMAGE

phasing harness

dipole element

mounting channel
or boom

Figure 14.1 Collinear antenna Model SD-22

Materials needed

Most of the materials needed to build the antenna model SD-22 are the same as those needed for the antenna model DP-2, except for the mounting channel. The mounting channel for the SD-22 is shorter, being only 8 inches long, and has slits on two sides instead of two holes.

Additionally, another long square channel is needed to mount the two dipole elements into a single mast. Also, a system of phasing harness made of coax cable is required to feed the two dipoles simultaneously. In short, the additional materials needed for the SD-22 are as follows:

Quantity	Description	Dimensions
2 pcs.	Square aluminum channel	1" x 1" x 8"
1 pc.	Square aluminum channel	1" x 1" x 115"
2 pcs.	Hose clamp	2 - 1/2" clamping capacity
6 pcs.	BNC VHF male connector	
2 pcs.	BNC 'T' connector	

Construction

Follow the procedures for constructing the dipole antenna model DP-2, and make two identical dipoles. The mounting channel for the model SD-22 is slightly different, and is described in the following illustration (see Figure 14.2 on the next page).

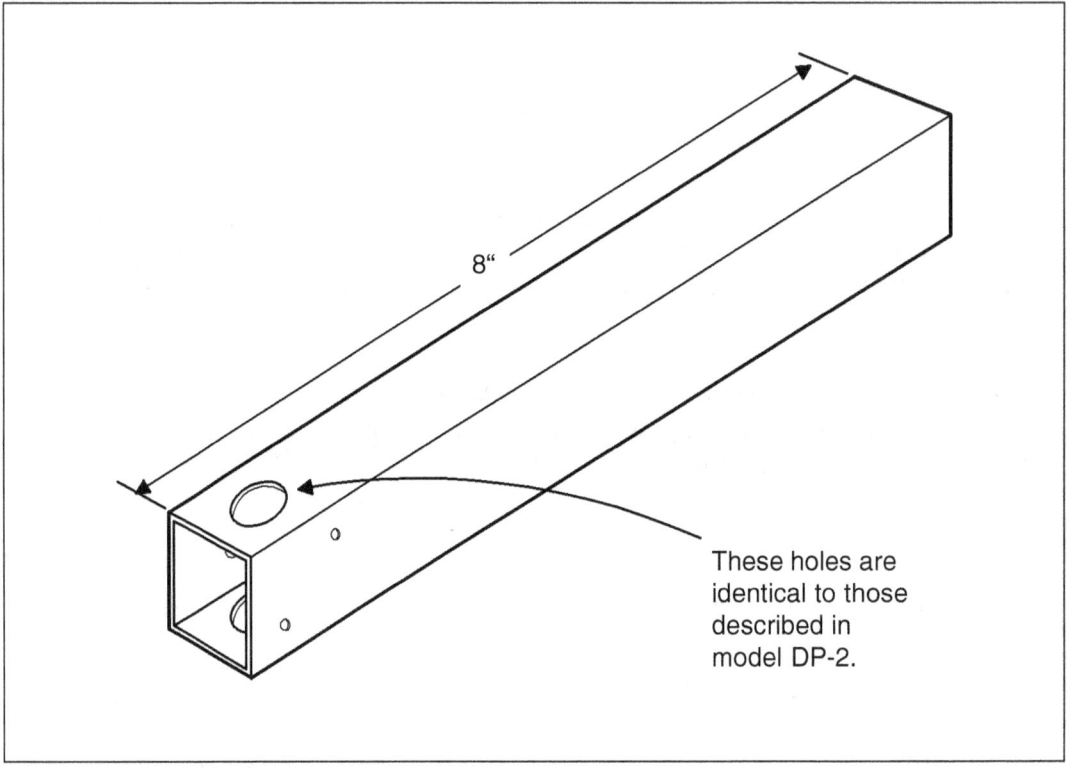

These holes are identical to those described in model DP-2.

Figure 14.2   Mounting channel dimensions.

Saw shallow slits at two sides of the channel using a hacksaw. A hose clamp will be inserted into these slits for the purpose of mounting the channel to the supporting mast (see Figure 14.3).

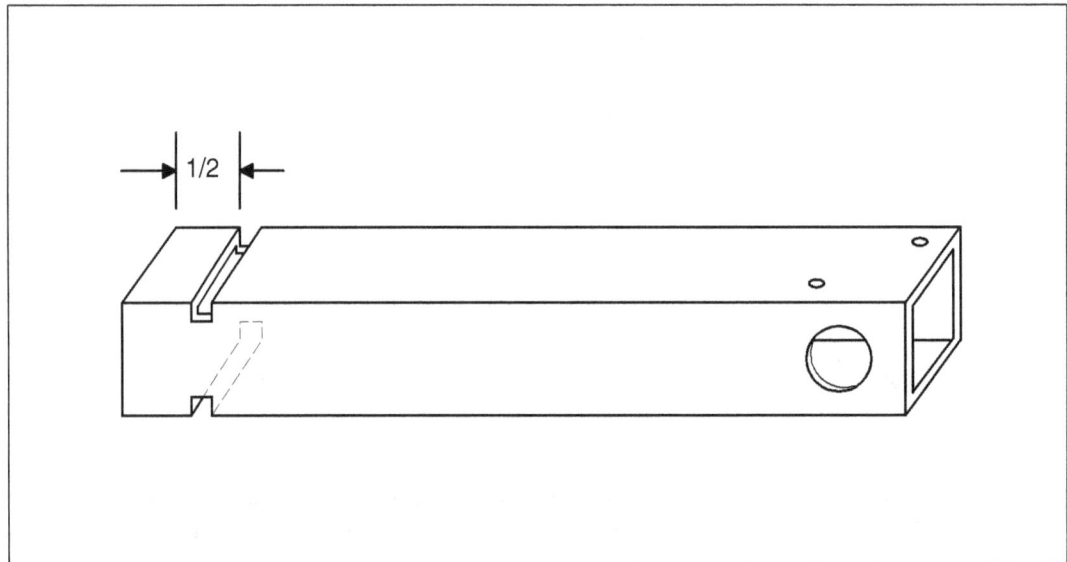

1/2

Figure 14.3   Saw slits at two sides of the channel.

Assemble the two dipole elements following the procedures described for the antenna model DP-2. After the dipoles are completed, insert the hose clamps through the slits in the channel (see Figure 14.4).

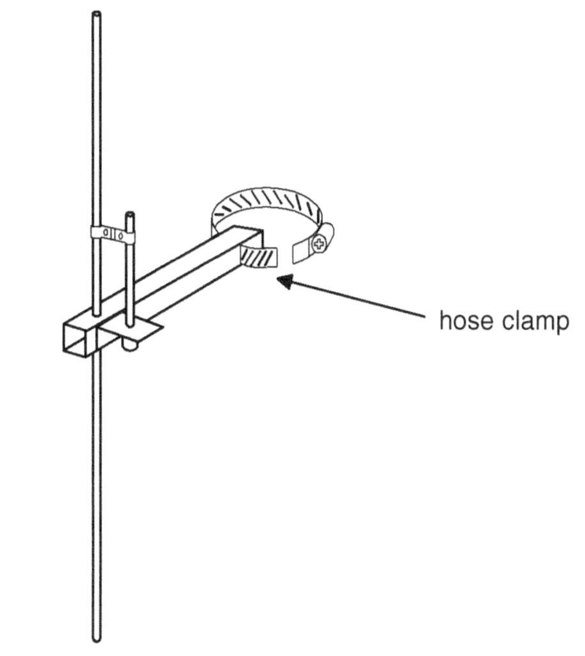

hose clamp

Figure 14.4   Assembled dipole element with hose clamp.

Mount the two dipoles to the aluminum supporting mast following the dimensions shown in Figure 14.5. Wrap the two hose clamps around the body of square channel mast, and tighten the clamps to hold the dipole elements rigidly.

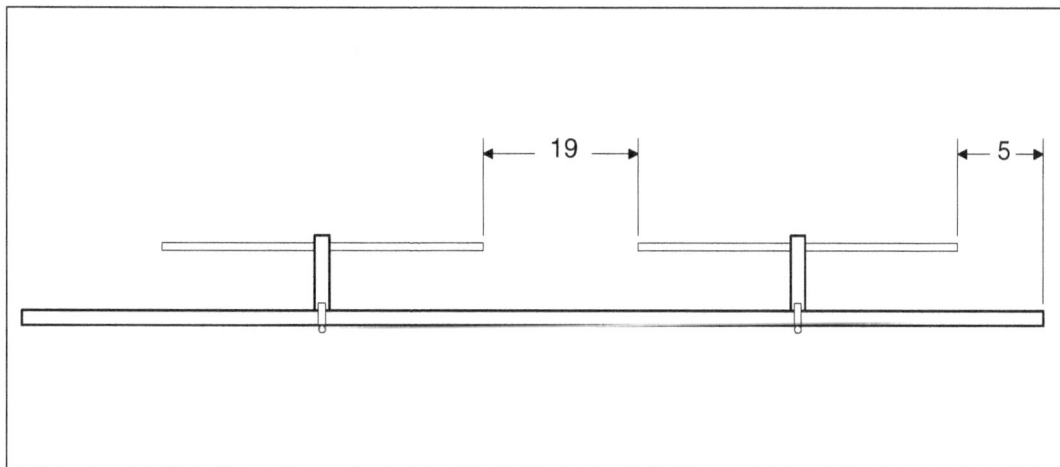

Figure 14.5   Mounting the dipoles to the aluminum mast.

Next, construct the phasing harness using RG-58/U coaxial cables and the appropriate connectors (see Figure 14.6).

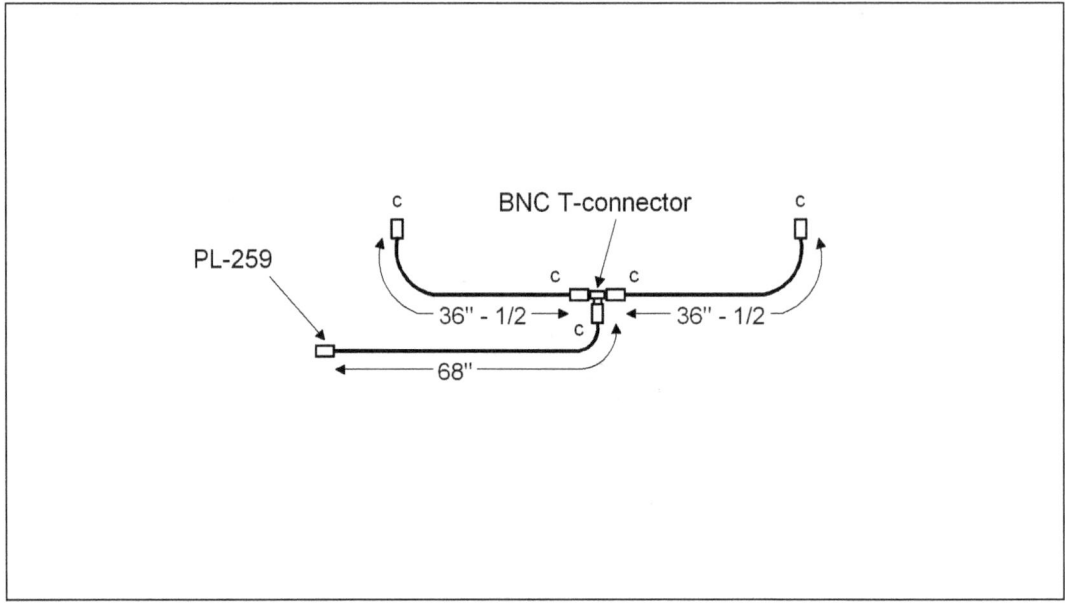

Figure 14.6   Constructing the phasing harness.

Finally connect the phasing harness to the two dipole elements, and secure it to the support mast with plastic binders (see Figure 14.7).

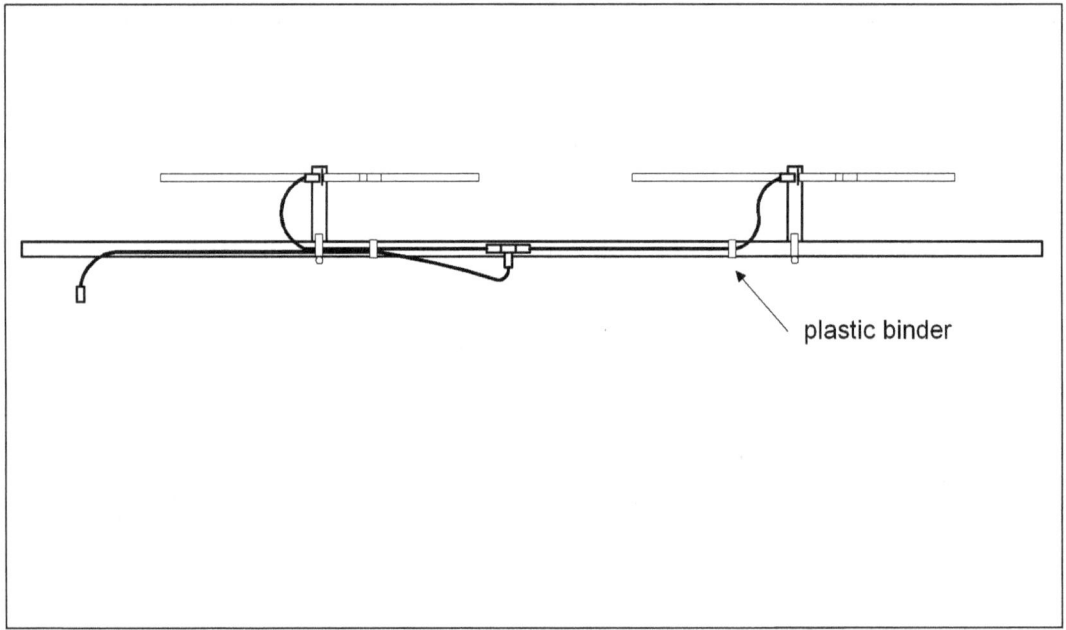

Figure 14.7   Connecting the phasing harness to the antenna.

Tuning the antenna

The tuning procedure for the antenna model SD-22 is similar to that of the antenna model DP-2. In tuning the SD-22, however, the two dipoles have to be tuned simultaneously. You have to do a lot of shuttling back and forth between the two dipoles before you can achieve a good match. If it is not practical to tune the antenna right in the main mast, then it can be tuned on the ground by placing it in a horizontal position with the dipoles facing upward. The antenna must be elevated to not less than 1 meter above the ground, supported by non-metallic materials, such as wooden benches.

REVIEW QUESTIONS

1. What is a collinear antenna?

2. What are the advantages of using a collinear antenna?

3. Why must the dipole elements be fed in phase?

4. By careful observation of the phasing harness, what does 'fed in phase' mean?

5. What is the optimum spacing between the tips of the dipole elements?

6. How is the collinear antenna tuned?

15 STACKED DIPOLE ARRAY

Model SD-24 (4 stacked dipoles)

This model demonstrates the capability of a simple dipole to attain high power gain by simply stacking identical units into a single structure, and feeding them all simultaneously with a phasing harness. This arrangement is also called a collinear array.

As stated earlier in chapter 14, power gain in a collinear array increases in direct proportion to the number of dipole elements used. However, in order to construct a practical phasing harness, the number of dipole elements installed cannot be simply dictated by personal choice. The correct method is to double the original number of dipole units. Another words, if the original array has two dipole elements installed, then the next array must have four dipoles, and the next must have eight dipoles, and so on. Every time the number of dipole elements used is doubled, the power ratio is also doubled.

NOTE:
The power ratio is not numerically the same with the dB figure. For accurate computations refer to Appendix.

The particular four-element array presented here has a power gain of 6 dB. A collinear array having eight dipole elements would have a power gain of 9 dB. An array with elements in excess of eight, is rarely constructed, because of the inherent mechanical problems encountered in erecting structures of this size. Most collinear antennas are mounted vertically, to effect an omnidirectional pattern of radiation.

The model SD-24 is specifically dimensioned to operate in the frequencies of the 140-150 MHz band. If properly tuned, this array exhibits an SWR of less then 1.5:1 over the entire band. The procedure for tuning this antenna to resonance is similar to the procedure for model SD-22.

SCALED DOWN IMAGE

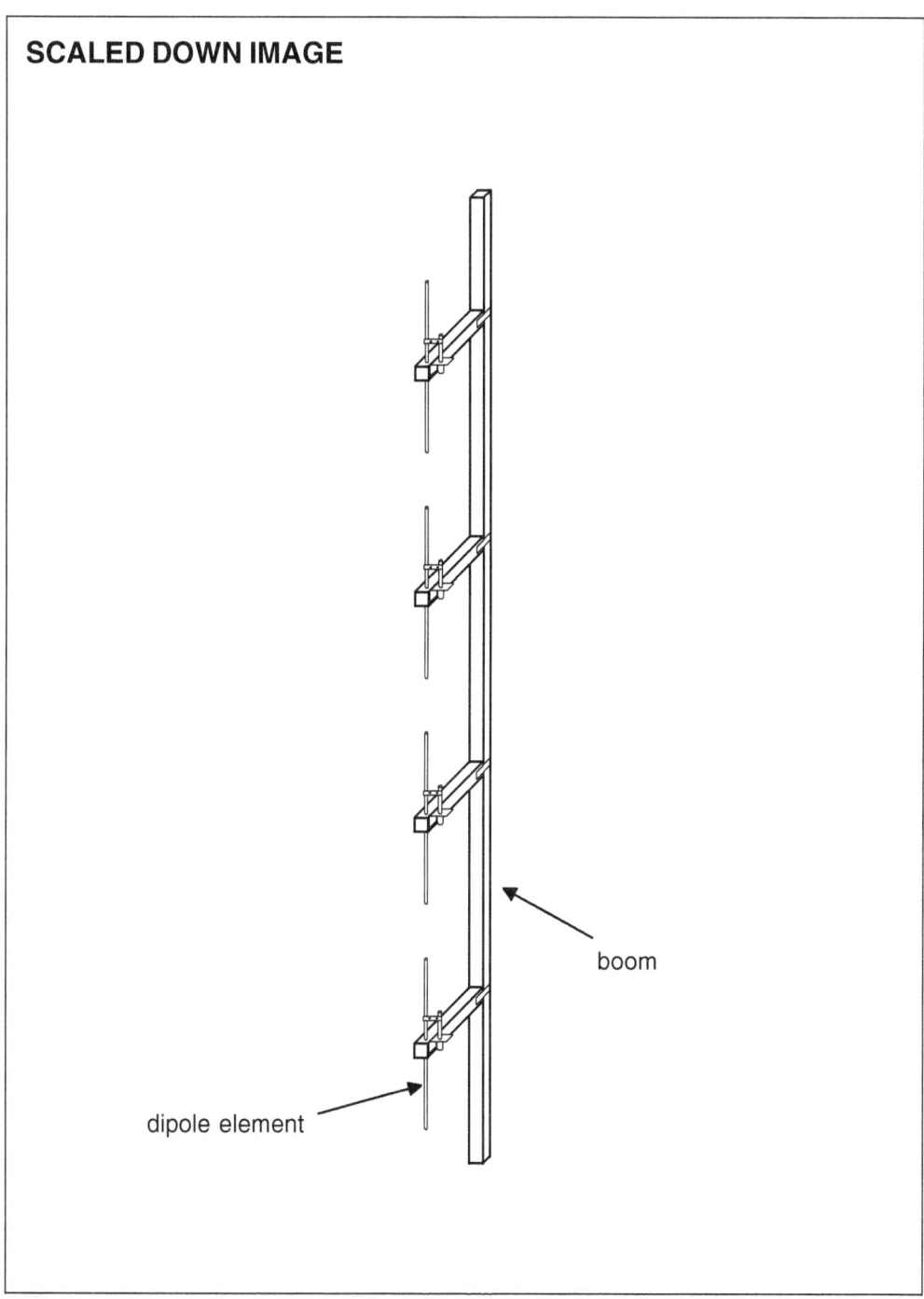

boom

dipole element

Figure 15.1   Stacked Dipole Array Model SD-24

Materials List

The necessary materials in building this antenna are the same as those needed for SD-22, being its extended version. The square channel used to mount the four dipoles is larger and twice longer than the one used for SD-22. An additional set of phasing harnesses is also needed to feed the four dipole elements simultaneously.

Additional materials for the Model SD-24 are as follows:

1 pc.	Square aluminum channel	1-1/2" x 1-1/2" x 235"
4 pcs.	BNC male connectors	
2 pcs.	BNC 'T' connectors	

Construction

Construct the four dipoles following the procedures described for models DP-2 and SD-22. Mount the four dipoles to the aluminum supporting channel by using hose clamps. The antenna elements must be attached to the mast separated by the proper distances from each other (see Figure 15.2).

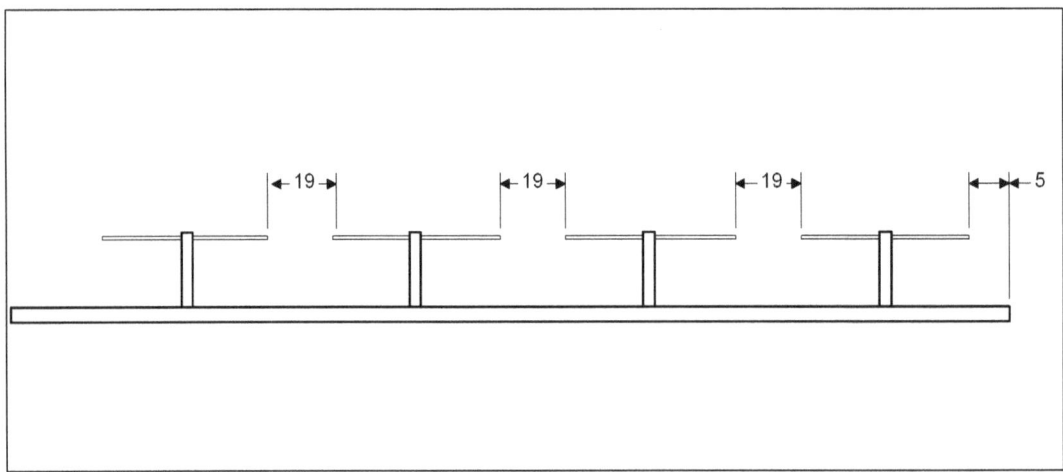

Figure 15.2   Mounting the four dipole elements on the mast.

Construct the phasing harness as shown below, and attach it to the four dipoles in similar fashion to the model SD-22 antenna (see Figure 15.3).

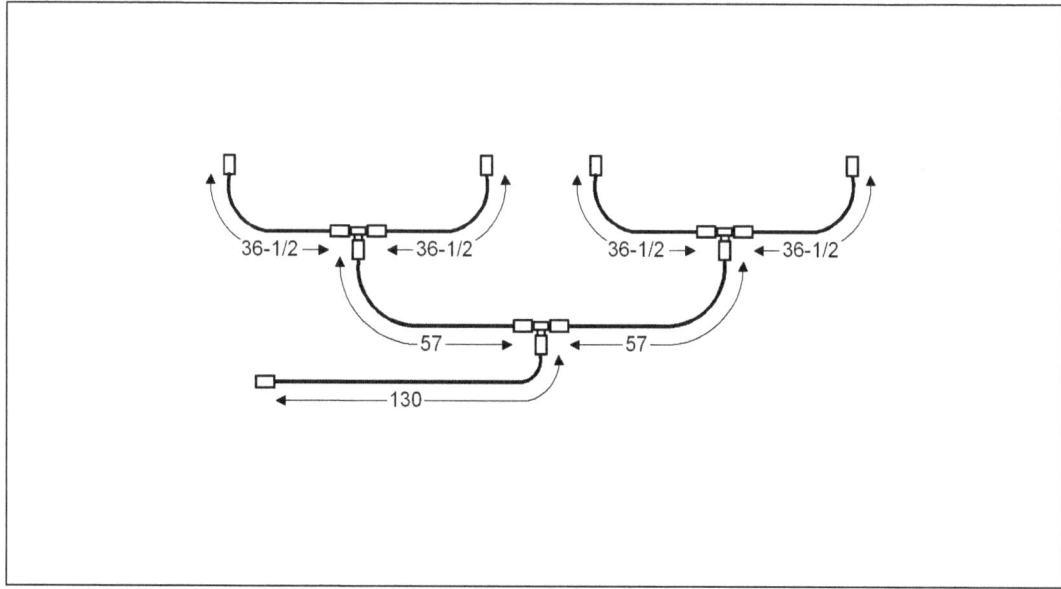

Figure 15.3 Constructing the phasing harness.

Tuning SD-24 to resonance

Tuning the antenna model SD-24 to get a good match is similar to procedures for tuning the antenna model SD-22.

REVIEW QUESTIONS

1. Why must a designer follow the method of doubling the number of collinear dipole elements used if he plans to expand the collinear antenna to obtain additional gain?

2. Why is an array with dipole elements in excess of eight rarely constructed?

3. If a 2-element collinear antenna has a power ratio of 2 and a gain of 3 dB, what then is the power ratio of a 4 element array?

4. What is the power ratio of a 16-element array? What is its gain in terms of decibels (dB)?

5. If you want to attain a power ratio of 64, how many dipole elements will be needed? What would be its gain in terms of decibels?

6. What are the advantages of a collinear antenna or array?

16 YAGI-UDA ANTENNA

Model YG-23 (3 element beam)

A Yagi-Uda antenna is a type of an array having one active dipole and two or more parasitic elements. It was named after the two Japanese physicists who invented it. The basic Yagi is one of the highest gain antennas yet developed. Several factors affect the performance of a Yagi. Among these are the number of elements used, their diameters, and the spacing between them.

A basic half wave dipole is cut to resonance at the center of the frequency band, and is utilized as driven element. High gain is attained by the addition of parasitic elements positioned either in front or behind the driven element. These parasitic elements are called 'directors' and 'reflectors', depending on their length and positioning with respect to the drive element. The reflector is longer by approximately 5%, and is positioned behind the driven element. The director, on the other hand, is cut shorter by approximately 5%, and is positioned at the front of the driven element. The combination of these elements produces the directivity of the radiated signal, thus resulting in higher power gain. However, the radiation pattern becomes uni-directional, and the much desired omni-pattern is completely lost.

Maximum radiation of signal is now concentrated at the front of the antenna, and there is only minimum radiation at the back. The ratio between the radiated signal at the front and the radiated signal behind it is called 'front to back ratio'. Radiation is weakest at the sides of the Yagi and these points are called 'null points'. The ratio between the radiated signal at the front and the radiated signal at the sides, is called 'front to side ratio'.

These highly directive and uni-directional characteristics of a Yagi antenna necessitate the use of a rotator device in order to beam it to the direction of the station in contact. If a rotator device is not used, then the high gain character of a Yagi becomes useless, unless the antenna is intended to be permanently beamed to a single direction, such as in the case of fixed point-to-point communication.

The dimensions of model YG-23 are specially designed to resonate in the frequency band of 140-150 MHz. If properly tuned, it exhibits an SWR of less than 1.5:1 over the entire band. It has a gain of approximately 7.3 dB compared to a standard dipole reference. This is only a basic configuration of a Yagi, and its gain and directivity can be increased by adding more directors at the front. Detailed information for the exact dimensions of additional director elements and their spacing is given in chapter 17. These Yagi dimensions are based on the information published by the National Bureau of Standards (NBS).

SCALED DOWN IMAGE

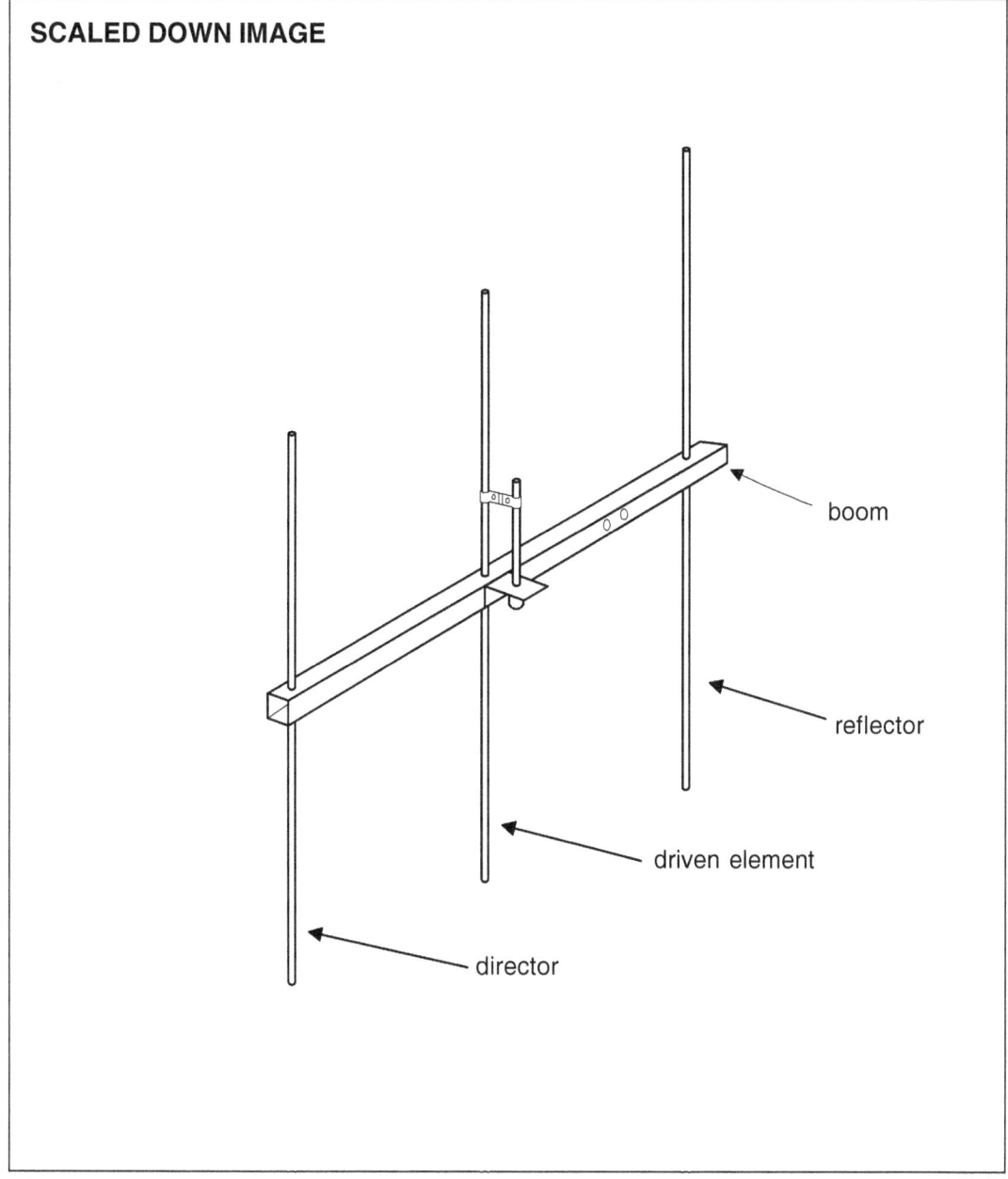

Figure 16.1　Yagi antenna model YG-23

Materials List		
Quantity	**Specification/Description**	**Dimensions**
1	Aluminum tube	3/8" od* 3 feet 4"
1	Aluminum tube	3/8" od* 3 feet 2-3/16"
1	Aluminum tube	3/8" od* 3 feet 7/8"
1	Aluminum square channel	1" x 1" x 2 feet and 32-1/4"

Other materials used in constructing the antenna DP-2 are also needed for this Yagi antenna, except for the mounting channel.

* od - outside diameter

Construction

Cut the three tubes to their exact lengths, and drill a hole (1/8" diameter) through and through at its middle length. The shortest tube will be used as a director element, the longest tube will the reflector element, and the medium length will be the driven element (see Figure 16.2).

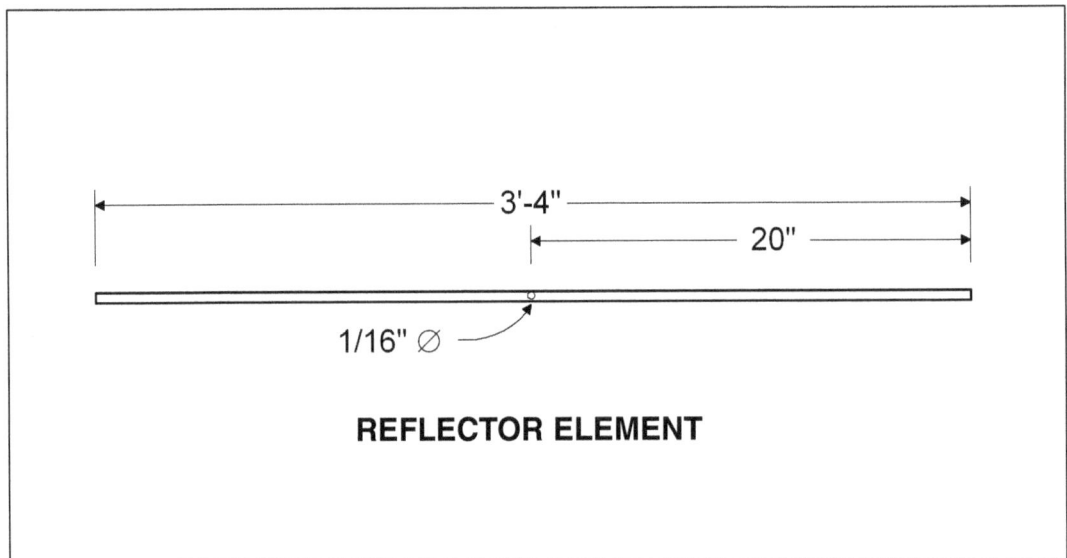

Figure 16.2 Preparing the reflector element.

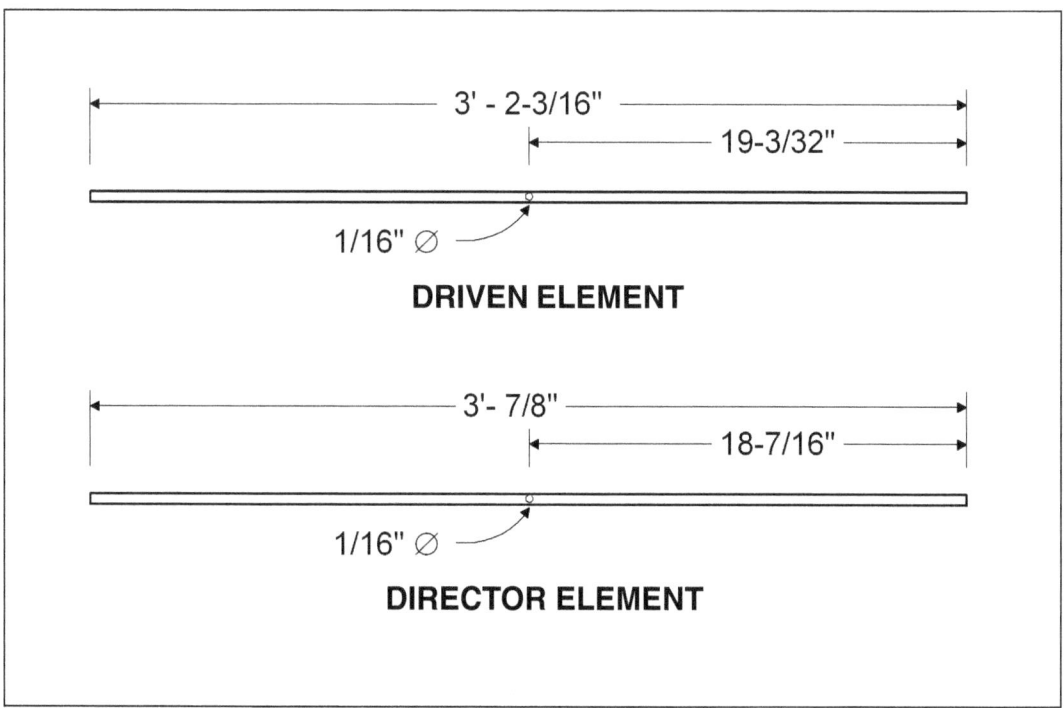

Figure 16.3   Preparing the driven and director elements.

Cut the aluminum mounting tube or boom to 2 feet and 11 inches long, and drill three holes through and through at one side. The holes must have a diameter of 3/8" or at least enough to accommodate the diameter of the tube that will be inserted into it. Follow the dimensions shown. Drill also three 3/8" diameter holes at the same point where the larger holes are, but at one side. The axis of the smaller holes must cross the axis of the larger holes (see Figure 16.4).

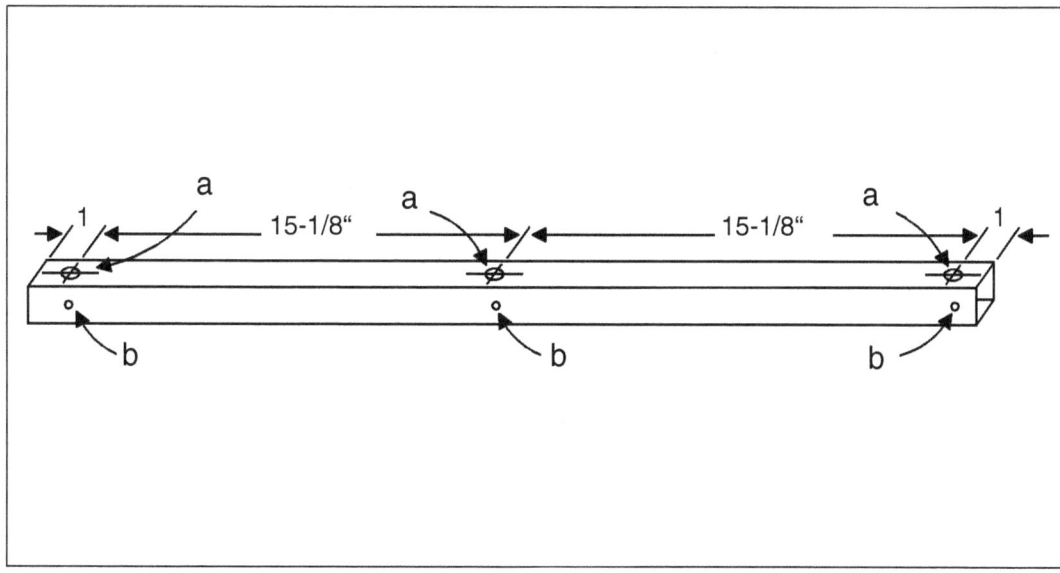

Figure 16.4   Preparing the boom.

Insert the aluminum tubes into the boom, following the illustration showing the proper arrangement of the elements. Secure the tube to the boom by placing the screws through the holes at the sides, similar to the method of attaching the dipole element of the antenna model DP-2 (see Figure 16.5).

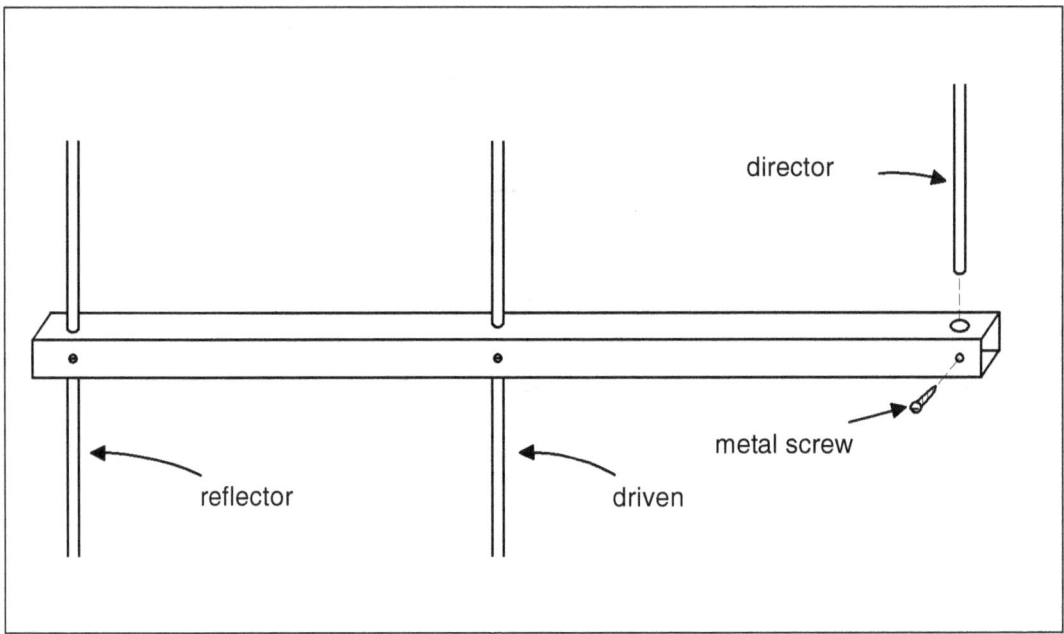

Figure 16.5   Assembling the antenna elements to the boom.

Complete the attachments of the *driven element* following the procedures described for the model DP-2. All other materials and dimensions (e.g. gamma, bracket, connector, clamp, etc.) are similar to those used for the dipole elements of the DP-2 (see Figure 16.6).

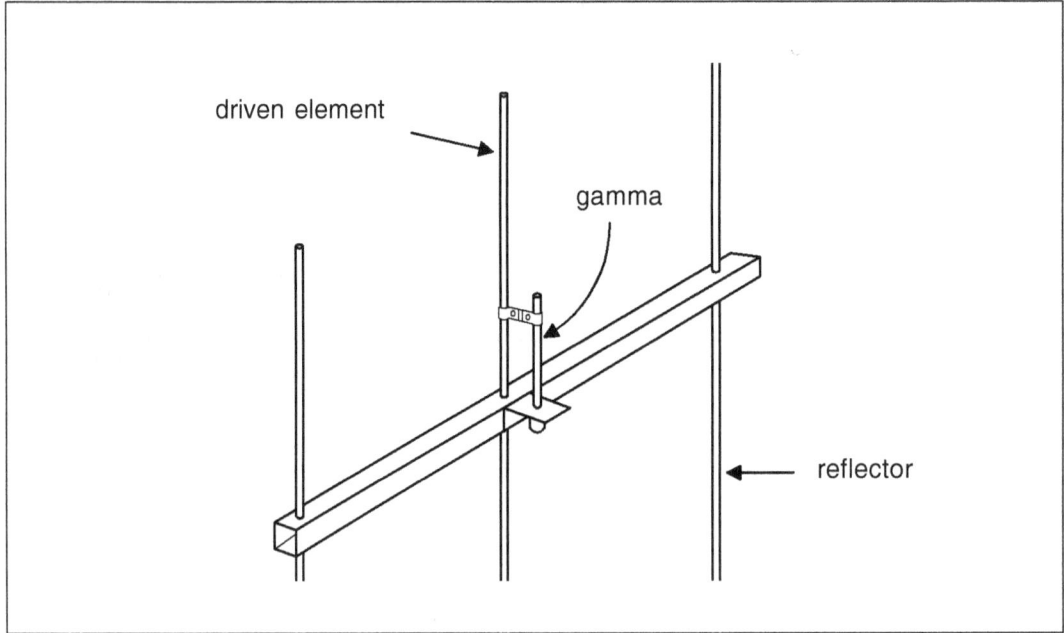

Figure 16.6   Complete assembly of the driven element.

Installation of YG-23

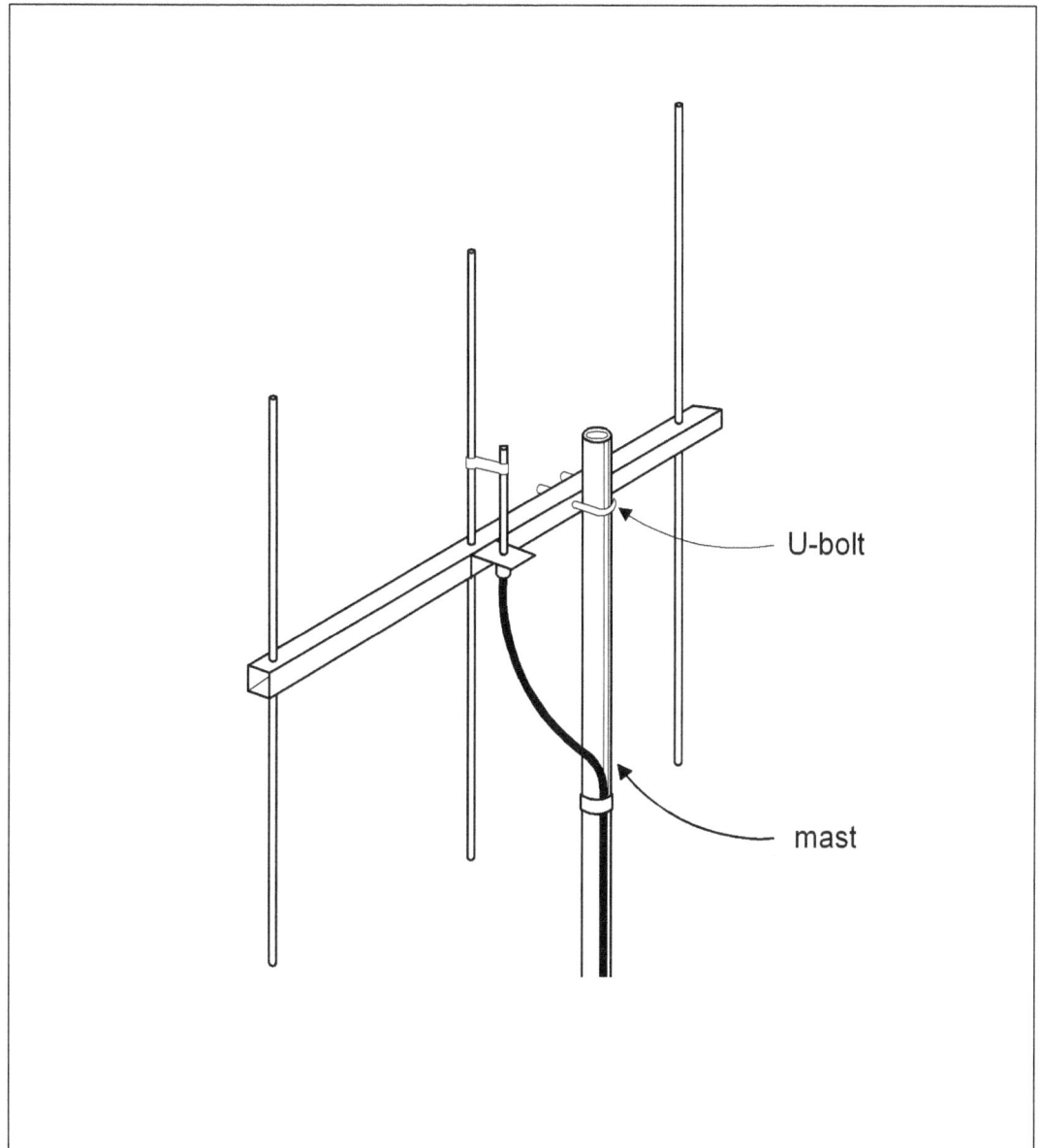

Figure 16.7   Installing the YG-23 to the mast.

Tuning YG-23 to resonance

The tuning procedure for the antenna model YG-23 is similar to the procedure for tuning antennas model DP-2, SD-22 or SD-24. The most important step is to tune the antenna while directly attached to the mast where it will be permanently installed whenever practical.

REVIEW QUESTIONS

1. What does 'directive array' mean?

2. What are the factors that affect the performance of a Yagi?

3. How is the reflector constructed with respect to the driven element?

4. How is the director element constructed with respect to the driven element?

5. What is the radiation pattern of a Yagi antenna?

6. What is 'front to back' ratio?

7. What is 'front to side' ratio?

8. What is 'null point'?

9. How can the directive characer of a Yagi antenna be maximized for communication?

10. In this particular model, what is the method employed to match the impedance of the antenna to the impedance of the transmission line?

17 MULTI-ELEMENT YAGI-UDA ANTENNA ARRAY

The Yagi-Uda antenna or simply "Yagi" model YG-23 described in the preceding chapter gives a fairly high gain figure in a very compact and easy-to-construct antenna. By adding more director elements at the front and extending the boom length of the Yagi, you can achieve a much higher gain figure from this type of antenna.

The following table shows the exact element lengths and dimensions for the various Yagi antennas based on the National Bureau of Standards (NBS) specifications. Two sets of dimensions are given. One set is for the type of Yagi antenna with elements that are insulated from the boom. Another set is for the Yagi antenna with elements directly attached to the metal boom. The latter set is widely popular among antenna constructors because it is easier to construct and eliminates the need for individual insulators.

The construction of the reflector, driven element, gamma match, director elements, assembly, and tuning procedures are basically similar to the model YG-23.

Boom Length	Boom Diameter	Element Diameter	Insulated Elements	Reflector	Driven	Dir.1	Dir.2	Dir.3	Dir.4	Dir.5
5'5-9/16" (0.8λ)	1"	3/8" od*	YES	3'4"	3'2-3/16"	3'7/8"	3'11/16"	3' 7/8"	–	–
			NO	3'4-5/8"	-do-	3'1-1/2"	3'1-3/8"	3'1-1/2"	–	–
8'2-5/16" (1.2λ)	1"	3/8" od*	YES	3'4"	-do-	3' 7/8"	3'7/16"	3'7/16"	3'7/8"	–
			NO	3'4-5/8"	-do-	3'1-1/2"	3'1-1/8"	3'1-1/8"	3'1-1/8"	–
15'1/4" (2.2l)	1-1/2"	3/8" od*	YES	3'4"	-do-	3'1-1/8"	3'5/16"	2'11-13/16"	2'11-1/4"	2'10-9/16"
			NO	3'4-13/16"	-do-	3'1-15/16"	3'1-1/8"	3'5/8"	3'	2'11-3/8"
21'10-1/16"(3.2λ)	1-1/2"	3/8" od*	YES	3'4"	-do-	3'7/8"	3'9/16"	2'11-3/4"	2'11-1/8"	2'10-7/8"
			NO	3'5-1/16"	-do-	3'1-15/16"	3'1-3/8"	3'13/16"	3'3/16"	3'
28'8-1/8" (4.2λ)	1-1/2"	3/8" od*	YES	3'3-3/8"	-do-	3'9/16"	3'9/16"	3'3/8"	2'11-5/8"	2'11-1/2"
			NO	3'4-1/2"	-do-	3'1-5/8"	3'1-5/8"	3'1-7/16"	3'11/16"	3'9/16"

Dir. 6	Dir. 7	Dir. 8	Dir. 9	Dir. 10	Dir. 11	Dir. 12	Dir. 13	Dir. 14	Dir. 15
–	–	–	–	–	–	–	–	–	–
–	–	–	–	–	–	–	–	–	–
–	–	–	–	–	–	–	–	–	–
–	–	–	–	–	–	–	–	–	–
2'10-9/16"	2'10-9/16"	2'10-9/16"	2'11-1/4"	2'11-13/16"	–	–	–	–	–
2'11-3/8"	2'11-3/8"	2'11-3/8"	3'	3'5/8"	–	–	–	–	–
2'10-9/16"	2'10-5/16"	2'10-5/16"	2'10-5/16"	2'10-5/16"	2'10-5/16"	2'10-5/16"	2'10-5/16"	2'10-5/16"	2'10-5/16"
2'11-5/8"	2'11-3/8"	2'11-3/8"	2'11-3/8"	2'11-3/8"	2'11-3/8"	2'11-3/8"	2'11-3/8"	2'11-3/8"	2'11-3/8"
2'11-1/8"	2'10-13/16"	2'10-9/16"	2'10-9/16"	2'10-9/16"	2'10-9/16"	2'10-9/16"	2'10-9/16"	–	–
3'3/16"	2'11-7/8"	2'11-5/8"	2'11-5/8"	2'11-5/8"	2'11-5/8"	2'11-5/8"	2'11-5/8"	–	–

ELEMENT SPACING WITH RESPECT TO BOOM LENGTH

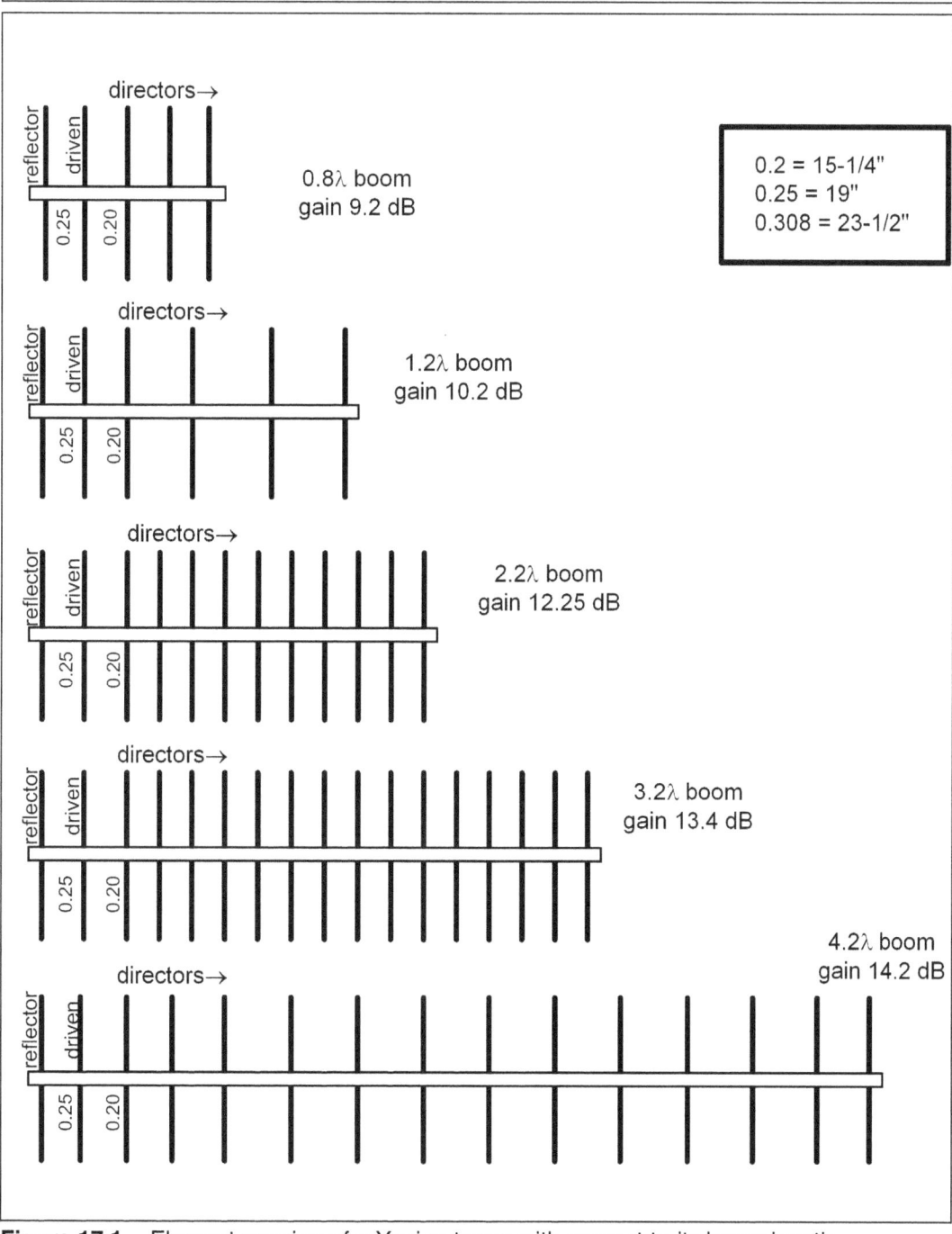

directors→

reflector
driven
0.25
0.20

0.8λ boom
gain 9.2 dB

0.2 = 15-1/4"
0.25 = 19"
0.308 = 23-1/2"

directors→

reflector
driven
0.25
0.20

1.2λ boom
gain 10.2 dB

directors→

reflector
driven
0.25
0.20

2.2λ boom
gain 12.25 dB

directors→

reflector
driven
0.25
0.20

3.2λ boom
gain 13.4 dB

4.2λ boom
gain 14.2 dB

directors→

reflector
driven
0.25
0.20

Figure 17.1   Element spacing of a Yagi antenna with respect to its boom length.

18 STACKING YAGI ANTENNAS

Stacking Yagi antennas means multiplying the number of Yagi antennas and feeding them all simultaneously. If the number of Yagi antennas is doubled, it will add an additional 3 dB to the original gain figure. For example, if you feed two identical 3-element Yagis which have a gain of 7.3 dB, it will give you a total gain of 10.3 dB. Similarly, a 17 element Yagi with a gain of 13.4 dB will give a whopping 19.4 dB if stacked to four identical pieces!

In stacking Yagis, the spacing between the antennas is very important. The distance between two Yagis stacked side by side must not be less than 1 wavelength, or it must be approximately 77 inches. The distance between the tips of the elements in vertically stacked Yagis must be not less than one-half wavelength, or it must be approximately 38 inches (see figures 18.1 and 18.2).

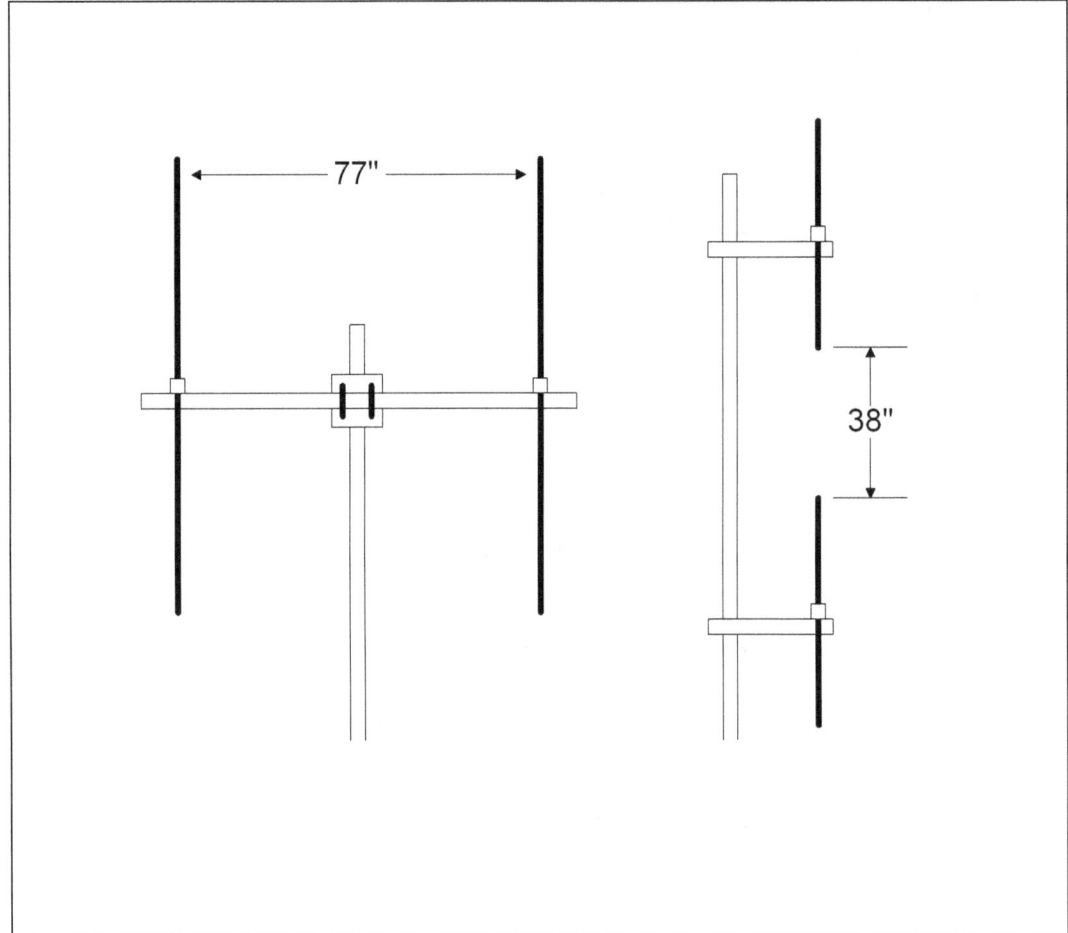

Figure 18.1   Two stacked Yagis viewed at their boom ends.

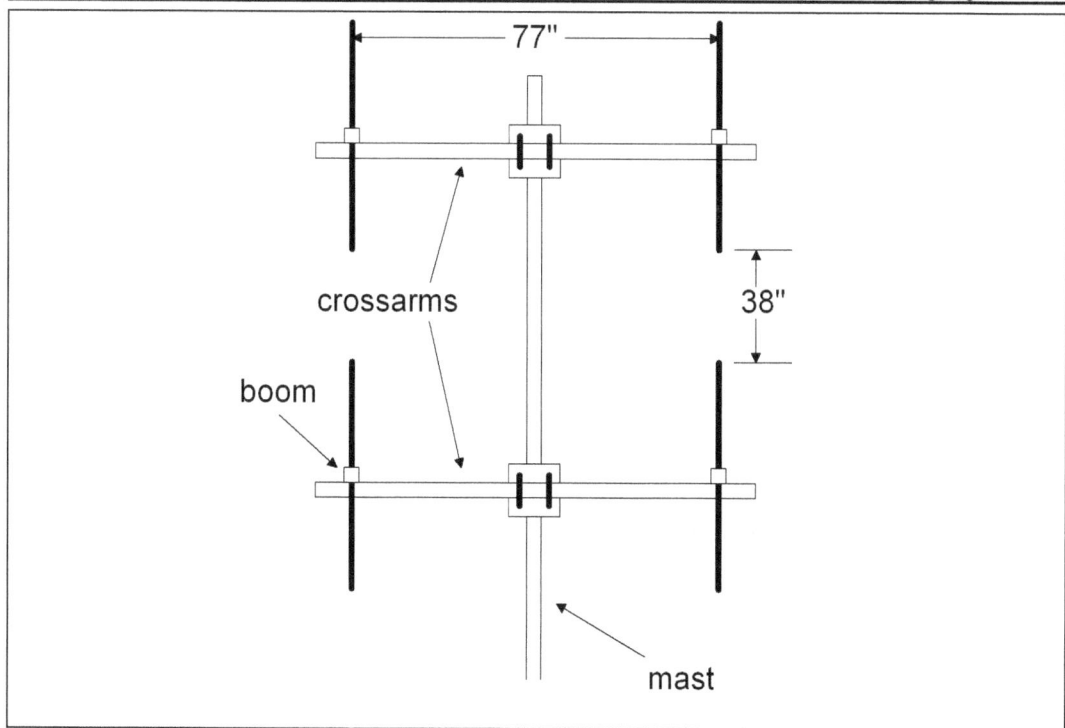

Figure 18.2   Four stacked Yagis viewed at their boom ends.

All the Yagis must be fed in phase with a phasing harness. For example, the configuration of a phasing harness for the 4 stacked Yagis shown above is described in the following illustration.

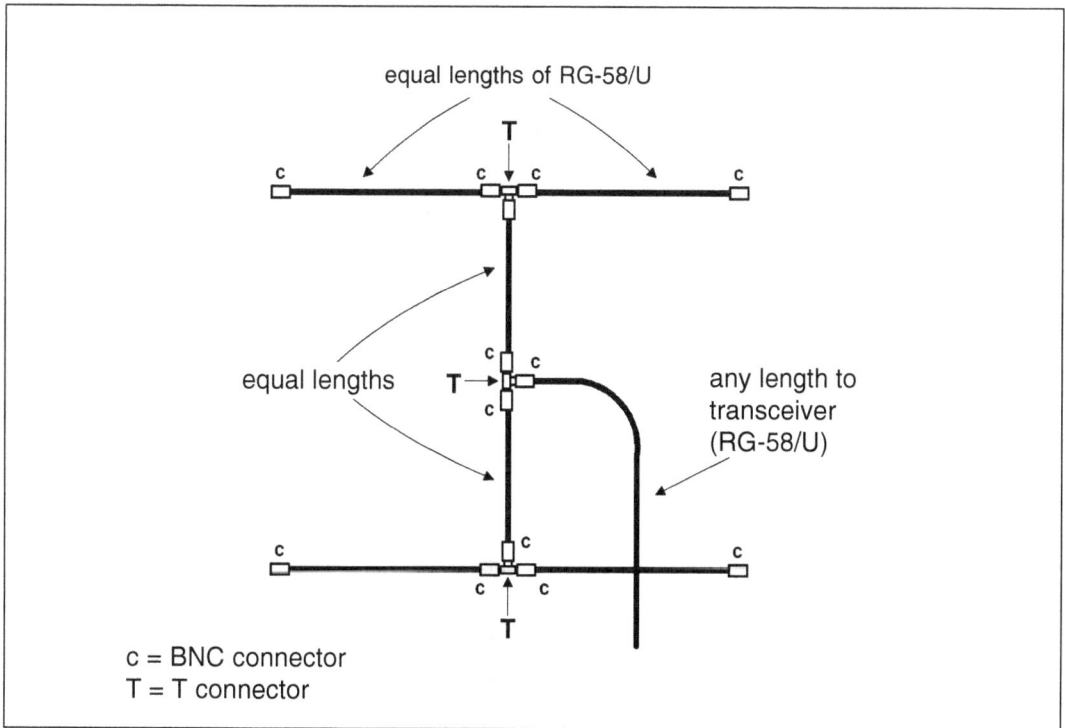

Figure 18.3   Phasing harness for a 4-stacked Yagi antenna.

Mechanical Construction

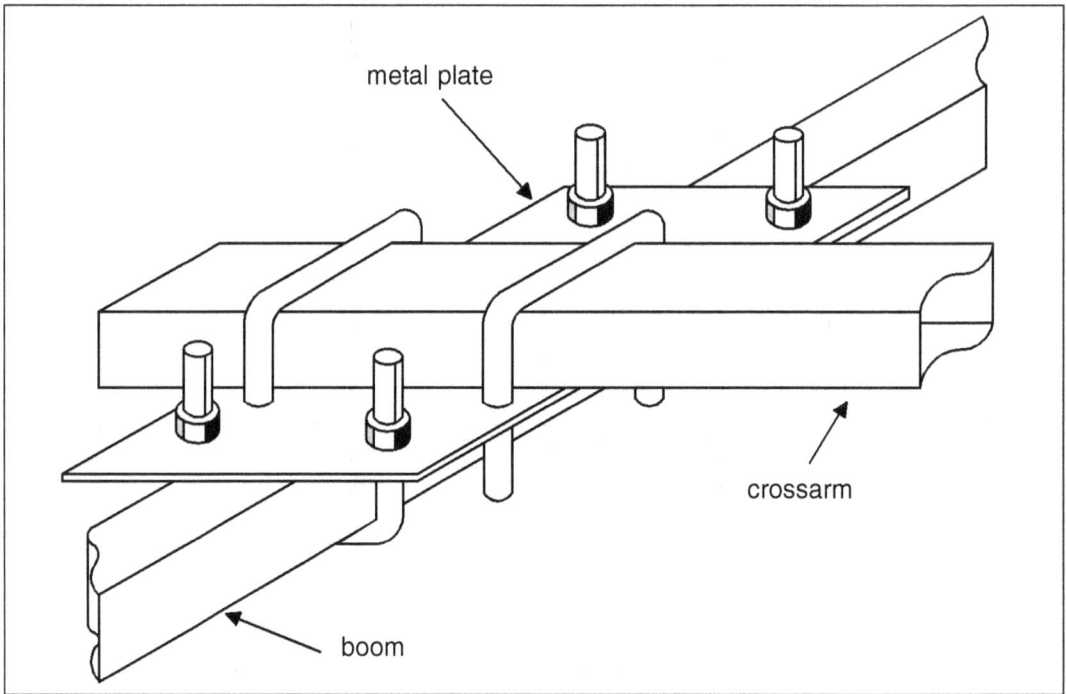

Figure 18.4   Attachment of the boom to the cross arm.

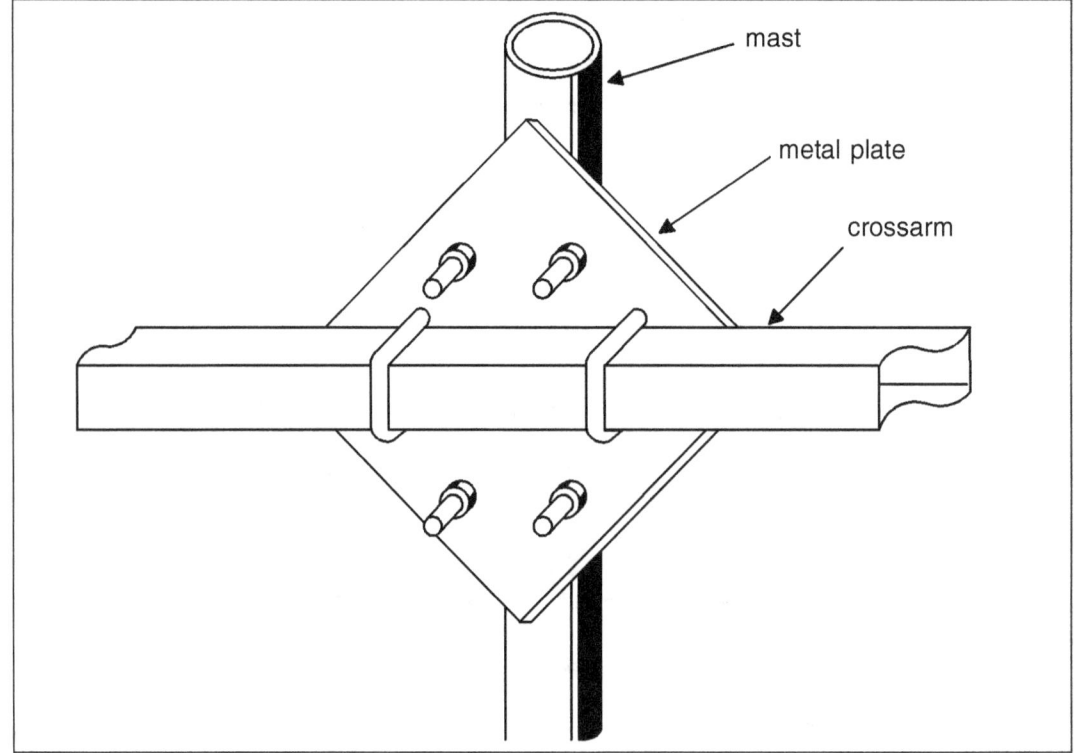

Figure 18.5   Attachment of the cross arm to the mast.

19

FORMULAS FOR CONVERTING ANTENNA DESIGNS FOR OTHER FREQUENCY BANDS

The dimensions of the antenna elements are generally derived from the antenna's electrical wavelength. *The electrical wavelength of a certain frequency is slightly different from its wavelength in free space*, where the former is the wavelength of the signal present in the physical conductor of the antenna, and is somewhat shorter

---

The formula to get the electrical wavelength of a frequency is:

In feet:

$$\frac{936}{Fc\ (MHz)} = \lambda\ (feet)$$

In meters:

$$\frac{286}{Fc\ (MHz)} = \lambda\ (meters)$$

---

where:

Fc is the center frequency of the band expressed in Megahertz.

Lambda λ is the symbol for the wavelength expressed either in feet or meters, depending on the particular units used.

* For example, the wavelength of 145 MHz is:

$$\frac{936}{145} = 6.46\ \text{feet or } 77.52\ \text{inches}$$

Other symbols of wavelength:

$\lambda/2$ or **(0.5)**λ = half wavelength
$\lambda/4$ or **(0.25)**λ = quarter wavelength

Groundplane elements

The formula to get the length of each element of a groundplane antenna is:

$$E \text{ (feet)} = \frac{\left(\dfrac{468}{Fc}\right)}{2} \qquad \text{or} \qquad E \text{ (feet)} = 0.25 \, \lambda$$

* To convert **E** (in feet) to inches, multiply it with 12.

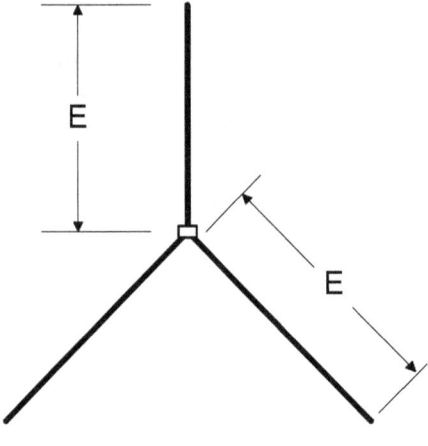

Example: Find the length of one ground plane element intended for 220 MHz.

Solution:

$$E \text{ (feet)} = \frac{\left(\dfrac{468}{220}\right)}{2} \qquad \textit{substituting the value of frequency}$$

$$E \text{ (feet)} = \frac{2.127}{2}$$

$$E \text{ (feet)} = 1.06 \qquad \textit{<<< this is the length of the element expressed in feet}$$

To convert the result to inches:

$$E \text{ (feet)} \times 12 = E \text{ (inches)}$$
$$1.06 \times 12 = 12.72$$

$$E = \textbf{12.72 inches} \quad \textit{<<< this is the length of the element expressed in inches}$$

Coaxial dipole elements

Formulas to find the elements of a coaxial dipole:

$$E \text{ (feet)} = \frac{\left(\dfrac{468}{Fc}\right)}{2} \qquad \text{or} \qquad E \text{ (feet)} = 0.25\ \lambda$$

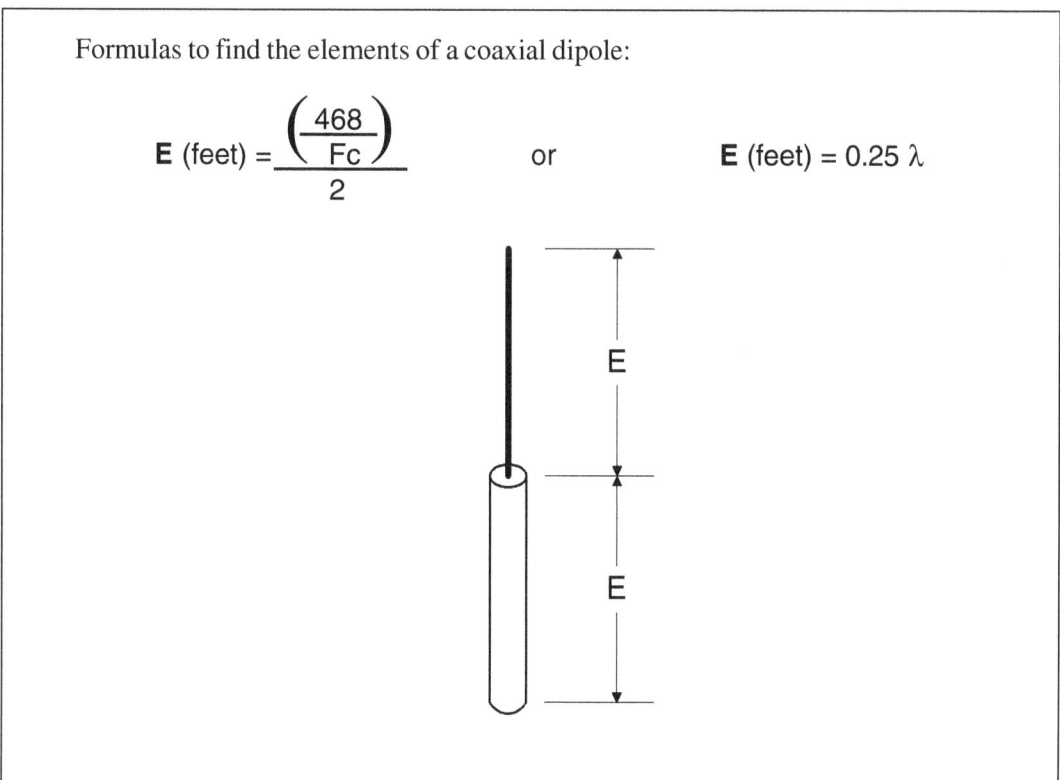

Example: Find the length of the element for a coaxial dipole intended for 110 MHz (also known as the aircraft band).

Solution:

$$E \text{ (feet)} = \frac{\left(\dfrac{468}{Fc}\right)}{2}$$

$$E \text{ (feet)} = \frac{\left(\dfrac{468}{110}\right)}{2} \qquad \textit{substituting the value of frequency}$$

$$E = 2.125 \text{ feet} \qquad <<< \quad \textit{this is the length of the coaxial element expressed in feet}$$

Convert the result to inches:

E (feet) x 12 = **E** (inches)
2.125 x 12 = 25.5

E = 25.5 inches <<< *this is the final value*
expressed in inches

Quad loop antenna element

The formula to get the length of each side of the loop element for the Quad loop antenna:

$$S\ (\text{feet}) = \frac{\left(\dfrac{468}{Fc}\right)}{2} \qquad \text{or} \qquad S\ (\text{feet}) = 0.25\ \lambda$$

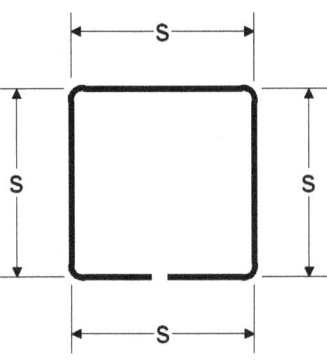

Example: Find the length of one side of the quad loop intended for 155 MHz (VHF commercial band).

Solution:

$$S\ (\text{feet}) = \frac{\left(\dfrac{468}{Fc}\right)}{2}$$

$$S \text{ (feet)} = \frac{\left(\frac{468}{155}\right)}{2}$$ *substituting the value of frequency*

$$S \text{ (feet)} = \frac{3.02}{2}$$

S   = 1.51 feet   <<< *this is the length of one side of the loop element expressed in feet*

Convert the result to inches:

S (feet) x 12 = S (inches)

1.51 x 12   = 18.12

S   = **18.12 inches** <<< *this is the final value expressed in inches*

Dipole element

The formula to get the length of the dipole element for the antennas DP-22, SD-22, and SD-24:

$$D \text{ (feet)} = \frac{486}{Fc} \qquad \text{or} \qquad D \text{ (feet)} = 0.5\,\lambda$$

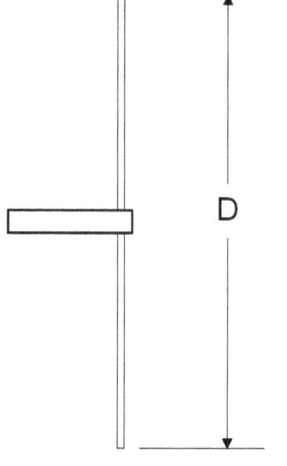

* **Example:** Find the length of the dipole element intended for 195 MHz.

Solution:

$$D \text{ (feet)} = \frac{468}{Fc}$$

$$D \text{ (feet)} = \frac{468}{195}$$

$$D \text{ (feet)} = 2.4 \text{ feet} <<< \textit{this is the length of the dipole element expressed in feet}$$

Convert the result to inches:

$$D \text{ (feet)} \times 12 = D \text{ (inches)}$$

$$2.4 \times 12 = 28.8$$

$$D = \textbf{28.8 inches} \quad <<< \textit{this is the final value expressed in inches}$$

Spacing between dipoles' ends

The formula to get the correct spacing between the ends of two dipoles in a collinear array such as the SD-22 or SD-24:

$$S \text{ (feet)} = \frac{\left(\dfrac{468}{Fc} \right)}{2} \qquad \text{or} \qquad S \text{ (feet)} = 0.25 \, \lambda$$

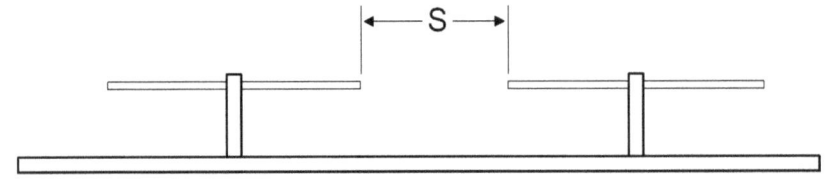

Example: Find the spacing between two dipoles designed for 220 MHz.

Solution:

$$S \text{ (feet)} = \frac{\left(\dfrac{468}{Fc}\right)}{2}$$

$$S \text{ (feet)} = \frac{\left(\dfrac{468}{220}\right)}{2}$$

$$S \text{ (feet)} = \frac{2.127}{2}$$

S = 1.053 feet <<< *this is the length of the dipole element expressed in feet*

Convert the result to inches:

S (feet) x 12 = S (inches)

1.053 x 12 = 12.45

S = **12.45 inches** <<< *this is the final value expressed in inches*

NOTE:

This spacing is only for the minimum allowable between the ends of the dipole elements in a collinear antenna. Optimum spacing is within 0.25l and 0.5l.

Discone dimensions

To get the exact lengths and dimensions of the discone antenna, first compute the wavelength of the lowest targeted frequency by using this formula:

$$\lambda \text{ (wavelength in feet)} = \frac{936}{Fc}$$

Formulas for dimensions:

* The disc element diameter is 0.19 of the wavelength.
* The cone element length is 0.29 of the wavelength.
* The spacing between the disc element and the apex of the cone element is 0.0077 of the wavelength.

* The diameter of the mounting tube is not critical.

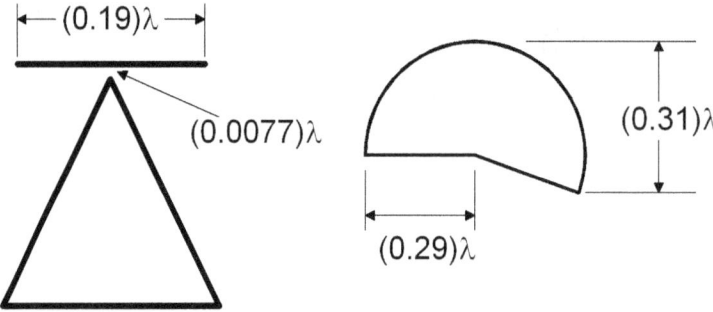

Example: Find the dimensions of a discone antenna with a cut-off frequency of 120 MHz (upper portion of the aircraft band).

Solution:

        First find the wavelength of the frequency by using the formula below and convert the result to inches.

$$\frac{936}{Fc} = \lambda \text{ (wavelength in feet)}$$

$$\frac{936}{120} = 7.8$$

$$7.8 \text{ feet} = \lambda \quad <<< \text{ \textit{this is the wavelength of the frequency in feet}}$$

Convert it to inches:

7.8 x 12 = 83.6

λ (inch) = **83.6 inches** <<< *wavelength of the frequency in inches*

From the result above you can proceed to compute the dimensions:

* DISC ELEMENT DIAMETER:

Formula: 0.19 x λ (inch) = disc diameter

(substitute value of λ)
        0.19 x 83.6   = 15.384

                15.382 inches   <<< *diameter of the disc*

* LENGTH OF CONE ELEMENT:

Formula: 0.29 x λ (inch) = length of cone element

(substitute value of λ)
        0.29 x 83.6   = 24.44

                24.44 inches   <<< *length of the cone element*

* SPACING BETWEEN DISC AND CONE'S APEX:

Formula: 0.0077 x λ (inch) = spacing

(substitute value of λ)
    0.0077 x 83.6 = 0.64

      0.64 inches <<< *spacing between*
                disc and cone's apex

> ## NOTE:
>
> *In calculating the dimensions of a discone, you must always use the value of the lowest frequency at which you intend to operate (the so-called 'cut-off frequency').*

5/8 Wave radiator element

The formula to get the length of the radiator element of a 5/8 element wavelength antenna such as the WA-2 and PF-2C:

  L (feet) = 0.65 x λ   or   **L** (feet) = (5 x λ)

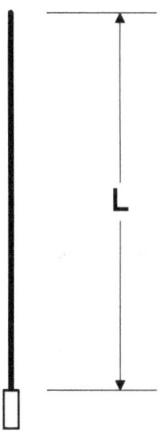

Example: Find the length of the radiator element intended for 160 MHz (VHF commercial band).

Solution:

First find the wavelength of the center frequency by using the formula below and convert the result to inches.

$$\frac{936}{Fc} = \lambda \quad \text{(wavelength in feet)}$$

$$\frac{936}{160} = 5.85$$

5.85 feet = λ <<<   *this is the wavelength of the frequency in feet*

Convert it to inches:

5.85 x 12 = 70.2

λ (inch)   = **70.2 inches** <<< *wavelength of the frequency in inches*

Finally, find the length of the 5/8 radiator element by using the following formula:

L (inch) = 0.65 x λ (inch)

L (inch) = 0.65 x 70.2

L = 45.6 inches <<<   *length of the radiator element*

REVIEW PROBLEMS

1. Find the wavelengths of the following frequencies both in feet and inches:

   a) 165 MHz

   b) 250 MHz

   c) 100 MHz

   d) 50 MHz

   e) 135 MHz

2. Find the length (in inches) of a dipole element intended for 185 MHz.

3. Find the length of a groundplane element intended for 175 MHz.

4. Find the length of one side of a Quad loop intended for 220 MHz.

5. Find the dimensions of a discone antenna intended to operate with a cut-off frequency of 35 MHz.

6. Find the actual distance between the ends of two collinear dipole elements intended for 220 MHz using 0.36l as spacing.

ANSWERS TO REVIEW QUESTIONS

Chapter 1 (Model FA-2)

1. The radiation pattern of an RF signal is important since it must conform to the needs of a particular situation. Omni-directional pattern is excellent in medium range multi-contact communications while directive pattern is very effective for long distance point-to-point communication links.

2. An antenna with an omni-directional pattern has the advantage of having the capability to communicate with any stations from any directions in the horizontal plane.

3.   Since antennas with directive patterns can only communicate effectively in one or two directions, it is practical when you know exactly the direction of the other station and you have the means to rotate the beam of the antenna.

4.   To overcome the limitations of line-of-sight communications. Raising the height of the antenna extends the horizon farther thereby expanding the area covered by its signal.

5.   The 45 degree bend of the radials lowers the impedance of the antenna to 50 ohms. Since the impedance of the coaxial cable is also 50 ohms, the entire sytem is matched.

6.   The radials simulate the ground needed by the antenna to attain electrical balance.

Chapter 2 (Model FQ-2)

1.   It has the advantage of being packed in a small size for easy transportation.

2.   It insulates the radiator element from the body of PL-259.

3.   Because the body of PL-259 is not part of the radiator element.

4.   To enable quick assembly and disassembly of the radials. It also allows one to construct the antenna without the use of drilling tools.

Chapter 3 (Model JF-2)

1.   The quarterwave matching section provides the return path for the RF current to maintain electrical balance in the antenna.

2.   It matches the impedance of the antenna to the impedance of the transmission line.

3.   This antenna is tuned by adjusting the position of the feeder clamps in the antenna element while monitoring the SWR on the transmission line. The antenna is said to be tuned when the SWR is or close to 1:1.

4.   Tuning the resonance is another term for antenna matching where the impedance of the antenna is made equal to the impedance of the transmission line so that a maximum transfer of energy is obtained.

5.   No. Metal plate would short the two sections of the antenna.

6.   Feedpoint is the point in the antenna element where the transmission or coaxial cable is connected.

Chapter 5 (Model CD-2)

1.   The coaxial dipole is best used in areas with strong winds.

2.   The outer tube acts as a groundplane element.

3.   The inner tube is for mounting purposes only.

Chapter 6 (Model DP-2)

1.   The antenna can be fine-tuned easily.

2.   Gamma match is a method of antenna impedance matching where the shield of the coax cable is connected to the center of the dipole element directly and the inner conductor is tapped into the element via a series capacitor.

3.   The coax inserted inside the gamma tube acts as the series capacitor.

4.   Because the balancing action is already performed by the gamma match.

5.   It is simple to construct, rugged, and the dipole element can be grounded.

6.   It taps the gamma tube to the dipole lement and serves as the fine tuning device.

7.   The shorting bar is adjusted while monitoring the SWR on the line.

Chapter 7 (Model QA-2F)

1.   A quad gives a gain of 2 dB than a dipole. Its pattern is directional which is desirable in some cases.

2.   Bidirectional.

3.   No. The lower plate would short the feed point of the loop element.

4.   Lower end. However, it seems that the other way works equally well.

5.   Perpendicular to the plane of the loop.

Chapter 8 (Model CD-2W)

1. It acts like a frequency independent transformer. Its dimensions are designed so that the impedance at the edges of the cone and disc elements matches the impedance of free space.

2. Because the SWR rises rapidly, if the antenna is operated outside its bandwidth.

3. The dimensions of the cone and disc elements.

4. It is the lowest frequency where the discone can operate effectively. Below this frequency, the SWR on the line increases rapidly.

5. It couples the low impedance transmission line to the higher impedance of free space.

6. To minimize the effect of wind.

7. For auto-scanning wideband monitors.

8. It is for mounting purposes only.

Chapter 9 (Model CD-2P)

1. Only on the mechanical aspect. The metal plate is more durable and rigid.

2. So that it will be grounded adequately to the mast of the tower.

3. The plate can be made of aluminum to make it corrosion resistant.

4. Aluminum is easy to machine as well as corrosion resistant.

Chapter 10 (Model CD-2T)

1. For reasons of portability. If the rods are retracted and collapsed, the resulting size of the antenna is very small.

2. When the telescopic rods are extended to full length.

3. Yes. In fact, the more rods used, the better.

4. Small size, lightweight, and fast assembly and disassembly.

Chapter 11 (Model WA-2)

1.   It gives a slight gain of 1.8 dB over a dipole.

2.   It functions as groundplane radials to simulate the ground needed for proper electrical balance of the antenna.

3.   It serves as a loading coil to match the impedance of the antenna to that of the transmission line.

4.   For mechanical reason only. It gives the copper coil some protection against corrosion.

5.   It displays a graphical representation of the SWR response over the entire band, and aids the constructor in speeding up the tuning process.

6.   By adjusting the tap of the hook-up wire while monitoring the SWR on the line. It is said to be tuned when the SWR is or near to 1:1.

Chapter 12 (Model WD-2)

1.   Unbalanced coupling results when an electrically balanced antenna is connected to a physically unbalanced transmission line such as coaxial cable.

2.   Stray RF current will flow along the outside surface of the line.

3.   It detunes the antenna from any unwanted RF current. It represents a very high impedance path for the stray RF outside the line.

4.   To minimize if not totally eliminate stray RF current that might overload or 'desensitize' the receiver unit.

5.   It means that the receiver unit of a repeater system is overloaded by its own transmitted signal so it cannot receive signals from distant stations.

6.   Yes. The more rods used, the better it simulates the function of the metal cone.

Chapter 13 (Model PF-2C)

1. It is very portable, practical for handheld transceivers, and it gives a slight gain of 1.8 dB over a dipole.

2. It functions as a loading coil to match the impedance of the antenna to the impedance of the transmission line. Also, it serves as a flexible spring to support the telescopic element.

3. It fine tunes the antenna impedance.

4. By raising the position of the antenna.

5. By adjusting the tap of the conductor inside the coil, while monitoring the SWR on the line. It is said to be tuned when the SWR is or close to 1:1.

6. No. It will function like a 'rubber ducky' then, and its gain is negative compared to a dipole.

Chapter 14 (SD-22)

1. It is an antenna made of identical elements (like dipole) assembled in a single structure with their axis arranged in a straight line and fed in phase.

2. It gives a higher power gain and exhibits an omni-directional radiation pattern.

3. To attain gain. If it is not fed in phase, it becomes a harmonic antenna, and its gain is reduced.

4. In-phase in simple terms means that the RF signal must arrive at the feed points of the driven elements at the same time.

5. Within 0.25 to 0.5 of the wavelength. The spacing used in this model is 0.25 of the wavelength, to save on the total length of the aluminum mast.

6. The driven elements in a collinear antenna are tuned at the same time, in a manner described in the instruction for matching a single element. For example, the design described here uses a gamma matching system, so each driven element is tuned, following the procedure for tuning a gamma match.

Chapter 15 (Model SD-24)

1.   To be able to make a practical phasing harness.

2.   Because of the problems in mechanical construction.

3.   Its power ratio is 4.

4.   Its power ratio is 16. Its gain is 12 dB.

5.   It needs 64 dipole elements. Its gain is 18 dB.

6.   High gain together with omni-directional radiation pattern.

Chapter 16 (Model YG-23)

1.   It is a multi-element antenna design having a maximum radiation pattern that is narrow and is concentrated in one or two directions only.

2.   Number of elements used, their diameter, and spacing between the elements.

3.   The reflector element is cut approximately 5% longer than the driven element and positioned behind it.

4.   The director element is cut approximately 5% shorter than the driven element and positioned in front of it.

5.   Uni-directional. Maximum radiation is in front of the array.

6.   The signal radiated to the front compared to the signal radiated towards the back.

7.   The signal radiated to the front compared to the signal radiated to the sides.

8.   Null point is the direction of the signal where it is weakest or minimum radiation compared to the major lobe or maximum radiation.

9.   By using a rotator device to beam the array towards the contact station.

10. Gamma matching system.

Chapter 19

1. a) 5.67 feet or 68.04 inches
    b) 3.74 feet or 44.88 inches
    c) 9.36 feet or 112.3 inches
    d) 18.72 feet or 222.64 inches
    e) 6.93 feet

2. 30.36 inches

3. 16.04 inches

4. 12.76 inches

5. Dimensions of a discone antenna with a cut-off frequency of 35 MHz:

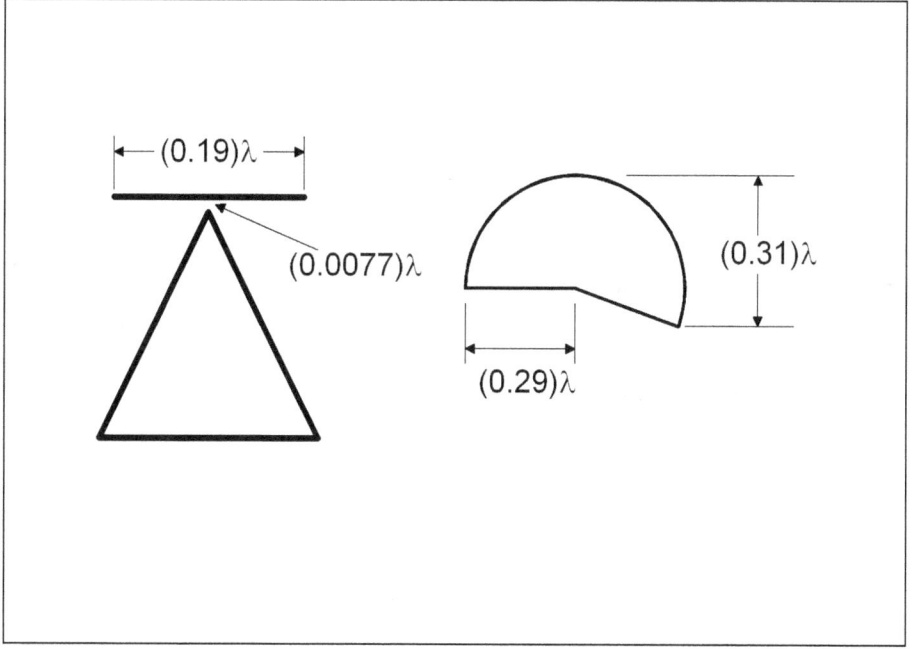

6. The spacing is 17.85 inches.

APPENDIX

POWER RATIO TO DECIBEL CONVERSION

Decimal Increments

Ratio	0.0	0.1	0.2	0.3	0.4	0.5	0.6	0.7	0.8	0.9
1	0.00	0.41	0.79	1.14	1.46	1.76	2.04	2.30	2.55	2.79
2	3.01	3.22	3.42	3.62	3.80	3.98	4.15	4.31	4.47	4.62
3	4.77	4.91	5.05	5.19	5.32	5.44	5.56	5.68	5.80	5.91
4	6.02	6.13	6.23	6.34	6.44	6.53	6.63	6.72	6.81	6.90
5	6.99	7.08	7.16	7.24	7.32	7.40	7.48	7.56	7.63	7.71
6	7.78	7.85	7.92	7.99	8.06	8.13	8.20	8.26	8.33	8.39
7	8.45	8.51	8.57	8.63	8.69	8.75	8.81	8.86	8.92	8.98
8	9.03	9.08	9.14	9.19	9.24	9.29	9.34	9.40	9.44	9.49
10	10.00	10.04	10.09	10.13	10.17	10.21	10.25	10.29	10.33	10.37
x10	+10									
x100	+20									
x1000	+30									
x10,000	+40									
x100,000	+50									

How to use this chart:

The decibel value is read from the body of the table for the desired ratio, including the decimal increment. For example, a power ratio of 1.8 is equivalent to 2.55 dB. Value from the table may be extended as indicated at the lower left in each section. For example, a power ratio of 16 which is the same as 10 x 1.6 is equivalent to 10 + 2.04 = 12.04 dB.

Power loss in relation to the SWR figure in the transmission line:

SWR	Power loss in %
1 :1	0 %
1,3 :1	2 %
1,5 :1	3 %
1,7 :1	6 %
2 :1	11 %
3 :1	25 %
4 :1	38 %
5 :1	48 %
6 :1	55 %
10 :1	70 %

METRIC EQUIVALENTS

Most of the antenna dimensions described in this book are in English units. If the constructor wants to use the metric system, he can convert all the dimensions by using the following conversion guide:

English to Metric

Inch = 25.4 millimeters
Inch = 2.54 centimeters
Foot = 0.305 meter
Yard = 0.914 meter

Metric to English

Centimeter = 0.3937 inches
Meter   = 39.37 inches
Meter   = 3.28 feet
Meter   = 1.094 yards

 # CONVERSION TABLES

Table 3.1   Conversion table - english foot to meter

engl. foot(')	0"	1"	2"	3"	4"	5"	6"	7"	8"	9"	10"	11"
0	0.000	0.0254	0.0508	0.0762	0.1016	0.1270	0.1524	0.1778	0.2032	0.2286	0.2540	0.2794m
1' (= 12")	0.305	0.330	0.356	0.381	0.406	0.432	0.457	0.483	0.508	0.533	0.559	0.584m
2'(= 24")	0.610	0.635	0.660	0.686	0.711	0.737	0.762	0.787	0.813	0.838	0.864	0.889m
3'(= 36")	0.914	0.940	0.965	0.991	1.016	1.041	1.067	1.092	1.118	1.143	1.168	1.194m
4'(= 48")	1.219	1.245	1.270	1.295	1.321	1.346	1.372	1.397	1.422	1.448	1.473	1.499m
5'(= 60")	1.524	1.549	1.575	1.600	1.626	1.651	1.676	1.702	1.727	1.753	1.778	1.803m
6'(= 72")	1.829	1.854	1.880	1.905	1.930	1.956	1.981	2.007	2.032	2.057	2.083	2.108m
7(= 84")	2.134	2.159	2.184	2.210	2.235	2.261	2.286	2.311	2.337	2.362	2.388	2.413m
8'(= 96")	2.438	2.464	2.489	2.515	2.540	2.565	2.591	2.616	2.642	2.667	2.692	2.717m
9' (= 108")	2.743	2.769	2.794	2.819	2.845	2.870	2.896	2.921	2.946	2.972	2.997	3.023m
10' (= 120")	3.048	3.073	3.099	3.124	3150	3.175	3.200	3.226	3.251	3.277	3.302	3.327m
11' (= 132")	3.353	3.378	3.404	3.429	3.454	3.480	3.505	3.531	3.556	3.581	3.607	3.632m
12' (= 144")	3.658	3.683	3.708	3.734	3.759	3.785	3.810	3.835	3.861	3.886	3.912	3.937m
13' (= 156")	3.962	3.988	4.013	4.039	4.064	4.089	4.115	4.140	4.166	4.191	4.216	4.242m
14' (= 168")	4.267	4.293	4.318	4.343	4.369	4.394	4.420	4.445	4.470	4.496	4.521	4.547m
15'(= 180")	4.572	4.597	4.623	4.648	4.674	4.699	4.724	4.750	4.775	4.801	4.826	4.851m
16'(= 192")	4.877	4.902	4.928	4.953	4.978	5.004	5.029	5.055	5.080	5.105	5.131	5.156m
17(=204")	5.182	5.207	5.232	5.258	5.283	5.309	5.334	5.359	5.385	5.410	5.436	5.461m
18'(=216")	5.486	5.512	5.537	5.563	5.588	5.613	5.639	5.664	5.690	5.715.	5.740	5.766m
19' (= 228")	5.791	5.817	5.842	5.867	5.893	5.918	5.944	5.969	5.994	6.020	6.045	6.071m
20' (= 240")	6.096	6.121	6.147	6.172	6198	6.223	6.248	6.274	6.299	6.325	6.350	6.375m
21' (= 252")	6.401	6.426	6.452	6.477	6.502	6.528	6.553	6.579	6.604	6.629	6.655	6.680m
22' (= 264")	6.706	6.731	6.756	6.782	6.807	6.833	6.858	6.883	6.909	6.934	6.960	6.985m
23' (= 276")	7.010	7.036	7.061	7.087	7.112	7.137	7.163	7.188	7.214	7.239	7.264	7.290m
24' (= 288")	7.315.	7.341	7.366	7.391	7.417	7.442	7.468	7.493	7.518	7.544	7.569	7.595m
25; (= 300")	7.620	7.645	7.671	7.696	7.722	7.747	7.722	7.798	7.823	7.849	7.874	7.899m
26' (= 312")	7.925	7.950	7.976	8.001	8.026	8.052	8.077	8.103	8.128	8.153	8.179	8.204m
27 (= 324")	8.230	8.255	8.280	8.306	8.331	8.357	8.382	8.407	8.433	8.458	8.484	8.509m
28' (= 336")	8.534	8.560	8.585	8.611	8.636	8.661	8.687	8.712	8.738	8.763	8.788	8.814m
29' (= 348")	8.839	8.865	8.890	8.915	8.941	8.966	8.992	9.017	9.042	9.068	9.093	9.119m
30' (= 360")	9.144	9.169	9.195	9.220	9.246	9.271	9.296	9.322	9.347	9.373	9.398	9.423m

1' = 0.3048 m;   1" = 0.0254 m;   1' = 12"

Table 4.1   Conversion Table: fraction and decimal of an inch to millimeter

in inch	in mm	in inch	in mm	in inch	in mm
1/64 = 0.015	0.396	23/64 = 0.359	9.127	45/64 = 0.703	17.858
1/32 = 0.031	0.793	3/8 = 0.375	9.525	23/32 = 0.719	18.255
3/64 = 0.047	1.190	25/64 = 0.391	9.921	47/64 = 0.734	18.652
1/16 = 0.063	1.587	13/32 = 0.406	10.318	3/4 = 0.750	19.050
5/64 = 0.078	1.984	27/64 = 0.422	10.715	49/64 = 0.766	19.446
3/32 = 0.094	2.381	7/16 = 0.438	11.112	25/32 = 0.781	19.842
7/64 = 0.109	2.778	29/64 = 0.453	11.508	51/64 = 0.797	20.239
1/8 = 0.125	3.175	15/32 = 0.469	11.905	13/16 = 0.813	20.637
9/64 = 0.141	3.571	31/64 = 0.484	12.302	53/64 = 0.828	21.033
5/32 = 0.156	3.968	1/2 = 0.500	12.700	27/32 = 0.844	21.429
11/64 = 0.172	4.365	33/64 = 0.516	13.096	55/64 = 0.859	21.827
3/16 = 0.188	4.762	17/32 = 0.531	13.492	7/8 = 0.875	22.225
13/64 = 0.203	5.159	35/64 = 0.547	13.890	57/64 = 0.891	22.621
7/32 = 0.219	5.556	9/16 = 0.563	14.287	29/32 = 0.906	23.017
15/64 = 0.234	5.952	37/64 = 0.578	14.683	59/64 = 0.922	23.414
1/4 = 0.250	6.350	19/32 = 0.594	15.080	15/16 = 0.938	23.812
17/64 = 0.266	6.746	39/64 = 0.609	15.477	61/64 = 0.953	24.208
9/32 = 0.281	7.143	5/8 = 0.625	15.875	31/32 = 0.969	24.604
19/64 = 0.297	7.540	41/64 = 0.641	16.271	63/64 = 0.984	25.002
5/15 = 0.313	7.937	21/32 = 0.656	16.667	1 = 1.000	25.400
21/64 = 0.328	8.334	43/64 = 0.672	17.064		
11/32 = 0.344	8.730	11/16 = 0.688	17.462		

Table 12.1   American and english units in relation to metric units

USA and UK	Abbbreviation	Metric unit	conversion factor
1 inch = 10 lines = 1000 mils	('') in	2.54 cm	0.3937
1 foot = 12 inches	(') ft	30.48 cm	3.281×10^{-2}
1 yard = 3 feet = 36 inches	yd	91.44 cm	1.094×10^{-2}
1 fathom = 6 feet	fath	1.8288 m	0.547
1 into nautical mile = 6076 feet	naut. mile	1.852 km	0.54
1 statute mile = 1760 yards = 5280feet	stat.mile	1.6093km	0.6214
1 mile per hour	MPH	1.6093 km/h	0.6214
1 square foot	sqft	0.0929 m^2	10.7643
1 pound	lb	0.4569 kg	2.2046

To convert a metric unit into an english unit, use the conversion factor listed at the last column. For example : 40,000 km = 0.54 x 40,000 = 21,600 naut. miles.

Table 7.1 American and English Wire gauges, diameter in inches and millimeter

The american standard wire gauge is based on the standards of the *Brown & Sharpe* company which uses numbers in identifying the wire size. In general, the abbreviation AWG (= *American Wire Gauge*) is used. In Great Britain, there are two standard wire gauges: *BWG* (= *Birmingham Wire Gauge*) and *ISWG* (= *Imperial Standard Wire Gauge*) or *SWG* (= *Standard Wire Gauge*). Both these standards also use numbers to identify the size of the wire.

Wire gauge Nr.	AWG diameter in inches	in mm	BWG diameter in inches	in mm.	ISWG(SWG) diameter in inches	in mm.
0000(4/0)	0.460	11.68	0.454	11.53	0.40	10.16
000(3/0)	0.409	10.41	0.425	10.80	0.372	9.45
00 *(210)*	0.365	9.27	0.380	9.65	0.348	8.84
0(110)	0.325	8.25	0.340	8.64	0.324	8.23
1	0.289	7.35	0.300	7.62	0.300	7.62
2	0.258	6.54	0.283	7.21	0.276	7.01
3	0.229	5.83	0.259	6.58	0.252	6.40
4	0.204	5.19	0.238	6.05	0.232	5.89
5	0.182	4.62	0.220	5.59	0.212	5.38
6	0.162	4.11	0.203	5.16	0.192	4.88
7	0.144	3.66	0.179	4.57	0.176	4.47
8	0.128	3.26	0.164	4.19	0.160	4.06
9	0.114	2.90	0.147	3.76	0.144	3.66
10	0.102	2.59	0.134	3.40	0.128	3.25
11	0.091	2.30	0.120	3.05	0.116	2.95
12	0.081	2.05	0.109	2.77	0.104	2.64
13	0.072	1.83	0.195	2.41	0.092	2.34
14	0.064	1.63	0.083	2.11	0.081	2.03
15	0.057	1.45	0.072	1.83	0.072	1.83
16	0.051	1.29	0.065	1.65	0.064	1.63
17	0.045	1.15	0.058	1.47	0.056	1.42
18	0.040	1.02	0.049	1.24	0.048	1.22
19	0.036	0.91	0.042	1.07	0.040	1.02
20	0.032	0.81	0.035	0.89	0.036	0.92

NOTE: Values in millimeter were rounded off. AWG 21 to 40 see Table 10.1 in page 119.

Table 8.1 Two-wire cable with plastic dielectric. Amphenol standard types.

	Impedance in W	Velocity factor		Wire diameter in mm	Attenuation in dB pro 1000 m long cable at: 7 MHz	150MHz	400MHz
14-080	75	0.68	7 x 0.32	68.62	137.24	311.82	467.30
14-023	75	0.71	7 x 0.7	16.50	49.50	160.70	
14-079	150	0.77	7 x 0.32	21.72	49.50	111.19	180.70
14-056	300	0.82	7 x 0.32	9.55	19.97	51.25	88.60
14-100	300	0.82	7 x 0.32	9.55	19.97	51.25	88.60
14-271	300	0.82	7 x 0.32	9.55	19.97	51.25	88.60
14-185	300	0.82	7 x 0.4	6.95	17.37	44.30	81.65
14-076	300	0.82	7 x 0.4	6.95	16.50	41.70	72.00
14-022	300	0.82	1.3	6.21	12.16	33.00	59.00

Table 9.1   Coax cable MIL-C-17E equivalent types of the obsolete MIL-C-17 types

Obsolete Type	Equivalent	Obsolete Type	Equivalent	Obsolete Type	Equivalent
RG5/U... B/U	RG212/U	RG22/U ... A/U	RG22B/U	RG 10810	RGI08A/U
RG6/U	RG64/U	RG23/U	RG23A/U	RG111/U	RGl11A/U
RG8/U ... A/U	RG213U	RG24/U	RG25A/U	RG115/U	RG 115A/U
RG9/U ... A/U	RG214U	RG29/U	RG58C/U	RG116/U	RG227/U
RGI0/U ... A/U	RG215/U	RG34/U... A/U	RG34B/U	RG133/U	RG 133A/U
RG 1110	RG 11 NU	RG35/U... A/U	RG35B/U	RGl42/U...A/U	RG142B/U
RG12/U	RG12NU	RG58/U... B/U	RG58C/U	RG159/U	RG142B/U
RG 13/U ... A/U	RG216/U	RG59 ... A/U	RG59B/U	RG174/U	RG174A/U
RG14/U ... A/U	RG217/U	RG62/U... C/U	RG62A/U	RG178/U...A/U	RG178B/U
RG15/U	RG 11 NU	RG63/U ... A/U	RG63B/U	RG179/U...A/U	RG179B/U
RG17/U... B/U	RG218/U	RG65/U	RG65A/U	RG180/U...A/U	RG180B/U
RG18/U ... A/U	RG219/U	RG71/U ... A/U	RG71B/U	RG211/U	RG211A/U
RG191U ... A/U	RG220/U	RG74/U... A/U	RG224/U	RG228/U	RG228NU
RG20/U ... A/U	RG221/U	RG79/U ... A/U	RG79B/U	RG307/U	RG307A/U
RG21/U ... A/U	RG222/U	RG87/U ... A/U	RG225/U		

Obsolete MIL-Coaxial Cable types without equivalents:

RG16/U; RG36/U; RG54/U ... A/U; RG55/U ... B/U; RG57/U... A/U; RG72/U; RG-78/U; RG86/U; RG94/U ... A/U: RG117/U... A/U; RG118/U... A/U; RG140/U; RG141/U ... A/U; RG143/U ... A/U; RG147/U; RG148/U; RGl49/U; RGl50/U; RG181/U; RG187/U... A/U; RG188/U... A/U; RG195/U...A/U: RG196/U... A/U: RG229/U; RG282/U; RG293/U... A/U; RG294/U ... A/U; RG295/U; RG324/U; RG325/U; RG326/U: RG366/U; RG388/U; RG389/U.

Table 10.1   American and English Wire gauges, diameter in inches and millimeter (Wire gauges 21 to 40)

The american standard wire gauge is based on the standards of the *Brown & Sharpe* company which uses numbers in identifying the wire size. In general, the abbreviation AWG (= *American Wire Gauge*) is used. In Great Britain, there are two standard wire gauges: *BWG* (= *Birmingham Wire Gauge*) and *ISWG* (= *Imperial Standard Wire Gauge*) or *SWG* (= *Standard Wire Gauge*). Both these standards also use numbers to identify the size of the wire.

Wire gauge Nr.	AWG diameter in inches	in mm	BWG diameter in inches	in mm	ISWG(SWG) diameter in inches	in mm
21	0.028	0.72	0.081	0.81	0.032	0.81
22	0.025	0.64	0.028	0.71	0.028	0.71
23	0.023	0.57	0.025	0.64	0.024	0.61
24	0.020	0.51	0.023	0.56	0.023	0.56
25	0.078	0.45	0.020	0.51	0.020	0.51
26	0.016	0.40	0.018	0.46	0.018	0.46
27	0.014	0.36	0.016	0.41	0.016	0.41
28	0.013	0.32	0.0135	0.356	0.014	0.36
29	0.011	0.29	0.013	0.33	0.013	0.33
30	9.010	0.25	0.012	0.305	0.012	0.305
31	0.09	0.23	0.010	0.254	0.011	0.29
32	0.008	0.20	0.009	0.229	0.0106	0.27
33	0.007	0.18	0.008	0.203	0.010	0.254
34	0.0063	0.16	0.007	0.178	0.009	0.229
35	0.0056	0.14	0.005	0.127	0.008	0.203
36	0.0050	0.13	0.004	0.102	0.007	0.178
37	0.0044	0.11	-	-	0.0067	0.17
38	0.0040	0.10	-	-	0.0060	0.15
39	0.0035	0.09	-	-	0.0050	0.127
40	0.0031	0.08	-	-	0.0047	0.12

NOTE: Values in millimeter were rounded off. AWG 0000 to 20 see Table 7.1 in page 88.

 # GLOSSARY OF ANTENNA TERMS

Actual ground   The point within the earth's surface where effective ground conductivity exists. The depth of this point varies with the frequency, the condition of the soil and the geographical region.

Antenna   An electrical conductor or array of conductors that radiates signal energy (transmitting) or collects signal energy (receiving).

Apex   The feedpoint region of a discone antenna.

Apex angle   The enclosed angle in degrees inside the cone element of a discone antenna and similar antennas.

Bandwidth   The group of frequencies where the antenna functions efficiently.

Band   A group of frequencies.

Coaxial cable   Any of the coaxial transmission lines that has the outer shield (either solid or braided) in the same axis as the inner or center conductor. The insulating material can be air, helium, or solid dielectric compounds.

Collinear array   A linear array of radiating elements (usually dipoles) with their axis arranged in a straight line. Popular in VHF and higher frequencies.

Conductor   A metal body such as tubing, rod or wires that permits current to travel continously along its length.

Counterpoise   A wire or group of wires mounted close to ground, but insulated from ground, to form a low impedance, high capacitance path to ground. Commonly used at medium frequency and high frequency to provide an effective ground for an antenna.

Dielectrics   Various insulating materials used in antenna systems, such as found in insulators and transmission lines.

Dipole   An antenna that is split exactly at the middle for connection to a feedline. Usually a halfwavelength in dimension. Also called a doublet.

Directivity   The property of an antenna that concentrates the radiated energy to form one or more major lobes.

Director A conductor placed in front of a driven element to cause directivity. Frequently used singly or in multiples with Yagi or cubical quad beam antennas.

Direct ray Transmitted signal energy that arrives at the receiving antenna directly rather than being reflected from the ionosphere, ground or man made reflector.

Doublet see Dipole

Driven array An array of antenna elements which are all driven or excited by means of a transmission line.

Driven element The radiator element of an antenna system. The element to which the transmission line is connected.

Efficiency The ratio of useful output power to input power, determined in antenna systems by losses in the system, including in nearby objects.

Feeders Transmission lines of assorted type that are used to route RF power from a transmitter to an antenna, or from an antenna to a receiver.

Feedline see Feeders

Front to back The ratio of radiated power off the front to the back of a directive antenna. A dipole would have a ratio of 1 for example.

Front to side The ratio of radiated power between the major lobe and the null side of a directive antenna.

Gain Increase in effective radiated power in the desired direction of the major lobe.

Gamma match A matching system used with driven antenna elements to effect a match between the transmission line and the feed-point of the antenna. It consists of an adjustable arm that is mounted close to the driven element and in parallel with it near the feedpoint.

Groundplane A man made system of conductors placed below an antenna to serve as an earth ground.

Groundscreen A wire mesh groundplane.

Impedance The ohmic value of an antenna feedpoint, matching section or a transmission line. An impedance may contain reactance as well as resistance components.

Lambda Greek symbol for L used to represent a wavelength with reference to electrical dimensions in antenna work.

Line loss The power lost in a transmission line, usually expressed in decibels.

Line of sight Transmission path of a wave that travels directly from the transmitting antenna to the receiving antenna.

Load The electrical entity to which the power is delivered. The antenna is a load for a transmitter. A dummy load is a nonradiating substitute for an antenna.

Loading The process of transferring power from its source to a load. The effect of a load has on a power source.

Lobe A defined field of energy that radiates from a directive antenna.

Matching The process of effecting an impedance match between two electrical circuits of unlike impedance. One example is matching a transmission line to the feedpoint of an antenna. Maximum power transfer to the load (antenna system) will occur when a matched condition exists.

Null A condition during which an electrical property is at minimum. The null in an antenna radiation pattern is that point in the 360 degree pattern where minimum field intensity is observed. An impedance bridge is said to be 'nulled' when it has been brought into balance.

Parasitic array A directive antenna that has a driven element and independent directors or reflectors or both. The directors and reflectors are not connected to the feedline. A yagi antenna is one example. See also driven array.

Phasing lines Sections of transmission line that are used to ensure correct phase relationship between the bays of an array of antenna. Also used to effect impedance transformations while maintaining the desired array phase.

Quad Rectangular or diamond shaped fullwave loop antenna. Most often used with a parasitic loop director and a parasitic loop reflector to provide approximately 8 dB of gain and good directivity. Often called the 'cubical quad'.

Radiation pattern   The radiation characteristics of an antenna as a function of space coordinates. Normally, the pattern is measured in the far field region and is represented graphically.

Radiator   A discrete conductor in an antenna system that radiates RF energy. The element to which the feedline is attached.

Reflector   A parasitic antenna element or a metal assembly that is located behind the driven element to enhance forward directivity. Large man made structures may reflect radio signals.

Source   The point of origination (transmitter or generator) of RF power supplied to an antenna system.

Stacking   The process of placing similar directive antennas atop or beside one another forming a 'stacked array'.

SWR   Standing wave ratio on a transmission line in an antenna system. More correctly, 'VSWR' or voltage standing wave ratio. The ratio of the forward to reflected voltage on the line and not the power ratio. A VSWR of 1:1 occurs when all parts of the antenna system are matched correctly to one another.

Velocity factor   That which affects the speed of radio waves in accordance to the dielectric medium they are in. A factor of 1 is applied to the speed of light and radio waves in free space, but the velocity is reduced in various dielectric mediums such as transmission lines. When cutting a transmission line to a specific electrical wavelength, the velocity factor of the particular line must be taken into account.

VSWR   Voltage standing wave ratio. See SWR.

Wave   A disturbance that is a function of time or space or both.
A radio wave for example.

Wave front   A continous surface that is the locus of points having the same phase at the same instant.

Yagi   A directive, gain type of antenna that utilizes a number of parasitic directors and a reflector. Named after one of the inventors (Yagi and Uda).

 # BIBLIOGRAPHY

Source materials for more advanced study of the antenna designs presented in this book can be found in the following references:

Lytel, Allan. *ABC's of Antennas*, 1973

"Antenna Fundamentals", *The ARRL Antenna Book*, 14th ed., chap.2

Bergren, A.L. 'The Multi-element Quad", *QST*, May 1963

Brown, "Directional Antennas", *Proc. I.R.E.*, January 1937

Brown, Lewis, and Epstein. "Ground Systems as a Factor in Antenna Efficiency", *Proc. I.R.E.*, June 1937

Carter. «Circuit Relations in Radiating Systems and Applications to Antenna Problems», *Proc. I.R.E.*, June 1932

Geiser, Dave. "The Discone - a VHF-UHF Tribander", *QST*, December 1978

Erhorn, P.C. "The Element Spacing in 3-element Beams", *QST*, October 1957

"Groundplane Antenna", *The ARRL Handbook*, 14th ed., chap.2

Jasik. *Antenna Engineering Handbook* (New York: Mcgraw Hill Book Co.)

Kraus. *Antenna* (New York: Mcgraw Hill Book Co.)

Laport. *Radio Antenna Engineering* , New York: Mcgraw Hill Book Co., 1952

"Omni-directional Antennas For VHF and UHF", *The ARRL Antenna Book*, 14th ed., chap. 11

"Portable and Mobile Antennas", *The ARRL Antenna Book*, 14th ed., chap. 13

Reynolds, F. "Simple Gamma Match Construction", *QST*, July 1957

Rumsey. *Frequency Independent Antennas*, New York: Academic Press, 1966

Terman. *Radio Engineering*, New York: Academic Press, 1966

INDEX

INDEX

INDEX

www.ingramcontent.com/pod-product-compliance
Lightning Source LLC
Chambersburg PA
CBHW081116170526
45165CB00008B/2467

* 9 7 8 1 4 4 0 4 5 1 0 5 8 *